Drug Discovery in Japan

Sadao Nagaoka
Editor

Drug Discovery in Japan

Investigating the Sources of Innovation

 Springer

Editor
Sadao Nagaoka
Department of Economics
Tokyo Keizai University
Kokubunji, Tokyo, Japan

ISBN 978-981-13-8908-5   ISBN 978-981-13-8906-1   (eBook)
https://doi.org/10.1007/978-981-13-8906-1

This Springer imprint is published by the registered company Springer Nature Singapore Pte Ltd.
The registered company address is: 152 Beach Road, #21-01/04 Gateway East, Singapore 189721,
Singapore

Preface

This book investigates the sources of drug-discovery innovation in Japan through detailed case studies of the 12 groups of 15 innovative drugs discovered in Japan. Japan is a major contributor to the global drug-discovery industry, as is illustrated by its breakthrough discoveries, such as the first statin and the recent immune checkpoint inhibitor against cancer. This book discusses the knowledge sources of the drug-discovery projects, the interaction between drug discovery and scientific progress, causes of unexpected difficulties, capturing serendipities, and uniqueness and competition in drug discovery, focusing on the sources of innovation.

Case studies of the following drugs are covered: compactin/pravastatin (Mevalotin, Pravachol), rosuvastatin (Crestor), leuprorelin (Leuplin, Lupron, Viadur), ofloxacin (Tarivid, Floxin) and levofloxacin (Cravit, Levaquin), tamsulosin (Harnal, Flomax), pranlukast (Onon), tacrolimus (Prograf), pioglitazone (Actos, Glustin), donepezil (Aricept), candesartan (Blopress, Atacand, Amitas), tocilizumab (Actemra), and nivolumab (Opdivo). Many of these drugs offered new treatments to diseases that had been previously intractable and significantly increased the patient length of life and/or quality of life. Most of these drugs are globally available, and their values are recognized internationally. The drugs discussed in this book covered not only those discovered through chemical synthesis, including peptides, but also those screened from natural products and antibody drugs.

Each case has unique features so that the book as a whole provides diverse and rich information on drug-discovery processes in Japan. Across the cases, one common finding is the pervasive high uncertainty that drug-discovery projects face. Many of the projects faced the dangers of discontinuation because of the emergence of unexpected difficulties. Serendipitous discoveries also played major roles in a number of drug-discovery projects. The major source of such uncertainty is the incompleteness of science. Over the past century, there has been significant scientific progress in the medical sciences, and some believed that such scientific progress would make drug discovery less random and more rational. However, the drug-discovery process continues to face significant uncertainty as demonstrated by the high failure rate of clinical trials. This is largely caused by the extensive list of unknowns with regard to the specific disease mechanisms, the potential targets, and the drug candidate's mechanism of action. As a result, even today, drug-discovery projects by pharmaceutical firms often start when the underlying science for such projects is incomplete. Such major uncertainty makes the progress of science and drug discovery by a firm mutually reinforcing. A drug discovered provides a tool for a scientist to analyze how the target molecule functions in a body, thereby increasing understanding of the disease mechanism.

High uncertainty makes the value of an entrepreneurial scientist high. Such scientists are the gap fillers as defined in the theories of entrepreneurship put forward by Leibenstein (1968).[1] The case studies described in this book highlight the role that entrepreneurial researchers have played in drug discovery. These scientists filled the knowledge gaps by absorbing and combining external scientific progress and by a relentless pursuit of possibilities through their own research, often including unauthorized research to overcome crises. Furthermore, high uncertainty and its resolution significantly characterize the evolution of competition. Competition is initially rare for a pioneer who engages in discovery projects for drugs with new mechanisms of action because uncertainty is very high. However, once the clinical trials by a successful pioneer credibly show that the new specific mechanism of action works, there emerges strong intra-mechanism competition (competition among drugs with the same mechanism discovered by the pioneer). Finally, the case studies described also highlight the contribution of policy and institutions to innovations, which are not often recognized. For example, the patent system promotes innovation under high uncertainty not only by enhancing appro-priability of such research and development investment but also by facilitating the combination of knowledge and capabilities between different organizations through disclosure.

As the editor of this book, I thank Dr. Yoshiyuki Ohsugi (Chairman, Ohsugi BioPharma Consulting Co., former global project leader for tocilizumab (Actemra)) and Dr. Hideo Kawabe (Patent Attorney at Hiraki & Associates/IP Counsel in Manufacturing Technology Association of Biologics) for their valuable professional advice on case studies in terms of medical science and patent law; and Yasushi Hara (Michelin Fellow, CEAFJP/EHESS, Paris, France/Adjunct Associate Professor, Faculty of Economics, Hitotsubashi University, Tokyo, Japan) who serves as a secretariat for the research projects described in this book and helped to complete the case studies. Dr. Akira Nagumo (Medical Science Liaison, MSD K.K.) provided useful comments for the introductory and final chapters of the English draft of the book. I also express my gratitude to the Institute of Innovation Research of Hitotsubashi University for assisting in many research tasks that lead to this book. Without the assistance of the center's research support staff, the implementation of such large-scale empirical projects would be difficult. I also thank Juno Kawakami of Springer for facilitating the publication of this book and thank Osami Kono of the Biotech Editorial Department of Nikkei BP (then) for enabling the compilation of the first edition of this volume in Japanese.

---

[1] The reference is indicated at the end of Chap. 1.

In the case-based investigations, it was indispensable to interview the principal researchers for each drug-discovery project. I express my deep gratitude to the following researchers for their time and support to this project (their titles appear in the corresponding chapters, while some researchers are not named in accordance with their company policy):

Akira Endo (compactin), Kazuo Nakamura (pravastatin; Mevalotin, Pravachol), Haruo Koike (rosuvastatin; Crestor), Hiroaki Okada (leuprorelin; Leuplin, Viadur), Isao Hayakawa (ofloxacin; Tarivid, Floxin; levofloxacin; Cravit, Levaquin), Toichi Takenaka (tamsulosin; Harnal, Flomax), Hisao Nakai (pranlukast; Onon), Toshio Goto (tacrolimus; Prograf), Hiroyuki Odaka (pioglitazone; Actos, Glustin), Hachiro Sugimoto (donepezil; Aricept), Youichi Iimura (donepezil; Aricept), Yoshiyuki Kawakami (donepezil; Aricept), Koji Shimizu (donepezil; Aricept), Toshio Hirano (tocilizumab; Actemra), Tetsuya Taga (tocilizumab; Actemra), Kiyoshi Yasukawa (tocilizumab; Actemra), Kazuyuki Yoshizaki (tocilizumab; Actemra), and Tasuku Honjo (nivolumab; Opdivo).

Finally, I acknowledge that the views expressed in this book are those of the authors and are published under the responsibility of the authors and the editor.

March 2019

Sadao Nagaoka
Professor, Faculty of Economics
Tokyo Keizai University
Kokubunji, Japan

Former Professor, Institute of Innovation Research
Hitotsubashi University
Kunitachi, Japan

Acknowledgements

The papers constituting this book are based on the findings from a research project: "Science Origins of Innovations and its Economic Effects" (Principal Investigator: Sadao Nagaoka), supported by the Research Institute of Science and Technology for Society of the Japan Science and Technology Agency (JST/ RISTEX). The authors thank the Japan Science and Technology Agency for its support. We also thank the Office of Pharmaceutical Industry Policy Research (OPIR) of the Japan Pharmaceutical Manufacturers Association for their cooperation in conducting the interviews with the principal researchers of the drug-discovery projects as well as the complementary large-scale surveys for the drug discovery and development projects in Japan. The editor thanks the Japan Society of the Promotion of Science for grants (26285055, 18H00854, and 17H00963) in advancing the research to the current version published in English. The editor also thanks the Research Institute of Economy, Trade and Industry (RIETI), where the editor has organized a number of research projects on innovation, which have contributed to this research project.

Contents

Editor and Contributors

About the Editor

Dr. Sadao Nagaoka is a professor of economics at Tokyo Keizai University (Japan). Prior to this position, he was a professor at the Institute of Innovation Research of Hitotsubashi University (Japan). He has been the program director for research on innovation at the Research Institute of Economy, Trade and Industry (RIETI), and he has worked at MITI, the World Bank, and the OECD. He has been an economic advisor for the JPO (Japan Patent Office) and a member of the Economic and Scientific Advisory Board of the EPO (European Patent Office). He holds a Ph.D. in economics from the Massachusetts Institute of Technology. His scientific interests are the economics of innovation, especially R&D, intellectual property, standards, and science. His recent publications have appeared in the *Journal of Economics & Management Strategy*, the *International Journal of Industrial Organization*, *Research Policy*, the *Economics of Innovation and New Technology*, the *Journal of Technology Transfer*, the *Journal of the Japanese and International Economies*, and from the US National Academies.

Contributors

Mr. Koichi Genda is an employee of Shionogi Career Development Center Co., Ltd. (Japan). Previously, he served as a research fellow at the Office of Pharmaceutical Industry Research (OPIR) of the Japan Pharmaceutical Manufacturers Association (JPMA) after working on the clinical development of antibiotics and antipsychotics at Shionogi & Co., Ltd. He has surveyed the current status of doctor's actions in response to clinical trials in Japan, the USA, and Korea and the clinical development and review for approval of new drugs in Japan.

Yasushi Hara is an adjunct associate professor in the Faculty of Economics, Hitotsubashi University. He was a CEAFJP/Michelin Fellow in the Fondation France-Japon de l'EHESS (the School of Advanced Studies in the Social Sciences) in France prior to his current position. Combining his IT capability as a former ICT infrastructure engineer and his academic training in economics and management, he has been conducting empirical studies relevant to policy making in the field of science, technology, and innovation. He worked in the Institute of Innovation Research, Hitotsubashi University, and then joined the SciREX Center (Center of Science for RE-designing Science, Technology and Innovation Policy) of the National Graduate Institute for Policy Studies (GRIPS). He recently published "20 Years of Human Pluripotent Stem Cell Research: It All Started with Five Lines" in *Cell Stem Cell*.

Dr. Yuji Honjo is a professor of commerce at Chuo University (Japan). He is a faculty fellow in the Research Institute of Economy, Trade and Industry (RIETI). He has a Ph.D. in economics from the University of Tsukuba (Japan). His scientific interests are business economics, entrepreneurship, and small business. His work has appeared in *Applied Economics*, *Industrial and Corporate Change*, the *International Journal of Industrial Organization*, *Japan, and the World Economy*, the *Journal of Evolutionary Economics*, the *Journal of Small Business Management*, *Research Policy*, the *Review of Industrial Organization*, *Small Business Economics*, and other publications.

Dr. Hideo Kawabe is a patent attorney specializing in the field of medicine and biotechnology and concurrently holds a position at Hiraki & Associates and is IP counsel of the MAB (LLC for Manufacturing Technology of Biologics). Previously, he was the director of the Strategic Planning Department, Japan Bioindustry Association (JBA). Prior to that position, he did medical research at biotech companies and worked on Dr. Osamu Hayaishi's projects, then was in charge of intellectual property in the IP department and in the R&D department of a biotech company, Kyowa Hakko Bio Co., Ltd. He had long been a bio and life science committee member of the Japan Patent Attorneys Association and became the deputy chairman in 2017. He received a Ph.D. in pharmaceutical science from Kyoto University in 1984. His specialty in natural sciences is the structure and function of proteins.

Dr. Akira Nagumo is medical science liaison in medical affairs at MSD K.K. Prior to this position, he was a research fellow at the Office of Pharmaceutical Industry Research (OPIR). He worked at the Tsukuba Research Institutes of Banyu Pharmaceutical Co., Ltd. as a biologist for basic research of new drug discovery in the metabolic disease area. He has a Ph.D. in pharmaceutical science from Tohoku University. His scientific interests are new technology and applications of innovative drugs such as personalized medicine. His recent publications appeared in the *European Journal of Pharmacology*, the *Journal of Medicinal Chemistry*, *Bioorganic & Medicinal Chemistry* and *Lipids*.

Dr. Kenta Nakamura is an associate professor in the Graduate School of Economics, Kobe University (Japan). He holds a Ph.D. in economics from Hitotsubashi University. From 2015 to 2017, he was a visiting professor at the Max Planck Institute for Innovation and Competition, Munich, Germany. His research interests center on the fields of R&D, innovation, and intellectual property rights (IPRs, including patents, design rights, and trademarks). He has been studying how IPR policies affect R&D incentives and efficiency of the technology market and how IPRs contribute to the growth of new technology-based firms. He has also worked on a database development project for patent statistics research.

Dr. Hajime Oda is an associate professor of business administration at Tohoku Gakuin University (Japan). Before this position, he was a specially appointed research assistant at the Institute of Innovation Research of Hitotsubashi University (Japan). He has a Ph.D. in commerce and management from Hitotsubashi University. His research interests are innovation studies, especially legalization and socially justification of new businesses. His recent publications appeared in *Organizational Science* (in Japanese).

Dr. Yoshiyuki Ohsugi is Chairman & CEO, Ohsugi BioPharma Consulting Co., Ltd. He initiated and led the research and development project for a drug against autoimmune diseases in Chugai Pharmaceutical Co., Ltd., in collaboration with Osaka University, resulting in the development of tocilizumab (Actemra/ RoActemra). He was the director of Exploratory Laboratories, the President of Chugai Molecular Medicine Co., Ltd., the global project leader for tocilizumab, a senior director of the Department of Drug Development, and a science adviser for Chugai Pharmaceutical Co., Ltd. He graduated from Osaka University (Post-graduate School). He also served as a postdoctoral fellow in the Department of Internal Medicine, University of California at Davis, and was an adjunct professor in the Institute for Innovation Research, Hitotsubashi University.

Dr. Naoki Takada is a specially appointed assistant professor at the Institute of Advanced Sciences of Yokohama National University (Japan). He obtained a Ph.D. in Commerce from Hitotsubashi University, with a doctoral thesis on the dynamics of innovation in the industry of scientific instruments, especially mass spectrometry. He specializes in the management of innovation, and he has studied R&D alliances, technological breakthroughs, and science/industry linkages.

Chapter 1
Introduction

Sadao Nagaoka

Abstract This book provides detailed case studies of 12 groups of 15 major innovative drugs discovered in Japan, and in doing so investigates the sources of their innovations. It discusses: (1) the knowledge sources of the drug-discovery projects; (2) the dynamic interaction between drug discovery and scientific progress; (3) the causes of unexpected difficulties; (4) the capture of serendipity and luck; and (5) the uniqueness and competition of discovery research. The case studies cover the following drugs (generic name of molecules followed by product names): compactin/pravastatin (Mevalotin, Pravachol), rosuvastatin (Crestor), leuprorelin (Leuplin, Leupron, Viadur), ofloxacin (Tarivid, Floxin) and levofloxacin (Cravit, Levaquin), tamsulosin (Harnal, Flomax), pranlukast (Onon), tacrolimus (Prograf), pioglitazone (Actos, Glustin), donepezil (Aricept), candesartan (Blopress, Atacand, Amitas), tocilizumab (Actemra), and nivolumab (Opdivo). It elucidates the concrete mechanism by which science becomes a source of innovation, especially given that science is often incomplete when the discovery project starts. It clarifies the sources and consequences of uncertainties in drug discovery. It expands our understanding of competitive mechanism of drug discovery.

1.1 Objectives

This book provides detailed case studies of 12 groups of 15 major innovative drugs discovered in Japan, and in doing so investigates the sources of their innovations. It discusses: (1) the knowledge sources of the drug-discovery projects; (2) the dynamic interaction between drug discovery and scientific progress; (3) the causes of unexpected difficulties; (4) the capture of serendipity and luck; and (5) the uniqueness and competition of discovery research. The case studies cover the following drugs (generic name of molecules followed by product names): compactin/pravastatin (Mevalotin, Pravachol), rosuvastatin (Crestor), leuprorelin (Leuplin, Leupron, Viadur), ofloxacin (Tarivid, Floxin) and levofloxacin (Cravit,

S. Nagaoka (✉)
Tokyo Keizai University, Tokyo, Japan
e-mail: sadao.nagaoka@nifty.com

© Springer Nature Singapore Pte Ltd. 2019
S. Nagaoka (ed.), *Drug Discovery in Japan*,
https://doi.org/10.1007/978-981-13-8906-1_1

Levaquin), tamsulosin (Harnal, Flomax), pranlukast (Onon), tacrolimus (Prograf), pioglitazone (Actos, Glustin), donepezil (Aricept), candesartan (Blopress, Atacand, Amitas), tocilizumab (Actemra), and nivolumab (Opdivo). Given that Leuplin is a drug delivery system (DDS) for sustained release of leuprorelin, and levofloxacin is obtained by the optical resolution of ofloxacin, we consolidated the case studies on these drugs, respectively. For compactin and pravastatin, given that each had its own discovery and development process (the development of compactin was discontinued), this book devotes a single chapter to each drug even though pravastatin is a metabolite of compactin. Thus, this book reports on 12 groups of 15 drugs, based on 13 case studies.

This book presents detailed accounts of the discovery processes for 15 innovative drugs; it covers the full project history, from the project beginnings to market launch. It is based on interviews with the principal researchers responsible for drug discovery, and on the analysis of the relevant literature describing the drug-discovery process, patent data, scientific papers, and drug sales. The final chapter presents a cross-sectional overview of the case studies discussed in this book following a unified framework and presents an analysis of the sources of innovation and their implications. In addition, we have included the following eight topic boxes, which highlight the important issues of innovation economics and management: (1) academic–industrial collaboration to understand the mechanism of the action of statin, (2) competition in statin development between Sankyo and Merck, (3) contribution of follow-on drugs to innovation, (4) optical activity and drug discovery, (5) effects of the innovations in diagnosing Alzheimer's disease by magnetic resonance imaging, (6) history of IL-6 discovery, (7) contribution of drug discovery to science: role of IL-6 revealed by Actemra, (8) the drug-development process exploiting frontier science: a comparison between nivolumab (Opdivo) and ipilimumab (Yervoy).

The drugs discussed in this book are all highly successful and represent only a small selection of Japanese drug-discovery projects. At the same time, they individually accounted for an average of 20% of the firm sales at peak share, indicating that these innovations significantly affected each drug company's performance and contribution. In addition, the major findings of the case studies are consistent with the results of a separate large-scale survey of drug-discovery projects in Japan with a focus on New Molecular Entity (NME) drugs, which included discontinued projects (see Chap. 15).

There are three major research motivations for this book. The first purpose of the case studies is to elucidate the concrete mechanism by which science becomes a source of innovation, especially given that science is often incomplete when the discovery project starts. Many studies have already pointed out the importance of science to industrial innovation, especially for drug discovery (Gambardella 1995; Klevorick et al. 1995; Mansfield 1998; Adams and Clemmons 2008; Cockburn and Henderson 1996, 2000; Pisano 2006). However, studies characterizing the specific mechanism from science to innovation are limited. Pisano's research on the biotechnology industry points to the difficulties of harnessing science in business (Pisano 2006), particularly the high uncertainty, while demonstrating the need to integrate information from diverse sources and the necessity of cumulative learning. This book aims to gain a deeper understanding of how science contributes to drug discovery at the project level. It is difficult to obtain an understanding of the mechanism of knowledge flow and creation unless we collect information at the project level; such understanding is critical for us to understand the interactions between the advance of science and pharmaceutical innovations.

In this book, each case study provides a comprehensive account of how the researchers applied external knowledge, such as recent scientific advancements, to drug discovery. Specifically, it aims to comprehensively review the contributions of science to drug discovery: the knowledge background leading to the discovery research, the availability of new research tools (animal models, imaging tools, etc.), and identification of the specific use of the new drug candidate.

This book provides an account of the dynamic relationship between drug discovery and scientific advancement. Specifically, advances in scientific understanding of disease mechanisms and potential targets create opportunities for drug discovery. However, discovery projects often start when the science is incomplete, meaning that drug discovery can create opportunities to deepen the scientific understanding of the disease. For example, the inhibition of a target molecule's activity by a drug as a research tool provides information on the target's specific function in the body, such as a disease mechanism (see Box 13.2).

This book also analyzes how corporate scientists absorbed and utilized scientific advancements for drug discoveries. Research by Zucker et al. (1998) on technology transfer in the early days of biotechnology in the United States identified that it is crucial to transfer the tacit knowledge embodied in university researchers, which is facilitated by the geographic clustering of these researchers and biotechnology companies capable of commercializing their findings. That is, tacit knowledge sharing is local. The case studies of this book clarify how direct human interactions (such as the placement of corporate researchers in university laboratories) were important for the creation of innovative drugs in Japan where the pharmaceutical start-ups played a minimal role.

The second purpose of this book is to analyze the sources and consequences of uncertainties in drug discovery. The history of drug discovery suggests that high uncertainty is its major feature (Kirsch and Ogas 2017). Some believe that the introduction of advanced scientific tools to drug discovery will contribute to a less random and more rational process (see Gambardella 1995). However, drug discovery continues to face significant uncertainties, as demonstrated by persistently high failure rates of discovery and clinical trials. There are many examples of discovery projects that failed to identify a target and a molecule with promising efficacy and safety in vitro. Moreover, even if such molecule is discovered and optimized, it may not succeed in preclinical or clinical testing. According to a study by the Japan Pharmaceutical Manufacturers Association (JPMA) in 2011, the probability of an optimized molecule advancing from the preclinical stage to clinical trials is 37%.[1] Further-more, among those molecules subjected to clinical trials, the probability that such a molecule will be approved as a drug is 28%. Thus, only one-tenth of the molecules entering preclinical testing survive as drugs that are launched on the market. This book analyzes the sources of such uncertainties, as well as how unexpected difficul-ties were overcome and how serendipities and instances of good luck were captured. High uncertainty makes the value of an entrepreneurial scientist high, as a gap filler for innovation (Leibenstein 1968).

The third motivation of this work is to improve our understanding of the com-petitive mechanism of drug discovery. This book finds that high uncertainty and its resolution significantly characterizes the evolution of competition. Competition is initially rare for a pioneer who engages in discovery projects for drugs with new mechanisms of action, because uncertainty is very high. However, once the clini-cal trials by a successful pioneer credibly show that the new specific mechanism of action works, there emerges strong intra-mechanism competition (competition among drugs with the same mechanism discovered by the pioneer). It is often under-stood that when a new drug is protected by patent protection, competition (with generic drugs) emerges only after the patent expires. Recent studies, however, indi-cate that significant competition occurs even during the protection period. DiMasi and Paquette (2004) demonstrated that shortly after the entry of the pioneer, follow-on drugs are introduced to the market, using the same mechanism of action and a similar molecule, and that these drugs are often evaluated as priority drugs by the US Food and Drug Administration (FDA). Lichtenberg and Philipson (2002) also divided competition into competition with generics (within-patent competition) and competition between patents, pointing out that the latter is at least as influential as the former. Consistent with these findings, Scherer (2010) pointed out that most of the gross margins of pharmaceutical companies are invested in research and devel-opment (R&D) because of such competition, meaning that there are no significant excess returns.

1.2 Drugs Discussed in the Case Studies

The basic characteristics of the 12 groups of 15 drugs included in the case studies are provided in Table 1.1, wherein each drug is listed with its internationally unique nonproprietary generic name as well as product names used for sales (with trademark protection in each country). This book uses these names interchangeably, although it uses the product name when referring to the market status of a drug in a specific country. The drugs in Table 1.1 are loosely ordered by the year in which the patent

---

[1] According to DATA BOOK 2011 by the Japan Pharmaceutical Manufacturers Association (JPMA), the number of lead compounds discovered by 20 member companies of the JPMA during the period from 2005 to 2009 was approximately 650,000, and the number of optimized lead compounds (those included in preclinical studies) was 203. The number of compounds that entered clinical studies was 75, and 21 were approved.

Table 1.1 Basic characteristics of drugs discussed in this book

	Generic and product names (Japan, USA, and Europe)	Substance patent priority year	Discovery company	Therapeutic domain	Drug type
1	Compactin (discontinued in Phase II)	1974	Sankyo (now Daiichi Sankyo)	High cholesterol	New mechanism of action, low molecular weight and from natural products (Mevalotin is a metabolite of compactin)
	Pravastatin (Mevalotin, Pravachol)	1980	Sankyo (now Daiichi Sankyo)		
2	Rosuvastatin (Crestor)	1991	Shionogi		Mechanism of action not new, low molecular weight and chemical synthesis
3	Leuprorelin (Leupron USA only)	1973	Takeda Pharmaceutical	Prostate Cancer	New mechanism of action, low molecular weight peptide and chemical synthesis (Leupurine is a DDS preparation)
	Leuprorelin (Leuplin, Viadur)	1983[a]	Takeda Pharmaceutical		
4	Ofloxacin (Tarivid, Floxin)	1980	Daiichi Pharmaceutical (now Daiichi Sankyo)	Broad-spectrum antibiotic	Mechanism of action not new, low molecular weight and chemical synthesis (levofloxacin is an optically active substance)
	Levofloxacin (Cravit, Levaquin)	1985	Daiichi Pharmaceutical (now Daiichi Sankyo)		
5	Tamsulosin (Harnal, Flomax)	1980	Yamanouchi Pharmaceutical (now Astellas Pharma)	Prostatic hyperplasia	New mechanism of action, low molecular weight and chemical synthesis

(continued)

Table 1.1 (continued)

	Generic and product names (Japan, USA, and Europe)	Substance patent priority year	Discovery company	Therapeutic domain	Drug type
6	Pranlukast (Onon)	1984	Ono Pharmaceutical	Bronchial asthma	New mechanism of action, low molecular weight and chemical synthesis
7	Tacrolimus (Prograf)	1984	Fujisawa Pharmaceuticals (now Astellas Pharma)	Organ rejection prophylaxis	Mechanism of action not new, low molecular weight and derived from natural products
8	Pioglitazone (Actos, Glustin)	1985	Takeda Pharmaceutical	Diabetes	New mechanism of action, low molecular weight and chemical synthesis
9	Donepezil (Aricept)	1987	Eisai	Alzheimer's disease	Mechanism of action not new, but the first effective drug, low molecular weight and chemical synthesis
10	Candesartan (Blopress, Atacand, Amitas)	1990	Takeda Pharmaceutical	High blood pressure and congestive heart failure	New mechanism of action, low molecular weight and chemical synthesis

(continued)

Table 1.1 (continued)

	Generic and product names (Japan, USA, and Europe)	Substance patent priority year	Discovery company	Therapeutic domain	Drug type
11	Tocilizumab (Actemra)	1992	Collaboration between Chugai Pharmaceutical, the UK Medical Research Council and Osaka University[b]	Rheumatoid arthritis	New mechanism of action, humanized antibody pharmaceuticals
12	Nivolumab (Opdivo)	2005[a]	Ono Pharmaceutical in cooperation with Medarex USA and Kyoto University [b]	Cancer	New mechanism of action, humanized antibody pharmaceuticals
	Frequency or average over case studies	1983			Eight drugs with new mechanisms of action

[a]Leuplin is a drug protected by the patent on its drug delivery system (DDS). A method-of-use patent for inhibiting PD1 signaling was applied for in 2002

[b]Tocilizumab (Actemra) and nivolumab (Opdivo) involved university and industry collaborations, based on the results of basic research

for the drug was applied for (the priority year),[2] but drugs with the same mechanism of action are consolidated. Here, "discovery" includes not only discovery through screening of natural products, but also chemical synthesis and preparation of antibody drugs. The actual discovery year may be considerably earlier than the patent filing year because it usually takes time to determine the structure and collect test data on efficacy after a new drug candidate is discovered. However, Table 1.1 gives the patent filing year as the discovery year. Compactin, the first drug in the list, was discovered in 1974, while the year of discovery of the last drug on the list, Nivolumab (Opdivo), was in 2005. Thus, the case studies in this book cover drug-discovery projects in Japan that were carried out over a period of approximately 30 years.

Diseases and therapeutic domains discussed in this book include high cholesterol, prostate cancer, bacterial infections, prostatic hyperplasia, bronchial asthma, organ rejection after transplantation, diabetes mellitus, Alzheimer's disease, hypertension, autoimmune diseases (rheumatoid arthritis) , and other cancers. Although these diseases are common, no effective treatments existed prior to those discussed in this

---

[2]The priority year is the year in which the invention was first applied for a patent.

book. They include some of the most common diseases in Japan: high blood pressure (10.1 million patients), diabetes (3.1 million patients), hyperlipidemia (2 million patients), heart disease (1.7 million patients, excluding hypertensive diseases), and malignant neoplasms (1.6 million patients, excluding dental diseases).[3]

Tocilizumab (1992) and nivolumab (2005), which were discovered relatively recently, are antibody drugs (biopharmaceuticals), while the others are low molecular weight compounds. Among the low molecular weight compounds, pravastatin is a metabolite of compactin, which was in turn identified through natural product screening. Similarly, tacrolimus was also derived from natural products; the remaining eight drugs were obtained by chemical synthesis.

As demonstrated in Table 1.1, eight of the drug groups have novel mechanisms of action, which is defined throughout this book as new, depending on whether the drug target was new. Often, a new drug must act on a new target to significantly enhance its therapeutic effects and reduce its side effects. For example, prior to the introduction of statins (see Chap. 2), which lower the level of cholesterol in the blood causing arteriosclerosis, cholesterol-lowering drugs acted by reducing the cholesterol ingestion or by inhibiting the absorption of ingested cholesterol. Exceptionally, there was a drug designed to inhibit the synthesis of cholesterol in the body; however, it was ineffective and complicated by serious side effects. In comparison, statins act by up-regulating low-density lipoprotein (LDL) receptors and thereby lower LDL cholesterol in the blood through competitively inhibiting HMG-CoA reductase (3-hydroxy-3-methylglutaryl-coenzyme A reductase), which is the rate-limiting enzyme in the body's synthesis of cholesterol. Statins are both highly effective and safe because of this new mechanism of action.

In the cases of tacrolimus, donepezil, ofloxacin/levofloxacin, and rosuvastatin, drugs with the same mechanism of action were already available. Nevertheless, these new drugs significantly increased the therapeutic efficacy and/or reduced harmful side effects relative to the pre-existing drugs. Tacrolimus, which inhibits the enzyme calcineurin (CaN) in T cells, leading to immune suppression, has the same mechanism of action as cyclosporine, a drug that was already commercially available. However, tacrolimus is a more effective immunosuppressant and has fewer side effects. The introduction of tacrolimus to medical practice thus significantly increased the success rate of transplantation. Similarly, donepezil replaced tacrine, which was approved for use in the treatment of Alzheimer's disease, yet was rarely used because of its high toxicity. Therefore, donepezil can be considered to be the first drug to effectively slow the progression of Alzheimer's disease.

As shown in Table 1.2, the development of each of these drugs required many years of R&D. On average, approximately 16 years elapsed from the start of the research (basic or discovery research) to the first approval or market sales of the drug. The R&D process for a drug can be divided into two stages: the search for new candidate molecules (New Molecular Entity) and preclinical testing, followed by clinical trials, which test whether the candidate molecules are efficacious in the treatment of a human disease and whether the side effects are tolerable to the patients. Each of

---

[3] According to a patient survey conducted by Ministry of Health, Labour and Welfare in 2014.

Table 1.2 Duration of R&D and sales data for drugs discussed in this book

Case	Generic and product names (Japan, USA, and Europe)	Approximate duration of R&D[a]	Share of company sales[b] (%)	Global expansion
1	Compactin (discontinued in Phase II)	17 years Discovery research began in 1972; phase II clinical trial of compactin discontinued in 1980; Pravastatin was discovered in 1980, its preclinical study began in 1981; first marketed in Japan in 1989		(Discontinued in Phase II clinical trial)
	Pravastatin (Mevalotin, Pravachol)		18	First statin in Japan, third statin in the world, sold in 115 countries (April 2010)
2	Rosuvastatin (Crestor)	11 years Research began in 1991 and approved in the Netherlands in 2002	38	Approved in more than 100 countries and sold in more than 80 countries (as of October 2009)
3	Leuprorelin (Leupron USA only)	14 years Research began in 1971, launched in USA in 1985		USA
	Leuprorelin (Leuplin[c], Viadur)	9 years Research began in 1980 and launched in USA in 1989	5	Approximately 80 countries
4	Ofloxacin (Tarivid, Floxin)	14 years Research for quinolones began in 1971 and launched in 1985 in West Germany		31 countries (October 2013)
	Levofloxacin[d] (Cravit, Levaquin)	8 years Research for optical resolution began after the launch of Tarivid in 1985, and launched in 1993	12	2008 and 2009 world's highest antimicrobial sales

(continued)

Table 1.2 (continued)

Case	Generic and product names (Japan, USA, and Europe)	Approximate duration of R&D[a]	Share of company sales[b] (%)	Global expansion
5	Tamsulosin (Harnal, Flomax)	17 years Research for developing new applications of the invention of Amoslarol began in 1976, launched in 1993	16	Approved and released in more than 60 countries worldwide
6	Pranlukast (Onon)	14 years Discovery research began in 1981, launched in 1995	16	Japan, South Korea, and Latin America
7	Tacrolimus (Prograf)	11 years Search began in 1982, launched in 1993	21	Approved in 104 countries worldwide, and sold in 96 countries (May 2011)
8	Pioglitazone (Actos, Glustin)	24 years Discovery of AL-321 in 1975, the launch of discovery research on drugs for the treatment of diabetes mellitus, approved in USA in 1999	27	90 countries
9	Donepezil (Aricept)	15 years Commenced in 1981 and approved in USA in 1996	40	Approved in 97 countries (November 2013)
10	Candesartan (Blopress, Atacand, Amitas)	18 years ARB research began in 1978 and launched in UK in 1996	10	In recent years, it has been sold in about 50 countries around the world

(continued)

Table 1.2 (continued)

Case	Generic and product names (Japan, USA, and Europe)	Approximate duration of R&D[a]	Share of company sales[b] (%)	Global expansion
11	Tocilizumab (Actemra)	24 years Start of basic research in 1984 to the launch in 2008 as a rheumatoid arthritis drug. 19 years Start of discovery research in 1986 (the start of industry–university collaborative research to explore IL-6 inhibitors with Osaka University) to launch as drug for Castleman's disease in 2005	17	Sales in over 130 countries (2014)
12	Nivolumab (Opdivo)	13 years 2002 application for the method-of-use patent with Prof. Honjo of Kyoto University, launched in Japan and USA in 2014	34	Malignant melanoma, non-small cell lung cancer, renal cell carcinoma, Hodgkins lymphoma, head and neck cancer approved in Japan, USA and Europe. Many other cancer disease areas in many countries are under clinical trials
	Average for case studies	16 years	21	

[a]From commencement of research to initial approval or launch
[b]Maximum value between FY2005 and FY2014, FY 2017 for Opdivo
[c]Leuplin is a new drug obtained by development of a sustained-release drug for Leuprorelin
[d]Levofloxacin is a drug obtained by optical resolution of Ofloxacin
Levofloxacin and Leuplin are considered for the calculation of the average duration of R&D

these processes may last many years. Furthermore, the transition period between the two stages is also an important determinant of the length of the R&D period. After the discovery project is completed, clinical development requires the organization of clinical trial protocols and the financial resources for large-scale investment, which can be time-consuming to prepare. In the case of antibody drugs, the development of a manufacturing process for a drug to be tested in clinical trials is also time-consuming. For example, tocilizumab and pioglitazone, among the case study examples, had the longest R&D periods, each of which lasted over 20 years. In the case of tocilizumab, the basic research began before the target was known; it took researchers 6 years to identify the target and to produce the humanized antibody. From there, 7 years lasted in pre-clinical research until the launch of the clinical trial, and an additional 11 years passed before the drug was approved for treatment of rheumatoid arthritis. Likewise, the drug candidate molecules that would eventually lead to pioglitazone were not discovered until 11 years after the identification of a lead compound. This was followed by an additional 5 years before clinical trials started in Japan and 8 years for approval of the drug.[4]

As Table 1.2 shows, each drug discussed in these case studies accounts for a large share of the entire company's sales. The average of the maximum yearly sales share between 2005 and 2014 is approximately 20% on a consolidated basis (2017 for the most recently approved nivolumab (Opdivo)). Donepezil (Aricept) and rosuvastatin (Crestor) each accounted for about 40% of Eisai's and Shionogi's sales, respectively, at peak share (including licensing income). The share of Leuplin is the smallest but still accounted for a maximum of 5% of Takeda's sales. Thus, the drugs described in this book are major contributors to the sales of each company. As highlighted in Table 1.2, these drugs have been developed and sold globally, including Europe and the United States (with the exception of Onon) , and their effects are recognized worldwide.

References

Adams, J.D., Clemmons, J.R. (2008). *The origins of industrial scientific discovery*. National Bureau of Economic Research Working Paper, No. 13823.

Cockburn, I. M., & Henderson, R. (1996). Public-private Interaction in Pharmaceutical Research. *Proceedings of the National Academy of Sciences of the United States of America, 93,* 12725–12730.

Cockburn, I. M., & Henderson, R. (2000). Publicly funded science and the productivity of the pharmaceutical industry. *Innovation Policy and the Economy, 1,* 1–34.

Zucker G. Z., Darby, M. R., & Brewer, M. B., (1998). Intellectual human capital and the birth of U.S. biotechnology enterprises. American Economic Review, 88(1).

DiMasi, J. A., & Paquette, C. (2004). The economics of follow-on drug research and development: Trends in entry rates and the timing of development. *Pharmacoeconomics, 22*(2), 1–14.

---

[4]It is important to note that the distinction between basic research on disease mechanisms and targeted discovery research on specific drugs is ambiguous and that there are uncertainties in deciding the first year of the discovery research.

Gambardella, A. (1995). *Science and innovation: The US pharmaceutical industry during the 1980s.* Cambridge University Press.

Kirsch, R. D., & Ogas, Ogi. (2017). *The drug hunters: The improbable quest to discover new medicines.* New York: Arcade Publishing.

Klevorick, A. K., Levin, R. C., Nelson, R., & Winter, S. (1995). On the sources and significance of inter-industry differences in technological opportunities. *Research Policy, 24*(2), 185–205.

Leibenstein, H. (1968). Entrepreneurship and development. *The American Economic Review, 56*(2), 155–177.

Lichtenberg, F. R., & Philipson, T. (2002). The dual effects of intellectual property regulations: Within- and between-patent competition in the US pharmaceuticals industry. Journal of Law & Economics, *45*, 643–672.

Mansfield, E. (1998). Academic research and industrial innovation: An update of empirical findings. *Research Policy, 26*, 773–776.

Pisano, G. (2006). *Science business: The promise, the reality and the future of biotech.* Harvard Business School Press.

Scherer F. M. (2010). Pharmaceutical innovation. In B. H. Hall & N. Rosenberg (Eds.), *Handbooks in economics of innovation* (Vol. 1), North Holland.

Chapter 2
Compactin

The Discovery of Statin, the "Penicillin" for Cholesterol

Sadao Nagaoka and Yasushi Hara

Abstract Compactin was a breakthrough discovery made by Dr. Akira Endo of Sankyo Co., Ltd. Its discovery enabled the creation of a class of drugs (statins) for hyperlipidemia that effectively and safely lowers cholesterol through competitive inhibition of HMG-CoA reductase. This chapter discusses the four major factors that contributed to this breakthrough discovery: (1) the uniqueness of the discovery program, (2) the adoption of a highly efficient research method that supported an uncertain discovery program, (3) the courage and persistence of Dr. Endo to overcome unexpected difficulties and the biases of conventional wisdom, and (4) a high degree of research freedom and a favorable research environment at Sakyo that supported his endeavor. This case has the following three implications; first, while the progress in underlying science creates opportunities for discovery projects for breakthrough drugs, such projects almost inevitably face unexpected difficulties in the early stages because of incomplete scientific knowledge. Second, the cooperation of industry and academia played key roles in solving these unexpected difficulties and concerns. Third, this case demonstrates significant knowledge spillover effects of pioneering drug discovery on subsequent drug discoveries.

2.1 Introduction

Compactin is regarded as the first statin, and was discovered in the project led by Dr. Akira Endo of Sankyo Co., Ltd. Statins are a class of drugs used against hyperlipidemia to effectively and safely lower cholesterol by competitively inhibiting HMG-CoA reductase. They greatly contribute to the prevention and treatment of fatal vascular diseases such as myocardial and cerebral infarctions.

S. Nagaoka
Tokyo Keizai University, Tokyo, Japan
e-mail: sadao.nagaoka@nifty.com

Y. Hara (✉)
CEAFJP/EHESS, Paris, France
e-mail: yasushi.hara@r.hit-u.ac.jp

Faculty of Economics, Hitotsubashi University, Tokyo, Japan

© Springer Nature Singapore Pte Ltd. 2019
S. Nagaoka (ed.), *Drug Discovery in Japan*,
https://doi.org/10.1007/978-981-13-8906-1_2

Although cholesterol inhibitors existed in the 1960s, most of these drugs controlled either the intake of cholesterol or the absorption of ingested cholesterol. As an exception, Triparanol was a commercially available drug that inhibited the body's synthesis of cholesterol. However, Triparanol had limited efficacy and caused severe side effects, which led to its withdrawal from the market in 1962. Statins were designed to safely lower the body's cholesterol levels by competitively inhibiting HMG-CoA reductase, the rate-limiting enzyme in the body's synthesis of cholesterol. The pharmaceutical industry's efforts to develop statins and the concurrent scientific advancement in uncovering its mechanism of action (Box 2.1) demonstrated that: (1) statins were drugs with a novel mechanism of action that up-regulates low-density lipoprotein (LDL) receptors and thereby lowers LDL cholesterol in the blood, and (2) their efficacy and safety are high, even with long-term administration. Today, statins continue to be administered to a very large number of patients worldwide. In the United States alone, 39 million people received statins between 2012 and 2013 (Salami et al. 2017). Such ubiquitous use has led Professors Michael Brown and Joseph Goldstein of the University of Texas to refer to statins as the "penicillin for cholesterol."

2.2 Timeline of the Discovery and Development of Compactin

Dr. Akira Endo and his team at Sankyo discovered compactin (ML-236B) in 1973, and through its development effort demonstrated that the inhibition of HMG-CoA reductase by compactin significantly lowered blood cholesterol levels. This was a discovery of a drug with a completely new mechanism of action, and it paved the way for the development of other statins, including lovastatin and pravastatin. Dr. Endo also contributed to the discovery of the mechanism of cholesterol metabolism mediated by LDL receptor in work with Professors Brown and Goldstein, who later received the Nobel Prize for such work.

The contributions made by Dr. Endo have been recognized through a number of globally recognized awards, including the Lasker Award, the Heinrich Wieland Prize, and the Japan Prize for the compactin discovery, which also led to his induction into the National Inventors Hall of Fame (USA), the first Japanese inventor to be awarded this honor (National Inventors Hall of Fame 2012). Although compactin was the first statin, its development was abandoned in Phase II clinical trials. The main processes from the start of the research to the discontinuation of the Phase II clinical study are described below:

April 1971: Endo launched the discovery program to search for inhibitors of cholesterol synthesis from physiologically active substances produced by microorganisms at Sankyo.
May 1971: Screening began.

July 1973: ML-236B (the code name of compactin) , a potent inhibitor of HMG-CoA reductase, was discovered from a blue mold. It was described in the August monthly laboratory report.

October 1973: The chemical structure of ML-236B was determined.

February 1974: The efficacy of ML-236B was tested in vivo using a rat model; it did not reduce the animal's cholesterol levels.

March 1974: Endo repeated the in vivo tests for efficacy using both rats and mice.

June 1974: The patent application for ML-236B was filed (published in Japan: December 1975).

October 1975: Endo requested three cell culture lines from Brown and Goldstein; homozygous FH (familial hypercholesteremia), heterozygous FH, and healthy cells; these were also used for their discovery of the LDL pathway (Brown and Goldstein 1976).

January 1976: Endo found that ML-236B strongly inhibited cholesterol synthesis in each of the cell lines.

July 1976: Endo confirmed the cholesterol-lowering effect of ML-236B in vivo using dogs.

August 1976: Preclinical testing of ML-236B was approved internally, and a drug-development project was launched with Endo as the leader.

December 1976: Japanese monkeys were used to confirm the hypocholesterolemic effect in vivo.

February 1977: A preclinical study of ML-236B was launched by Sankyo.

February 1977: Brown and Goldstein requested a sample of compactin from Endo for the treatment of patients with familial hypercholesterolemia.

April 1977: A joint study with Brown and Goldstein was initiated, and a sample of compactin crystals (not enough for clinical use) was sent to their laboratory.

April 1977: Suspected hepatotoxicity in rats.

August 1977: Professor Yamamoto of Osaka University requested a sample of compactin from Endo for the treatment of patients with familial hypercholesterolemia.

May 1977: Long-term toxicology studies in dogs began.

February 1978: Compactin was administered to patients with familial hypercholesterolemia at Osaka University.

November 1978: Sankyo commenced Phase I clinical trials of compactin.

End of December 1978: Endo retired from Sankyo and moved to the Tokyo University of Agriculture and Technology.

January 1979: Compactin Phase I clinical trial completed.

August 1979: Compactin Phase II clinical trial began.

August 1980: Sankyo discontinued the development of compactin based on the results of long-term toxicity studies in dogs.

2.3 Statin Discovery Program

When Endo joined Sankyo in 1957, the company did not have a laboratory dedicated to R&D. Thus, he began his R&D career in the factory laboratory. In 1962, Sankyo established its Independent Central Research Laboratory (Sankyo 2000), and this was where Endo initiated the statin discovery program in the early 1970s. Of note, Japanese patent protection did not cover product (substance) patents until 1975. Consistently, up to this point, the business focus of Japanese pharmaceutical companies had been to conduct clinical trials in Japan of drugs licensed from foreign firms and to improve the manufacturing process of these drugs. Japanese pharmaceutical companies had limited experience in full-scale drug discovery at the time.

In this context, there was no in-house research experience at Sankyo directly related to cholesterol-lowering drug discovery. However, Endo had extensive experience in discovery projects that utilized microorganisms, and he applied similar biochemical techniques to the discovery of compactin. In his research at the factory laboratory, Endo isolated pectinase, an enzyme that breaks down pectin, from white rot fungus; this research contributed to his doctorate in agriculture, which he received in 1966 as well as a JSBBA (Japan Society for Bioscience, Biotechnology, and Agrochemistry) Award for Young Scientists. It was quite common in Japan for a corporate scientist to receive a doctorate based on research accomplishment within the firm without attending a graduate school. Endo then took the opportunity of a 2-year research stay in the USA from 1966 and worked on a project that aimed to identify the role of phospholipid in enzymatic reactions at the Albert Einstein College of Medicine (Endo 2006a). He played a pivotal role in the successful completion of this project.

At the end of his stay in the United States, Endo returned to Japan with the idea of conducting a discovery project on cholesterol-lowering drugs by screening physiologically active substances from molds and mushrooms. He launched this discovery project in Sankyo's newly established Fermentation Laboratory, the primary goal of which was the discovery of new antibiotics. The project was started following the appointment of a new Research Director for the Fermentation Laboratory, who granted a high degree of freedom to the laboratory's scientists. Such liberty was the key for Endo to launch a highly novel and uncertain discovery program in April 1971. It aimed at searching for a drug candidate based on his hypothesis that inhibiting cholesterol synthesis would be more effective than inhibiting cholesterol absorption and that some fungi and mushrooms naturally produce inhibitors of cholesterol synthesis. This final part of his hypothesis assumed that inhibition of cholesterol synthesis would serve as a mechanism of survival competition between microorganisms. Endo recognized that his project was associated with a high level of uncertainty, so he set the time framework in advance that he would terminate the project if he could not obtain promising results in two years.

The concept that "an inhibitor of cholesterol synthesis would be more effective than an inhibitor of its absorption" was based on the scientific understanding of the biosynthesis pathway of cholesterol, which was gained in the 1950s and 1960s. In particular, researchers identified that internally synthesized cholesterol accounts for two-thirds of the total cholesterol supply in humans. Furthermore, research into the cholesterol synthesis pathway led to the discovery of HMG-CoA reductase, which then allowed it to be used as the target molecule in the Endo discovery program.

Endo also had several other reasons for implementing the discovery project based on inhibition of cholesterol synthesis. First, according to his view, many physiologically active substances produced by microorganisms, particularly molds and mushrooms, have long been used for human consumption; thus, many such compounds could be considered safe. Second, Endo had extensive experience in screening for compounds from molds and mushrooms, and third, he considered research opportunities in the context of competition with other research groups. From Endo's perspective, "Conducting a research project that directly competed with American researchers would be absolutely unsuccessful. Although many Japanese researchers who had completed projects in the United States pursued a research topic with the same or similar research theme as that they had pursued in the United States, they often ended picking up the remaining threads, serving only as an ornament to the work done by US researchers. Even if it was going to become painstaking work, we needed to conduct novel research." (Endo 2006a) Endo also considered that "on the basis of the scientific work of Dr. Konrad Bloch and colleagues, a number of US researchers would start to look for cholesterol synthesis inhibitors using synthetic compound screens, but it was unlikely that they bet on the use of fungi and mushrooms."[1]

2.4 Novelty of the Endo Discovery Program

The discovery program designed and led by Endo was novel for a number of reasons. First, it was based on the recently characterized mechanism of cholesterol biosynthesis, which had been extensively studied during the 1950s and 1960s, and set up HMG-CoA reductase as an optimal target molecule.

Second, in screening substances produced by fungi and mushrooms as the method of discovery, the program capitalized on Endo's previous experience. Indeed, the Japanese drug-discovery community had extensive experience in screening natural products, and was considered a strength from an international viewpoint. According to Brown and Goldstein (2004) , "no random chemical library would ever yield an HMG-CoA reductase inhibitor as potent as the natural statins." Although fully artificial synthetic statins were subsequently developed, all of them imitate the chemical structure of natural statins. In particular, synthetic statins mimic the "head" of molecule ML-236B.

---

[1] From an interview with Prof. Akira Endo.

In October 1973, Endo's team successfully determined the crystal structure of ML-236B,[2] which led to a patent application being filed with the Japan Patent Office in June 1974 and publication of the patent in December 1975. The US patent was filed in May 1975 and registered in September 1976. Of note, the Japanese patent (application: 49-64823) was a process patent, while both product and process patents were submitted in the US (3,983,140 and 4,049,495). An article discussing the basic science of ML-236B was published in December 1976 (Endo et al. 1976).

At the start of the study (April 1971), the research team consisted of four individuals: Akira Endo, Masao Kuroda, and two research assistants. However, the research team expanded to six researchers in April 1973. The compactin patent listed five inventors: Akira Endo, Masao Kuroda, Akira Terahara, Yoshio Tsujita, and Chihiro Tamura.

2.5   Collaboration with Academia

Although the discovery program itself was implemented single-handedly by the Endo team at Sankyo, the team also cooperated extensively with university researchers in Japan and the United States to promote scientific understanding of the mechanism of action of statins and the promotion of clinical research. Such collaborations made significant contributions to the understanding of the efficacy and safety of statins on the basis of the mechanism of action and accelerated the clinical development of statins. In fact, statins were an important research tool for the study of the mechanism of cholesterol metabolism. Notably, the mechanism of action of statins through the LDL pathway was established by Brown and Goldstein of the University of Texas in collaboration with Endo (Box 2.1). Their research clarified the mechanism by which statins selectively lower LDL cholesterol in the blood while maintaining the normal levels of cholesterol in cells, confirming that the high degree of safety of statins is based on their mechanism of action.

Endo also began working with Prof. Akira Yamamoto of Osaka University and Dr. Hiroshi Mabuchi of Kanazawa University following the publication of the compactin patent (December 1975 in Japan). Joint research with Yamamoto was initiated by Endo's 1976 review article in the Japanese journal *Biochemistry* (Endo 1976). Yamamoto's clinical study, which included patients with FH, showed that compactin was very effective at lowering blood cholesterol levels in patients that were FH heterozygotes without side effects, while it was ineffective in FH homozygotes (Yamamoto et al. 1980). Later, it was found that there was no LDL receptor in FH homozygotes; thus, the results supported evidence showing that compactin acts through the LDL receptor.

---

[2]The molecular structure and weight of the ML-236B was determined, and the fact that it was a new molecular entity was established in September 1972.

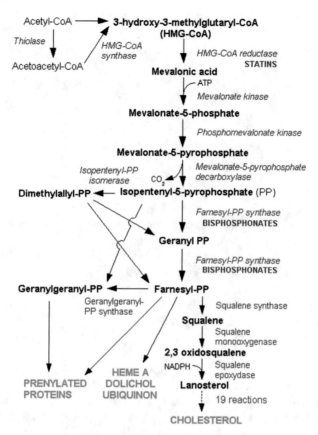

Fig. 2.1 Pathway of cholesterol biosynthesis. *Source* Wikipedia, https://en.wikipedia.org/wiki/
File:HMG-CoA_reductase_pathway.png [CC BY-SA]

**Box 2.1 Academic–Industrial Collaboration to Understand the Statin
Mechanism of action**

Cholesterol biosynthesis begins with acetyl CoA and involves a few dozen steps
until the final production of cholesterol. The rate-limiting enzyme is HMG-
CoA reductase, which catalyzes the conversion of HMG-CoA to mevalonate.
Statins are HMG-CoA reductase inhibitors that are primarily routed to the
liver where they bind and inhibit HMG-CoA reductase, reducing its enzyme
activity and subsequent cholesterol synthesis. The reduction of cholesterol
synthesis is compensated for by the increase in the expression of intracellular
LDL receptors, which in turn promote the uptake of LDL cholesterol from the
bloodstream into the liver. As a result, the level of LDL cholesterol in the blood
decreases, while the levels of cholesterol in the liver remain normal (Goldstein
and Brown 2009; Nakaya 1988) (Fig. 2.1).

This mechanism of action of statins was established by Brown and Goldstein at the University of Texas Southwestern Medical School in the United States, and was supported by the collaborations with Endo and others. They found the LDL receptor in 1973, the same year that Endo discovered compactin, and began investigating the mechanism of cholesterol metabolism mediated by the LDL receptor. Their research ultimately clarified that statins exert their effects through the action of the LDL receptor. After learning about the study of the mechanisms of cholesterol metabolism in the Brown and Goldstein laboratory, Endo reached out to them to request cell lines of normal, heterozygous, and homozygous familial hypercholesterolemia (FH) patient fibroblasts. Endo also indicated that he had discovered a potent inhibitor of HMG-CoA reductase. Brown and Goldstein responded to Endo a week later explaining how to obtain the cell lines, and offering to start a collaborative study on the effects of statins using their already established experimental system.

According to Brown and Goldstein (2004), they made a list of scientists who had cited their work. Through such monitoring, they found Endo's 1976 paper, which reported his discovery of compactin, and they "were amazed to find that Endo had discovered a molecule, which he called ML-236B, that was a potent competitive inhibitor of HMG-CoA reductase." For them, Endo's paper was very convincing in three respects: (1) that ML-236B was very potent and acts within 10 nM; (2) they performed kinetic experiments showing that ML-236B acted competitively with respect to the substrate HMG-CoA; and (3) that ML-236B had a five-member lactone ring that resembled mevalonate.

Endo provided compactin in response to a request from Brown and Goldstein, and they confirmed that the compound was a potent inhibitor of cholesterol synthesis. They also found that there was a large increase in the amount of enzyme activity. In addition to the feedback regulation of the LDL receptor by cholesterol, HMG-CoA reductase was also under feedback control from cholesterol levels. This research result was published as a joint paper, which became the first research paper published about statins outside of Japan (Brown et al. 1978).

Following this, Brown and Goldstein (2004) also succeeded in demonstrating that lowering of cholesterol by statin (using lovastatin supplied by Merck) increases the number of LDL receptors and results in a lowering of LDL in the blood. This research was inspired by Yamamoto et al. (1980) who showed that FH heterozygotes responded well to compactin, but two FH homozygotes showed marked resistance to the drug. Thus, statins served as a useful research tool for the study of the mechanism of regulation of cholesterol metabolism by the LDL receptor. Brown and Goldstein's research clarified the mechanism by which statins could selectively lower LDL cholesterol in the blood while maintaining normal levels of cholesterol for vital activities, showing that the high degree of safety of statins was caused by its mechanism of action. They received the Nobel Prize in Physiology or Medicine in 1985 for their "discoveries concerning the regulation of cholesterol metabolism."

2.6 Crises and Discontinuations

2.6.1 First Crisis: Failure in Rats

To move from the discovery stage to preclinical testing, a study of compactin in rats was conducted by a specialized unit of Sankyo's Central Research Laboratory in January 1974. It was reported in the following month that compactin passed the acute toxicity and general pharmacological tests, but that compactin failed to lower cholesterol. At the time, it was common practice to screen new drugs using rat models of disease and this result could have caused the discontinuation of the research and development of compactin. However, Endo reviewed the literature and found that cholestyramine, which was developed in the mid-1960s, had been validated in chickens and dogs (Huff et al. 1963), but had failed to reduce cholesterol in rats (Gallo et al. 1966).

Given these findings, Endo initiated a new animal study with the assistance of seven researchers in March 1974. By the end of 1975, their group had demonstrated that while compactin inhibited HMG-CoA reductase in rats, it also significantly increased production of hepatic HMG-CoA reductase, effectively counteracting the benefits gained by HMG-CoA reductase inhibition. In 1976, Endo conducted a pathological study of cholesterol in chickens and dogs in collaboration with Dr. Noritoshi Kitano of the Central Research Laboratory's Pathology Division. This study confirmed the efficacy of compactin in these models, thereby resolving Endo's first crisis. Later, when the compactin mechanism of action was identified, the reason compactin was ineffective in rats became apparent, because LDL is largely absent in rats (Brown and Goldstein 2004).

2.6.2 Second Crisis: Suspected Hepatotoxicity

In August 1976, a preclinical study of ML-236B was initiated, and the results from the mid-term toxicity studies in dogs and rats were reported in July 1977. In dogs, no toxicity concerns arose following the administration of a high dose of compactin (250 times the effective human dose). However, in rats, microcrystals were found in hepatocytes, raising concerns about drug safety. This was the second crisis faced by researchers during the development of compactin.

Three research efforts were critical to overcoming this crisis: (1) Endo showed that the microcrystals had a high probability of being cholesterol esters and characterized their formation routes, (2) Prof. Yamamoto conducted clinical studies in critically ill patients, the results of which indicated compactin was effective in humans, and (3) the in-house safety testing center demonstrated similar microcrystals in control rats in December 1977 (Endo 2006a; Renneberg 2014).[3] Thus, four and a half years had passed after the compactin patent application was filed (June 1974) before clinical trials were initiated (November 1978).

2.6.3   Third Crisis: Discontinuation of Clinical Development

In November 1978, a Phase I clinical trial was initiated and completed without a problem. During this time, Endo left Sankyo. However, a Phase II clinical trial that started in March 1979 was discontinued in August 1980 because of the results from a long-term canine toxicity study. Notably, symptoms of lymphoma had been observed in dogs administered an extremely high dose of compactin, raising serious safety concerns. Merck & Co., Inc. had concurrently begun a clinical trial of lovastatin, which the company discontinued in September 1980 because of the Sankyo decision. However, Merck subsequently resumed the lovastatin clinical research for critically ill patients in 1982, following a recommendation by clinical researchers, and the clinical trial itself was resumed in November 1983. In the following March, Sankyo began a Phase I clinical trial of pravastatin, a metabolite of compactin.

2.7   Science Sources that Supported the Discovery of Compactin

Endo points to the following three scientific advancements as being pivotal to his discovery of compactin:

(1) The discovery and elucidation of the mechanism of cholesterol biosynthesis,
(2) Discovery of the cholesterol feedback mechanism by HMG-CoA reductase,
(3) Drug screening method innovation.

The discovery and elucidation of the mechanism of cholesterol biosynthesis was a prerequisite for Endo's statin discovery project because this allowed for the identification of a target molecule. Cholesterol is synthesized from acetic acid through a number of complex processes, which were characterized by several teams of researchers, among them were the scholars Konrad Bloch and Feodor Lynen (Endo 2006a).[4]

---

[3]Interview with Prof. Akira Endo.
[4]Bloch and Lynen won the Nobel Prize in 1964 for their discoveries "concerning the mechanism and regulation of the cholesterol and fatty acid metabolism."

This major scientific advancement was essential for the design of drug-discovery programs targeting cardiovascular diseases such as arteriosclerosis, of which Endo's program is one example.[5] It is also important to note the major contribution that isotope measurement technology, which became available in the 1950s, made to this research field.

By the late 1960s, it had been determined that cholesterol was sourced both exogenously and endogenously. Notably, researchers demonstrated that endogenous cholesterol accounted for two-thirds of the cholesterol in the human body, and this was regulated by a feedback control system, such that when the supply of exogenous cholesterol decreased, the supply of endogenous cholesterol increased.[6] It was also established that HMG-CoA reductase, one of the rate-limiting enzymes in cholesterol biosynthesis, is responsible for this feedback control.

Endo also used a new experimental method to test for cholesterol synthesis. Prior to these studies, experiments had primarily been conducted with rat liver slices, while Endo opted for a new method that used cell-free extracts from ground liver. This method was based on work using murine liver enzymes that had been established by Knauss et al. (1959). Endo also exploited a screening technique designed by Heller and Gould (1973) that improved the selectivity of HMG-CoA reductase.[7]

Based on these scientific advancements, Endo scaled down and improved the efficiency of the existing methods, which allowed him to screen over 6000 naturally occurring substances for inhibitory effects against cholesterol synthesis in 2 years. Endo's original goal had been to complete the discovery program within 2 years; thus, it was necessary to perform a large amount of screening in a short time. To achieve this goal and efficiently measure inhibition of cholesterol synthesis, Endo exploited the latest research tools and drastically improved existing experimental methods.

The availability of radioactive HMG-CoA and radioactivity measuring devices were also of great importance for the statin discovery program. Radioactivity measuring equipment (like that from Packard Co., Ltd.) quickly became popular research tools in Japan; two sets were purchased for the Sankyo laboratory, and Endo's research team had access to these facilities. It was also essential for the research team to purchase radioactive HMG-CoA from companies in the United States. The discovery project would not have been possible without access to such equipment

---

[5]"A detailed knowledge of the mechanisms of lipid metabolism was necessary to deal with these medical problems in a rational manner. The importance of the work of Bloch and Lynen lies in the fact that we now know the reactions which have to be studied in relation to inherited and other factors." In Nobel Prize Award Ceremony Speech (1964) Nobelprize.org. http://www.nobelprize.org/nobel_prizes/medicine/laureates/1964/press.html. Accessed: 16 Sep 2012.

[6]Much of the endogenous cholesterol is produced in the liver. See review by Dietschy and Wilson (1970).

[7]In an interview with Dr. Endo he said, "The grinded liver tissues contain at least 30 enzymes that are needed for cholesterol synthesis. It was very difficult to extract only HMG-CoA from such mixture. Heller and Gould showed that such a task could be almost done, although the extraction of pure HMG-CoA had been impossible at that time."

and materials. Thus, Sankyo's ability to provide Endo's team with modern research facilities and a flexible budget was another major contributor to the success of his discovery research in the 2-year period.

2.8   Responses to Discontinuation of the Clinical Trial

As noted above, the clinical trial of compactin was permanently discontinued in the middle of the Phase II study in August 1980 because of suspicions of carcinogenicity, which arose in canine long-term toxicity tests after the administration of extremely high dozes. As a repercussion of this Sankyo decision, Merck also suspended the development of lovastatin in the following month (September 1980). However, after an interruption of almost 4 years, Merck resumed the development of lovastatin. The scientific community played a major role in encouraging Merck to resume this clinical trial through the implementation of university-based clinical research for critically ill patients and an improved understanding of the mechanism of action.

According to Dr. P. Roy Vagelos, Merck's director of research at the time, after discontinuing the clinical trials of lovastatin, the Merck team focused on a discovery project for a new drug targeting HMG-CoA reductase that had a different structure from compactin (Vagelos and Galambos 2004). However, this was very difficult work and only led to the discovery of simvastatin, a modified version of lovastatin (patent filed: August 1980).

In September 1981, Prof. Hiroshi Mabuchi at Kanazawa University reported the results of a clinical study of compactin in FH-heterozygote patients in the *New England Journal of Medicine* (Mabuchi et al. 1981). The study demonstrated that the drug was highly effective with minimal side effects, which drew worldwide attention. Subsequently, Merck and the US Food and Drug Administration (FDA) approved the resumption of a clinical research by researchers in Oregon and Texas who wished to treat patients with high-risk familial cholesterolemia in July 1982. The study was completed in collaboration with Brown and Goldstein and took approximately 1 year to complete; it demonstrated the cholesterol-lowering effects of lovastatin and the absence of serious side effects.

In addition, Brown and Goldstein's characterization of the mechanism of action of statin was critical for the resumption of the clinical development of lovastatin by Merck. According to their accounts, Brown and Goldstein persuaded Merck to resume the clinical development of lovastatin by explaining the mechanism-based theory behind statin's selective reduction of LDL levels in blood without causing toxicity.

2.9 "Invisible" Discovery Competition

Researchers at Beecham Laboratories independently discovered ML-236B from a different blue mold (*Penicillium brevicompactum*) in their discovery program for antibiotics, which was published as an academic paper in July 1976 (Brown et al. 1976). The first statin was thus known as "compactin" because the paper was published earlier than Endo's work. However, the patent application for ML-236B (compactin) was filed in June 1974, more than 2 years earlier than the Beecham paper, and the patent application (Japanese and Belgian) was published in December 1975. For this reason, Endo's invention did not lose priority to Beecham's work, allowing Sankyo to acquire the patents globally, including the product patent in the United States. Notably, Beecham also evaluated the cholesterol-lowering effect of the compound and concluded that there was no such effect because there was no lowering of cholesterol in rats.

2.10 Compactin Is the Basis for All Statins

After the discovery of compactin, drugs based on the compactin structure were developed worldwide and functioned as competitive inhibitors of HMG-CoA reductase using the same mechanism; these drugs are collectively referred to as "statins" (Fig. 2.2). Statins as a group include compounds that were discovered from natural sources (compactin and lovastatin), compounds obtained by modifying naturally occurring statins (pravastatin and simvastatin), and fully synthesized compounds (fluvastatin, atorvastatin, pitavastatin, and rosuvastatin).

As the pioneer of statins, compactin has had a major knowledge spillover effect on the discovery and synthesis of the subsequent statins. Although the first-in-class statin was Merck's lovastatin, its discovery was later than the discovery of compactin by approximately 5 years. According to Vagelos and Galambos (2004), their laboratory screened both a synthesized chemical and microbiological library in the search for inhibitors of HMG-CoA reductase; however, priority was given to the former. Following the publication of the compactin patent application (December 1975), Merck offered to collaborate with Sankyo on the assessment of compactin. As a result of this collaboration, Merck received crystals of compactin from Sankyo in January 1976,[8] and also obtained the critical information that compactin was ineffective in rats but effective in dogs. The following year (1977), Merck developed a new screening technique for its microbial exploration program, which led to the

---

[8] Sankyo cooperated with Merck in the hope that Merck would become a development partner in the United States, but Merck continued their search of statins in-house. Sankyo also provided compactin crystals to Sandoz (Novartis), Warner-Lambert (Pfizer), and Eli Lilly, but these companies did not show interest in the drug because of the lack of efficacy in rats (Endo 2006b).

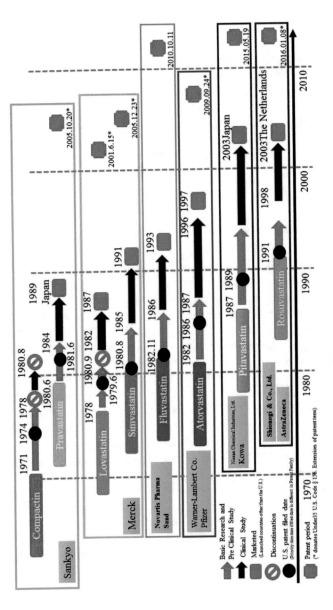

Fig. 2.2 Research and development timeline for statins. *Source* authors

discovery of lovastatin from microbes in Spanish soils in November 1978.[9] The new compound (lovastatin) was isolated in February 1979, and a patent application was made in June 1979.

Merck's simvastatin in turn was a derivative of lovastatin and succeeded lovastatin. While pravastatin (Mevalotin) was the first statin marketed in Japan and became the second statin in the world, it is a metabolite of compactin (direct derivative) and was discovered by Sankyo (see Chap. 3 for details). Thus, compactin was the invention essential for the discovery of these two highly successful statins.

In November 1982, 6 years after the publication of the compactin patent, Sandoz (currently Novartis) filed a patent application for the first synthetic statin: fluvastatin. The development of fully synthetic statins continued with the release of atorvastatin (Lipitor), pitavastatin, and rosuvastatin. All of these chemically synthesized statins were designed to include the fundamental part of the chemical structure of compactin as well as its mechanism of action. When the research team at Warner-Lambert (Pfizer), who were responsible for Lipitor, began their research and development program in 1982, they modeled their drug after both compactin and lovastatin. According to their accounts (Shook 2007), "The first thing we did was to study the published data on Sankyo's compactin. By this time, Merck had also filed its patent on Mevacor... we carefully studied all of that information, and then we set off to see if we could come up with something that would be different but would have similar kinds of activity in the body."

Thus, the discovery of compactin, and the efforts that went into its development laid the groundwork for the discovery and development projects of subsequent statins. The discovery of compactin revealed that there was a chemical structure that competitively inhibits HMG-CoA reductase, and that there were significant species-specific differences in the inhibitory effects of statins. All of the subsequent statin research and development efforts exploited such information.

Merck resumed clinical development of statin earlier than Sankyo, and lovastatin was the first statin to receive FDA approval. In the process, Merck thoroughly examined the drug's toxicity, including its potential carcinogenicity, and the safety of lovastatin, and the results were published in academic journals (MacDonald et al. 1988). Therefore, Merck's research efforts also had a positive spillover effect on the development of subsequent statins.

2.11   Statin Sales

Globally, many pharmaceutical companies entered the R&D competition for new statin development, following Sankyo and Merck. By 2014, a total of eight statins had been introduced to the global market, although cerivastatin was discontinued in

---

[9]Around the same time (January 1979), Prof. Endo discovered the same compound and he named it monacolin K. The patent application was filed earlier by Endo, and in the first to file countries (most countries) Endo's patent was established.

2001 because of safety issues. In terms of global statin sales, the long-term trends are shown in Fig. 2.3. Note that only non-generic statin sales are shown; data from generic drug manufacturers and pharmaceutical companies in developing countries where patent rights are not well protected are not included. In addition, nominal dollar sales were converted into real sales using the Consumer Price Index of the USA. Notably, global sales grew rapidly over the 18 years from the introduction of the first statin to the market in 1987–2005. During this period, drugs produced by Merck (lovastatin and simvastatin) and Sankyo (pravastatin) , which had the earliest market entry during the first decade, occupied most of the market. With the introduction of simvastatin, Merck overcame the constraints on the patent coverage of lovastatin, which was limited to a few countries including the United States, and avoided competitive disadvantages with pravastatin in its global development.

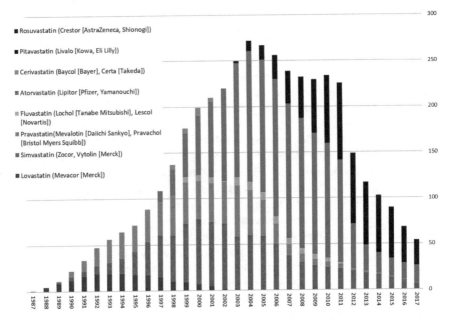

Fig. 2.3 Trends in real statin sales (1987–2017; 1M USD per unit). *Sources* Pharma Future (Uto Brain Division, Cegedim Strategic Data) [1999–2012], excerpted from company history and press releases, etc. [1987–1998, 2013–2017]; based on US Consumer Price Index [medical care] Annual Value, prices were adjusted as 1987 = 100 (For lovastatin, sales by Merck. For simvastatin, the sales of the following products by Merck: Zocor, Lipovas, and Vitrin, a combination drug with ezetimibe. For pravastatin, the sum of Mevalotin from Daiichi Sankyo and pravastatin from Bristol-Myers Squibb. The figures for fluvastatin are the sum of Lochol from Mitsubishi Tanabe Pharma and Lescol from Novartis. For atorvastatin, Pfizer, Astellas Pharma and Yamanouchi Pharmaceutical provided sales of Lipitor, Rivazo Alumina Inc. and Caduet, a combination of Amlodipine by Pfizer. For cerivastatin, Bayer's Baycol sales are summed with Takeda Chemical's Certa sales. For pitavastatin, sales of Livalo by Kowa Pharmaceutical, Daiichi Sankyo and Eli Lilly are shown. Rosuvastatin is the sum of sales of Crestor by AstraZeneca and Shionogi)

Subsequent growth in the non-generic market for statins was supported by Pfizer's marketing of atorvastatin (Lipitor). Atorvastatin was fully synthesized and is much stronger than conventional statins; thus, it is often referred to as a "super statin." In the late 2000s, another super statin, rosuvastatin (Crestor), was synthesized by Shionogi and developed by AstraZeneca. Of note, these super statins were discovered and developed by companies other than the two early entrants (Sankyo and Merck) . This suggests that the early disclosure of discoveries through patent system and the involvement of diverse R&D capabilities have enhanced the performance of industry innovation.

2.12 Conclusion

Within this chapter, we identified the following four major factors that contributed to the discovery of compactin: uniqueness of the discovery program, adoption of a highly efficient research method that supported an uncertain discovery program, the courage and persistence to overcome unexpected difficulties and the biases of conventional wisdom, and a high degree of research freedom. First, the discovery project for compactin was unique in that it screened for inhibitors from natural products, such as molds and mushrooms, while targeting HMG-CoA reductase, an enzyme that had recently been identified in the scientific research on the mechanism of cholesterol biosynthesis. Notably, this was unlike the strategy employed by Merck, which focused on a chemical library screen. This strategy was conceived by Endo and was based on his awareness that he would only be able to compete with US researchers by developing a novel approach.

Second, Endo significantly improved the screening methods, which enabled him to screen over 6000 active substances of microbial origin within 2 years. This new technique allowed Endo to undertake a discovery program with high uncertainty, for which Endo himself had set a 2-year completion deadline at the start of the discovery project. While Endo held only a college degree when he joined Sankyo (he would later earn a Ph.D. based on his research at Sankyo), he built up the capability to conduct cutting-edge research through the experience of in-house research and independent learning.

Third, Dr. Endo had the courage and persistence to overcome the unexpected difficulties and biases of conventional wisdom that could have killed his project. Most importantly, the apparent absence of an inhibitory effect in rats did not lead him to abandon the project; instead, he embarked on the preclinical research by his team, which ultimately enabled him to discover that there exist important differences in the effects of compactin among species. As discussed within this chapter, other pharmaceutical firms had not shown interest in compactin or had discontinued their own research, because they followed the conventional wisdom that the results of the rat trials meant the drug was ineffective in vivo. Endo did not follow this conventional thinking.

Fourth, Sankyo gave Endo significant freedom of research to engage in a highly uncertain research project and provided him with a research environment that had substantial resources, including the use of modern research instruments and the purchase of novel research reagents. Notably, this coincided with Sankyo's initiation of full-scale R&D activities through the establishment of the central laboratory, although Endo was not a member of the main research group focusing on antibiotics.

There are three important implications from the compactin discovery project. First, it demonstrates that while the progress in underlying science creates opportunities for discovery projects for breakthrough drugs, such projects almost inevitably face unexpected difficulties in the early stages because of the absence of a complete scientific understanding. Brown and Goldstein discovered LDL receptor in the same year (1973) that compactin was discovered, indicating that the Endo's discovery program began while the scientific understanding was still significantly incomplete with respect to the control mechanism of cholesterol metabolism. Therefore, Endo encountered a number of unexpected difficulties, in particular, the absence of an inhibitory effect in rats. There was also the risk of discontinuation at Sankyo in the face of hepatotoxicity concerns. Furthermore, there were widely held concerns about the safety of lowering cholesterol because the mechanism of action was initially unknown. Merck had similar concerns.

Second, cooperation of industry and academia played key roles in solving these unexpected difficulties and concerns. There were two major scientific advances that were the keys to the development of statins. One was the discovery by Brown and Goldstein of a regulatory mechanism of cholesterol metabolism mediated by the LDL receptor. This led to the discovery that statins can lower LDL levels selectively in the blood without affecting the liver cholesterol levels. The second advance was the result from a clinical study of critically ill patients with familial hypercholesterolemia that was conducted by Yamamoto and Mabuchi. The discovery of compactin itself also made significant contributions to the scientific community, because the drug could be used as a research tool. This cooperation was mediated through Endo's decision to publish his results in academic journals.

Third, the case of compactin demonstrates the existence of large knowledge spillover effects as a result of drug discovery, as well as the important contributions made by subsequent discoveries. As a pioneer, the discovery of compactin and the efforts to clinically develop the drug formed the basis for the discovery and development of all subsequent statins. The discovery of compactin revealed that a chemical substance could competitively inhibit HMG-CoA reductase, and that there were significant species-specific differences in cholesterol reduction. All subsequent statin discovery and development projects were conducted using this knowledge. However, because these statins did not infringe on the compactin patents, Sankyo was unable to benefit from the significant knowledge spillover effects. At the same time, subsequent statins also contributed new knowledge to the field, which furthered the innovation of statins, including the discovery and development of "super statins" with enhanced medicinal efficacy (see Chap. 4 on Crestor).

Finally, this case highlights the limitations faced by Japanese firms with regard to global clinical development, which significantly constrained their discovery-related commercial benefits. In addition to the pioneering discovery of compactin, three of the seven statins on the market were successfully discovered by Japanese pharmaceutical companies: pravastatin (Sankyo), pitavastatin (Kowa), and rosuvastatin (Shionogi). However, the first-in-class was lovastatin by Merck, and two of the best-in-class are atorvastatin (Lipitor) by Pfizer, and rosuvastatin (Crestor), which AstraZeneca developed after licensing it from Shionogi.

References

Brown, M. S., Faust, J. R., Goldstein, J. L., Kaneko, I., & Endo, A. (1978). Induction of 3-hydroxy-3-methylglutaryl coenzyme A reductase activity in human fibroblasts incubated with compactin (ML-236B), a competitive inhibitor of the reductase. *Journal of Biological Chemistry, 253*(4), 1121–1128.

Brown, M. S., & Goldstein, J. L. (1976). Receptor-mediated control of cholesterol metabolism. *Science New Series, 191*(4223), 150–154.

Brown, M. S., & Goldstein, J. L. (2004). A tribute to Akira Endo, discoverer of a "penicillin" for cholesterol. *Atherosclerosis. Supplements, 5,* 13–16.

Brown, A. G., Smale, T. C., King, T. J., Hasenkamp, R., Thompson, R. H. (1976). Crystal and molecular structure of compactin, a new antifungal metabolite from Penicillium brevicompactum. Journal of the Chemical Society. Perkin Transactions, *1*(11), 1165–1170.

Dietschy, J. M., & Wilson, J. D. (1970) Regulation of cholesterol metabolism. New England Journal of Medicin, *282,* 1241–1249.

Endo, A. (1976). Cholesterol-metabolic regulation in human cultured cells. *Biochemistry, 48*(6), 301–307.

Endo, A. (2006a). Gift from nature (in Japanese). Medical Review.

Endo, A. (2006b). Discovering a new drug statin-challenging cholesterol (in Japanese). Iwanami Library.

Endo, A., Kuroda M., & Tanzawa, K. (1976) Competitive inhibition of 3-hydroxy-3-methylglutaryl coenzyme A reductase by ML-236A and ML-236B fungal metabolites, having hypocholesterolemic activity. FEBS letters, *72,* 323–326.

Gallo, D. G., Harkins, R. W., Sheffner, A. L., Sarett, H. P., & Cox, W. M. (1966). The species specificity of cholestyramine in its effect on synthesis of liver lipids and level of serum cholesterol. *Proceedings of the Society for Experimental Biology and Medicine, 122,* 328–334.

Goldstein, J. L., & Brown, M. S. (2009). History of discovery: The LDL receptor. *Arteriosclerosis, Thrombosis, and Vascular Biology, 29*(4), 431–438.

Heller, R. A., & Gould, R. G. (1973). Solubilization and partial purification of hepatic 3-hydroxy-3-methylglutaryl coenzyme A reductase. *Biochemical and Biophysical Research Communications, 50*(3), 859–865.

Huff, J. W., Gilfillan, J. L., & Hunt, V. M. (1963). Effect of cholestyramine, a bile acid-binding polymer on plasma cholesterol and fecal bile acid excretion in the rat. *Proceedings of the Society for Experimental Biology and Medicine, 114,* 352–355.

Knauss, H. J., Porter, J. W., & Wasson, G. W. (1959). The biosynthesis of mevalonic acid from 1-C14-acetate by a rat liver enzyme system. *Journal of Biological Chemistry, 234,* 2835–2840.

Mabuchi, H., Haba, T., Tatami, R., Miyamoto, S., Sakai, Y., Wakasugi, T., et al. (1981). Effects of an inhibitor of 3-hydroxy-3-methylglutaryl coenzyme a reductase on serum lipoproteins and ubiquinone-10 levels in patients with familial hypercholesterolemia. *New England Journal of Medicin., 305,* 478–482.

MacDonald, J. S., Gerson, R. J., Kornbrust, D. J., Kloss, M. W., Prahalada, S., Berry, P. H., Alberts, A. W., Bokelman, D. L. (1988). Preclinical evaluation of lovastatin. American Journal of Cardiology, 62(5), 26–27.

Nakaya, Noriaki. (1988). HMG-CoA reductase inhibitor. Farumashia, 24(12), 1217–1219.

National Inventors Hall of Fame. (2012). ENDO AKIRA. http://www.invent.org/hall_of_fame/461. html. Accessed: January 29, 2014.

Renneberg, R. (2014). Statin drug, how did Dr. Endo change the lives of million heart attack patients? Textbooks on biotechnology (Blue Back). Kodansha. pp. 130–149.

Salami, J. A., Warraich, H., & Valero-Elizondo, J. (2017). National trends in statin use and expenditures in the US adult population from 2002 to 2013: Insights from the medical expenditure panel survey. JAMA Cardiol, 2(1), 56–65. https://doi.org/10.1001/jamacardio.2016.4700.

Sankyo (2000) History of Sankyo 100 year. Sankyo Co., Ltd.

Shook, L. R. (2007). Miracle medicines: Seven lifesaving drugs and the people who created them. Penguin Group: Portfolio.

Vagelos, R.P., & Galambos, L. (2004). Medicine, science and Merck. Cambridge University Press.

Yamamoto, A., Sudo, H., & Endo, A. (1980). Therapeutic effects of ML-236B in primary hypercholesterolemia. Atherosclerosis, 35, 259–266.

Chapter 3
Pravastatin (Pravachol, Mevalotin)

"Blockbuster" Statin Discovered from the Metabolites of Compactin

Yasushi Hara and Sadao Nagaoka

Abstract Pravastatin was a serendipitous discovery by Sankyo; its discovery occurred while the company was in the process of investigating compactin metabolism during its development process. Pravastatin became the second statin in the world in terms of commercialization (1989). Although the launch of pravastatin was delayed, it demonstrated hydrophilicity and organ selectivity unlike lovastatin and simvastatin, which enabled its global success. Indeed, in 2010, pravastatin was used clinically in 115 countries. However, there were a number of challenges in the commercialization process of pravastatin. In particular, it required the establishment of a new two-step fermentation production process. Pravastatin also promoted cooperative research between industry and academia, which played an important role in the elucidation of the action mechanism of statin, including the effect of the hydrophilic nature of pravastatin, as well as its therapeutic value. Despite the pioneer status of Sankyo, lovastatin by Merck became the first statin in the world (1987). Although Merck also suspended the clinical development of lovastatin following the Sankyo decision to discontinue the compactin project, it resumed the development sooner because of the higher willingness of the US market to try new innovative drugs. The resumption of the suspended clinical trial of lovastatin in the USA occurred at the request of clinical researchers. An educational program by the NIH pointing out the importance of lowering cholesterol to prevent heart disease began in 1985.

Y. Hara (✉)
CEAFJP/EHESS, Paris, France
e-mail: yasushi.hara@r.hit-u.ac.jp

Faculty of Economics, Hitotsubashi University, Tokyo, Japan

S. Nagaoka
Tokyo Keizai University, Tokyo, Japan

© Springer Nature Singapore Pte Ltd. 2019
S. Nagaoka (ed.), *Drug Discovery in Japan*,
https://doi.org/10.1007/978-981-13-8906-1_3

3.1 Introduction

Mevalotin (pravastatin) was identified by Sankyo Co., Ltd. (now Daiichi Sankyo) as an active compactin metabolite in canine urine during the compactin drug development phase. As with most statins, pravastatin is indicated for the treatment of hyperlipidemia and familial hypercholesterolemia (FH). It was the first statin to be launched in Japan (1989) and the second in the United States (1991), where it followed Merck's lovastatin (1987). Pravastatin quickly became a global blockbuster drug with annual sales of approximately 5 billion USD in 2003.

In 1980, Sankyo filed a patent application for pravastatin after selecting it as a drug development candidate. Preclinical studies confirmed its efficacy in 1981, the Phase I clinical trial was conducted in 1984, and was subsequently followed by the Phase II clinical trial in 1985, which was completed in 1986. The Phase III clinical trial began in December 1986, and it was completed in September 1987. Around this time, a two-step fermentation process was developed for the efficient mass production of pravastatin. In 1989, pravastatin was launched in Japan with the product name "Mevalotin," and in the United States in 1991 by Bristol-Myers Squibb as "Pravachol."

3.2 Characteristics of Pravastatin

Pravastatin has the following two characteristics that differentiate it from Merck's lovastatin and simvastatin, which were also discovered and developed during the 1980s and early 1990s: (1) high water solubility, and (2) tissue specificity for the hepatic system. Of the three earliest statins, pravastatin has the lowest octanol–water partition coefficient.[1] Given its high water solubility, drug interactions via cytochrome P450s (CYPs) are unlikely to occur (Ishigami and Yamazoe 1998). Pravastatin's hydrophilicity is considered to be a significant contributor to its global success. Indeed, Mevalotin is sold in 115 countries across the globe.

Pravastatin demonstrates liver tissue specificity, which is critical for statins to effectively reduce blood cholesterol and is a reflection of its hydrophilicity. In liver cells, statins are taken up by both active transport and passive diffusion mechanisms, whereas other organ tissues only take statins up using passive diffusion mechanisms. Given that pravastatin contains a hydroxyl group, it is highly hydrophilic and its uptake via passive diffusion is extremely low. Lovastatin and simvastatin, on the other hand, are highly hydrophobic and are taken up by the cells of other organs through passive diffusion (Sankyo 1991). Accordingly, lovastatin and simvastatin are highly potent inhibitors of cholesterol synthesis not only in the liver and small intestine, but also in other organs. In contrast, pravastatin strongly inhibits cholesterol synthesis in the liver and small intestine, but only very weakly in other organs. In

---

[1] See Table 4.1

addition, high hydrophilicity implies that pharmacokinetic fluctuations are unlikely to be caused by the use of pravastatin in combination with other drugs (Tsujita 2000).

3.3 Timeline of Research and Development for Pravastatin

The major events of the research and development (R&D) process for pravastatin were:

1972: Sankyo began research on statins (Sankyo 2002).

1973: Sankyo isolated ML-236B (compactin).

January 1979: Assistant Prof. Yoshio Watanabe, Kobe University, developed WHHL (Watanabe Heritable Hyperlipidemic) rabbits as an animal model for the study of arteriosclerosis.

November 1979: Sankyo presented the results of compactin (ML-236B) metabolism studies in rats and dogs at the "11th Drug Metabolism and Medicinal Toxicity Symposium" in Huston, TX in the United States.

June 1980: Pravastatin, a metabolite of compactin, was isolated and adopted as a development candidate (Sankyo 2002) and a patent application for pravastatin was submitted (Patent Publication No. 57-2240).

August 1980: Sankyo discontinued the development of compactin.

November 1980: Sankyo isolated fungi from Australian soil, which was later used for the two-stage fermentation for pravastatin.

1981: Sankyo initiated preclinical studies of pravastatin in dogs, cynomolgus monkeys, and Japanese white rabbits. The WHHL rabbits showed that pravastatin inhibits the progression of coronary arteriosclerosis.

1981: Professor Hiroshi Mabuchi, Kanazawa University, published a clinical trial of compactin in the *New England Journal of Medicine* (Mabuchi et al. 1981).

1982: The pravastatin patent was published.

March 1984: Sankyo initiated the Phase I clinical trial of pravastatin.

May 1984: Sankyo completed the Phase I clinical study of pravastatin.

1985: Sankyo decided to use SANK 62585 (a fungus isolated in Australia) for the two-step fermentative process for pravastatin manufacture.

April 1985: Sankyo started the pravastatin Phase II clinical study.

May 1985: Sankyo and Bristol-Myers Squibb signed a licensing agreement for the joint development of pravastatin.

April 1986: An open-label Phase II clinical trial for pravastatin was completed and was followed by a blinded trial in September 1987.

December 1986: Sankyo started the Phase III clinical trial of pravastatin.

September 1987: Sankyo completed the Phase III clinical trial of pravastatin and began production.

December 1987: The application for approval of pravastatin was filed.

February 1989: The WOSCOPS (West of Scotland Coronary Prevention Study) began; it would be completed in May 1995.

March 1989: Pravastatin received manufacturing approval.

August 1989: The official price of pravastatin was set in Japan.
October 1989: Pravastatin was launched as Mevalotin in Japan.
1990: Pravastatin was launched in Europe.
October 1991: Pravastatin was launched in the United States as Pravachol by Bristol-Myers Squibb.

3.4   Discovery of Pravastatin: From the Metabolites of Compactin

Prior to the development of pravastatin, Sankyo had been conducting clinical trials of compactin (see Chap. 1), the first cholesterol synthesis inhibitor. The search for inhibitors of cholesterol synthesis by Sankyo began in 1971, and, in 1973, the HMG-CoA reductase inhibitor ML-236B (compactin) was discovered. Through the preclinical investigations on compactin, the research team demonstrated that: (1) there are clear species differences in the hypocholesterolemic effect of compactin, and (2) drug efficacy testing in vitro and in vivo using a rat animal model is not consistent for this reason (Sankyo 1991).

In 1979, pravastatin, which is more powerful and demonstrates better target organ specificity than compactin, was discovered during the process of investigating drug metabolism for the development of compactin. Namely, pravastatin was an active urinary metabolite of compactin that was first detected in canine urine (Tsujita 2001). The results of compactin metabolism studies in rats and dogs were presented at the 11th Drug Metabolism and Medicinal Toxicity Symposium in November 1979. Subsequently, pravastatin was isolated and purified from canine urinary metabolites, and a patent application was filed in June 1980 (Tanaka 2008). Structural analysis revealed that pravastatin featured a hydroxyl group at the C6β position of the compactin decalin skeleton, and further characterization showed that the inhibitory activity of pravastatin was ten times stronger than that of compactin in vitro (Naito 2000).

Clinical trials of compactin were initiated in 1978, but its development was halted in 1980 after lymphoma-like symptoms were observed in the dogs that received extremely high doses of compactin in long-term toxicity studies. Accordingly, Sankyo selected pravastatin, which exhibited stronger activity and hepatocyte specificity, as a candidate for drug development (Naito 2000). Furthermore, the high water solubility and hepatocyte specificity were considered to contribute to the optimal safety profile of pravastatin.

3.5 Preclinical Studies Using WHHL Rabbits

Preclinical studies began in 1981 using WHHL rabbits to demonstrate the efficacy of pravastatin against atherosclerosis (Okuda 1991; Sankyo 1991). The WHHL rabbit model of arteriosclerosis was created by Assistant Prof. Yoshio Watanabe at the Faculty of Animal Experimentation, Kobe University School of Medicine. This model is characterized by blood cholesterol levels that are 10–20 times higher than normal rabbits from birth, ensuring all WHHL rabbits develop arteriosclerosis upon reaching maturity at the age of 5–6 months. The development of this animal model was fortuitous and dependent on Prof. Watanabe discovering a mutant rabbit with hyperlipidemia in 1973. Watanabe then began a breeding program with this rabbit, which led to the establishment of the WHHL rabbit model by 1979. Subsequently, a WHHL rabbit model that frequently suffers coronary atherosclerosis was established in 1985 (Shiomi and Ito 2009).

Incidentally, as Sankyo searched for an animal model that could be used for the development of pravastatin, Watanabe founded the first colony of WHHL rabbits. Thus, a collaboration began between Sankyo and Kobe University to use WHHL rabbits to investigate the anti-atherosclerotic effects of pravastatin. The researchers investigated whether 6 months of oral pravastatin at a dose of 50 mg/kg to WHHL rabbits aged 2–3 months with mild arteriosclerosis would inhibit the progression of arteriosclerosis. The results showed that the serum cholesterol levels were approximately 25% lower in the treatment group relative to controls, and significant inhibition of the progression of coronary artery lesions was also observed. In addition, the onset of xanthoma in the limb joints was significantly suppressed compared to the control group (Sankyo 1991). Furthermore, in 1988, pravastatin was shown to reduce cholesterol levels and slow the progression of coronary atherosclerosis in WHHL rabbits (Watanabe et al. 1988).

In the absence of the technology to create genetically modified animals, WHHL rabbits significantly contributed to the development of pravastatin as: (1) a model for spontaneous atherosclerosis, and (2) a model for FH. Throughout the 1980s, which corresponds to the period pravastatin trials were occurring, 115 papers pertaining to WHHL rabbits were published.[2] As shown in Table 3.1, WHHL rabbits were used not only by Kobe University and Sankyo, which were directly involved in the clinical research of pravastatin, but also by the other research institutions across the globe, including the University of Texas, where Brown and Goldstein were conducting research.

---

[2]Data source: Web of Science, Query: TS (topic name) = "WHHL rabbit."

Table 3.1 Author affiliations for all WHHL rabbit papers published between 1980 and 1990

Author-affiliated organizations	Number of records	Percentage of all records (%)
Kobe University	26	22.6
University of Texas	17	14.8
Kyoto University	15	13.0
National Cerebral and Cardiovascular Center	7	6.1
Sankyo Co., Ltd.	5	4.4
University of California, San Diego	5	4.4
Fukuoka University	4	3.5
Shinshu University	4	3.5
Washington University	4	3.5
Radboud University	3	2.6

Source Web of Science. Query: TS = "WHHL rabbit"

3.6 Development of a Two-Step Fermentation Production Process

The next step in the development of pravastatin was to establish an efficient production method for commercialization. Initially, a chemical synthesis method was considered; however, the yield rate was insufficiently low (Kuroda 1994). Therefore, Sankyo began researching the ability of microbes to produce pravastatin from compactin; this research theme was designated as "high priority," as Sankyo sought success quickly (Naito 2000). As a result of this R&D, a two-step fermentation process was adopted in which: (1) compactin was produced by the blue mold *Penicillium citrinum* SANK 11480, and (2) a hydroxyl group was added to the 6β position of compactin sodium salt using a CYP from the actinomycete *Streptomyces carbophilus* SANK 62585.

During the first fermentation process, the reaction conditions were controlled to achieve productivity on a factory scale. These conditions included: (1) maintenance of a productive and fine pellet form throughout the entire fermentation process, (2) control of the feed rate of sugar sources through the use of fuzzy computer control, which mimics the practices of highly skilled workers, and (3) increasing the oxygen uptake rate by removing compactin covering the surface of the mold. By implementing these controls, it became possible to reliably produce compactin using *Penicillium citrinum* SANK 18767.

In comparison, the establishment of the second fermentation process took considerably more effort to develop. To begin with, researchers screened Sankyo's fermentation laboratory microbial strain collection for the potential to produce pravastatin. This led to the initial selection of the mold *Mucor hiemalis* SANK 36372, and the actinomycete *Nocardia autotrophica* SANK 62781. However, these strains were not suitable for the production process because of problems with concentration levels of the conversion substrate, as well as the conversion rate. Therefore, an additional search was conducted with the aim of isolating *Streptomyces* sp. from oligotrophic soils, and this was based on Sankyo's experience with the microbial conversion of steroids and alkaloids. Accordingly, *S. carbophilus* SANK 62585 was isolated from a sample of Australian desert soil in November 1980 (Okazaki et al. 1989). Given that Australian soil has approximately only 25% of the moisture and 50% of the organic matter of Japanese soil, and is of neutral pH, the isolated strain was suitable for fermentation and had excellent conversion activity (Okazaki and Naito 1989).

Sankyo's final step in the development of the pravastatin two-step fermentation process was the construction of the production facility. To do this, Sankyo acquired land in Onahama, Fukushima Prefecture, and began factory construction in June 1984. The factory opened in March 1987, with the second stage of construction beginning in April 1987. The plant, which performed the fermentation and extraction of pravastatin, was equipped with state-of-the-art computer facilities and was completed in September 1988 (Sankyo 2002).

3.7 Five Years from Initiating Clinical Trials to the Approval for Human Use

The Phase I clinical trial began in March 1984, more than 5 years later than that of compactin (November 1978), and was completed in May 1984. A second Phase I study, this one double-blind and in hyperlipidemic patients, ran from September 1984 to January 1985 and was designed to determine the optimal dosage. This study was conducted at four sites: the University of Tokyo, Tokai University Hospital, Tokai University Oiso Hospital, and Keiyo University Hospital (Nakaya et al. 1988).

Following the success of the Phase I trials, open-label[3] and blinded[4] Phase II clinical trials were initiated in April 1985. The open-label Phase II clinical study was completed in April 1986 (CS-514 Study Group 1988a), while the blinded trial ended in September 1987 (CS-514 Study Group 1988b). Based on these results, a Phase III clinical trial was initiated in December 1986 and completed in September 1987. During this time, Sankyo began production of pravastatin.

Sankyo subsequently submitted an application for pravastatin approval in December 1987 and obtained the manufacturing approval in March 1989 (Goshima et al. 1988). Given the novelty of pravastatin, the economic evaluation was high, and the National Health Insurance (NHI) price was set at more than twice the price of existing drugs (Sankyo 2002).

---

[3] A clinical trial in which the patient's test group is known by the doctor, subject, and staff.

[4] A clinical trial in which the patient is randomly assigned to the treatment or placebo group. In a double-blinded study, neither the patient nor the physician is aware of the patient's group.

3.8   Global, Large-Scale and Long-Term Clinical Studies

To demonstrate the long-term therapeutic effect of pravastatin, large-scale clinical trials were conducted in ten countries around the world. Table 3.2 lists the characteristics of pravastatin clinical trials that were conducted outside of Japan up to 1998. Notably, a number of different clinical outcomes were observed, including the prevention of arteriosclerosis progression, coronary artery disease onset, and coronary artery disease recurrence, with a total sample size of 20,000 (Sankyo 2002).

Bristol-Myers Squibb conducted large-scale clinical trials abroad, including the WOS (West of Scotland coronary prevention) and LIPID (Long-term intervention with pravastatin in ischemic disease) studies. In Japan, Sankyo conducted a large-scale trial, the MEGA (management of elevated cholesterol in adults) study, which began in February 1994 and ended in March 2004 (Nakamura 2009). A total of 8214 patients were randomized to the MEGA study, 3966 in the diet group, and 3866 in the group that combined dietary intervention and pravastatin treatment. The results showed a 33.3% decrease in the cholesterol levels of patients treated with pravastatin relative to patients that only received the dietary intervention; thus, pravastatin was effective for the management of elevated cholesterol (Nakamura et al. 2006; Hamada 2009).

Table 3.2   Large-scale clinical studies outside Japan after the launch of pravastatin

Clinical trial name	Year of publication	Number of patients	Country
Arteriosclerosis progression			
PLAC-I	1995	408	United States
PLAC-II	1995	151	United States
REGRESS	1995	885	Netherlands
KAPS	1995	447	Finland
Coronary artery disease			
WOS	1995	6595	United Kingdom
Coronary angiopathy			
PMNS	1993	1062	United States
CARE	1996	4159	United States
LIPID	1998	9014	Australia

Source (Sankyo 2002)

3.9 Clinical Development Through Collaboration Between Industry and Academia

Cooperation between industry and academia was crucial to the preclinical development of pravastatin. Of note, the preclinical research conducted using WHHL rabbits showed that pravastatin inhibits the progression of xanthoma and is effective against coronary artery diseases. Furthermore, this work was useful for preclinical safety studies and characterizing pravastatin's efficacy relative to existing statins (Sankyo 2002).

During development of the two-step fermentation process, the collective experience and knowledge of microorganism metabolism that had been obtained by Sankyo from industry–academia joint research projects also played an important role. Essentially, the knowledge that Sankyo had accumulated internally was indispensable for establishing the strain collection, selecting the optimal strain, and designing the production process mechanism, including the computer control system that mimicked human skills. This accumulation of scientific knowledge began in 1956 when Sankyo collaborated with Prof. Kyosuke Tsuda of the University of Tokyo's Institute of Applied Microbiology on microbial transformation of steroids (Ikegawa 2000). Indeed, prior to the production of pravastatin, Sankyo had utilized the knowledge and the findings from these earlier studies to the microbial conversion of siccanin, an antifungal.

Regarding clinical studies, trials on Sankyo's compactin and Merck's lovastatin focused on patients with severe FH, and this work contributed to the development of a better understanding of the efficacy and safety of HMG-CoA reductase inhibitors. In particular, statin treatment of FH-heterozygotes, which was performed by Prof. Mabuchi of Kanazawa University, showed convincingly that statins (compactin) reduced plasma LDL (Mabuchi et al. 1981). In the same issue of the *New England Journal of Medicine* (1981), which published the above study, Professors Brown and Goldstein published an editorial, "Lowering plasma cholesterol by raising LDL receptors," which declared that statins were useful for the treatment of hyperlipidemia (Brown and Goldstein 1981) . This sentiment stimulated the resumption of lovastatin clinical trials by Merck, as well as the clinical development and the eventual launch of Mevalotin by Sankyo.

Given the hydrophilic nature of pravastatin relative to other statins, it less likely to cross the lipid bilayer of the plasma membrane, rendering it less likely to be taken up by cells. This means that pravastatin has weak capability of inhibiting cholesterol synthesis in organs other than the liver. However, hepatocytes actively take up pravastatin through organic anion transporters, allowing pravastatin to potently inhibit cholesterol biosynthesis in hepatocytes, which leads to an increase in LDL receptors in the liver and a resulting decrease in serum cholesterol. Understanding of this process was only possible through collaboration with Prof. Sugiyama (University of Tokyo), who, after the market launch, characterized the transportation of pravastatin through a specific transporter, and linked this to the tissue specificity of pravastatin.

3.10   Toward a Blockbuster Drug

The sale trends of pravastatin by Daiichi Sankyo and Bristol-Myers Squibb are shown in Fig. 3.1. In this figure, Sankyo's sales of Mevalotin are combined with Bristol-Myers Squibb's sales of Pravachol, which recorded net sales of 4.75 billion USD in 2003, just before the basic patent expired. Since the sale of generic drugs was approved in July 2003, the sales have declined. In particular, the sales of Pravachol declined sharply from 2.3 billion USD in 2005, to 1.20 billion USD in 2006. The launch of Lipitor and Crestor have also affected the sales of pravastatin (Tanabe et al. 2008). Notably, although sales in the Japanese market have also been decreasing since 2004, the decline has been slower than that in the United States market.

As stated above, pravastatin and other statins have been shown to significantly reduce mortality in studies of atherosclerotic progression, coronary artery disease onset, and coronary artery disease recurrence. In Japanese clinical trials, statins have also been shown to significantly reduce the incidence and mortality of cardiac accidents (fatal, nonfatal reinfarction, sudden cardiac death, and heart failure death) in patients with myocardial infarction (Furuta et al. 2002).

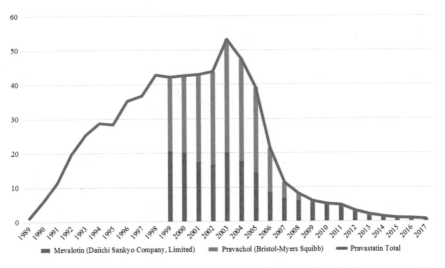

Fig. 3.1 Pravastatin sales between 1989 and 2012. The total pravastatin real sales are presented as the sum of the values of Pravachol and Mevalotin (unit 1M USD). *Sources* Cegedim Strategic Data K.K., Uto Brain Division "Pharma Future," Financial Report by Bristol-Myers Squibb from 1999 to 2008 and Daiichi Sankyo from 1999 to 2017

3.11   Partnership with Bristol-Myers Squibb

In 1985, Sankyo signed an agreement with Squibb (now Bristol-Myers Squibb) granting the company exclusive marketing rights in countries around the world except for Japan, Korea, Taiwan, and Thailand. Pravastatin was subsequently sold by Bristol-Myers Squibb as Pravachol. Pravachol was approved in Europe immediately after pravastatin gained approval in Japan (March 1989). In December 1989, Pravachol was approved in Iceland, and the United States followed suit in August 1991. As of 1992, pravastatin was sold in 26 countries across the globe (Sankyo 1991); this figure jumped to 115 countries as of 2011.[5]

Box 3.1 Statin Development Competition: Sankyo Versus Merck

Lovastatin is an HMG-CoA reductase inhibitor that Merck discovered and developed. Merck initiated a search for HMG-CoA reductase inhibitors from microorganisms, following the discovery of compactin by Sankyo, and their search led to the discovery of lovastatin in November 1978. Subsequently, crystals of lovastatin were isolated in February 1979, and a patent application was filed in June 1979 (Vagelos 1991). Interestingly, lovastatin was the same substance as Sankyo's patented monacolin K. The two firms discovered the compound almost simultaneously, and Sankyo had the patent to market monacolin K in a first to file country, while Merck had only the patent to market lovastatin in a first to invent country, such as the United States. Merck later discovered and developed simvastatin from lovastatin as a semi-synthetic statin, which was sold globally.

Merck established crystallization and fermentation production technology for statins in August 1979, and also started toxicity tests. Clinical trials were initiated in April 1980, but were discontinued in September 1980, after Sankyo discontinued the development of compactin because of suspicions of carcinogenicity, which arose in dogs from very high dose administrations in the long-term toxicity test.

One year after the discontinuation of lovastatin and compactin development in 1981, Prof. Mabuchi at Kanazawa University published research results on compactin in the *New England Journal of Medicine* (Mabuchi et al. 1981). His results indicated that the level of LDL cholesterol in a patient with familial hypercholesterolemia was lowered by the administration of compactin and that there were no serious side effects. Merck and the FDA approved the resumption of the clinical study in July 1982, under the request of clinical researchers from the University of Oregon and the University of Texas who wished to treat patients with high-risk familial hypercholesterolemia. Merck not only resumed the clinical trials earlier but also implemented comprehensive clinical studies in

---

[5]Source: Drug Interview Form: Pravastatin Sodium tablets Mevalotin tablets, a drug for treatment of hyperlipidemia, HMG-CoA reductase inhibitor.

a short period of time. Through a collaborative study with Professors Goldstein and Brown, Merck spent about a year demonstrating the cholesterol-lowering effects of lovastatin and the absence of serious side effects. Subsequently, a clinical trial of lovastatin by Merck was resumed in November 1983. The Phase II clinical studies were resumed in May 1984, and long-term toxicity studies in dogs demonstrated that lovastatin was not neoplastic in October 1986. Subsequently, a new drug application was filed with the FDA in November of the same year and approved in August 1987. The drug was launched as Mevacor.

Comparison of the development process of Sankyo's pravastatin and Merck's lovastatin show that after the discontinuation of compactin in 1980, Merck resumed the clinical trials of lovastatin in November 1983, while Sankyo started the Phase I clinical trial of pravastatin in March 1984. Afterwards, the approval application for lovastatin in the United States was submitted in November 1986, and that for pravastatin in Japan was submitted in December 1987. Pravastatin was delayed by 4 months in the initialization or resumption of Phase I clinical trial, and 13 months in approval applications compared with lovastatin.

The resumption of the clinical trial of lovastatin took place at the request of clinical researchers in the United States, as noted above, indicating the existence of strong unmet medical need for cholesterol-lowering drugs in the United States. In this context, one of the reasons that the market launch of statin occurred significantly earlier in the US than in Japan was likely the introduction of the Cholesterol Education Program in the United States. The program, introduced by the NIH in 1985, recommended lowering cholesterol levels to prevent heart disease (NIH 1984). Lovastatin, which was filed with the FDA in 1986, was thus positioned as a very important drug of interest in US healthcare policies. On the other hand, in Japan, clinicians were strongly concerned about the risk of adverse side effects from a new drug, given their past experiences with drugs such as thalidomide and chloroquine. They tended to follow the philosophy that "*strong powerful drugs are dangerous like sharp edges.*" Also, the number of familial hypercholesterolemia patients who needed statins in Japan was limited (Nakamura 2004). Thus, given the environment in Japan, Sankyo may have felt encouraged to shift to the development of pravastatin, abandoning compactin in an effort to seek a statin that was both safer and more effective. On the other hand, the development of lovastatin by Merck was supported by a nationwide policy program, which allowed it to pioneer the sale of statins.

Merck also developed simvastatin, which led to its global market expansion. Simvastatin, a semi-synthetic statin synthesized from lovastatin in 1979, demonstrated 2.5 times stronger inhibitory activity than lovastatin (Li 2009). In Japan, the Phase I clinical trial began in 1986, and the Phase III clinical trial began in 1988. Afterwards, simvastatin was approved under the brand name Zocor in Japan and the United States in 1991.

Because of the Sankyo patent, lovastatin could not be marketed in Japan, and on 1 August 1989, pravastatin became the first statin to be launched in Japan. Subsequently, simvastatin was approved as Lipovas in Japan in October 1991. Pravastatin was approved by the FDA on 31 August 1991 as Mevalotin, significantly later than lovastatin (Mevacor), which had been approved in 1987. However, pravastatin was approved 4 months earlier than simvastatin (Zocor), the follow-on product of lovastatin.

Simvastatin was first approved in Sweden in 1988, which was the first approved statin in Europe. It 1989 it was approved in 5 more countries: UK, France, Italy, the Netherlands, and New Zealand. In 1990, additional approval was given by 11 countries, including Australia, Germany, Switzerland, and Singapore; 6 more countries followed in 1991, and the USA in 1992.

3.12   Conclusion

The major characteristics of the pravastatin R&D process are summarized by the following four points. First, while Sankyo had led the world in statin development by the discovery of compactin, it became the second runner behind Merck in the United States and Europe after the discontinuation of compactin. Based on the experience of the discontinuation of the clinical development project of compactin, the clinical development of pravastatin emphasized the safety of statins, and the importance of large-scale and long-term clinical studies. Although its launch was delayed, pravastatin, which has hydrophilicity and organ specificity properties unlike lovastatin and simvastatin, became a highly successful medicine that is used all over the world.

Second, pravastatin was discovered unexpectedly in the process of investigating compactin metabolism for the purpose of drug development. Thus, its discovery was serendipitous.

Third, a new two-step fermentation process was established in a short period of time. Sankyo utilized internal knowledge that had accumulated during collaborative studies with the Institute of Applied Microbiology at the University of Tokyo.

Fourth, given that the mechanism of action of statins and their therapeutic value were not yet fully understood, Sankyo's industry–academia collaborations played an important role in promoting these scientific advancements. Indeed, industry–academia cooperative research, which included both basic scientists and clinical researchers, played an important role in elucidating the action mechanism of statins through LDL receptors, including the effect of the hydrophilic nature of pravastatin. Furthermore, animal models of hyperlipidemia (WHHL rabbits) were also crucial to the preclinical tests of pravastatin and characterization of the mechanism of action.

Lovastatin became the first statin in the world when it was marketed in the United States and Merck also launched the first statin (simvastatin) in Europe. Why did Sankyo, a pioneer in the field and with strong patent positions, end up being second? First, there were significant differences in the willingness to try new innovative drugs such as statins, between the US and Japanese markets at that time. In the USA, the resumption of the suspended lovastatin clinical trial occurred at the request of clinical researchers, which coincided with an NIH educational program that stressed the importance of lowering cholesterol to prevent heart disease. On the other hand, Japanese medical professionals, at the time, were hesitant to introduce a new powerful drug. This difference of the two markets mattered because Merck and Sankyo developed statins in their home markets first (because of limited clinical development capabilities of Sankyo). Second, Merck not only resumed the clinical trials earlier but also implemented comprehensive clinical studies in a short period of time. Furthermore, it synthesized and developed simvastatin early to capture the non-US market. In comparison, Sankyo switched from compactin to pravastatin, which it first launched only in Japan, while depending on Bristol-Myers Squibb for clinical development in overseas markets.

References

Sankyo Co., Ltd. (1991). *Development of pravastatin for treatment of hyperlipidemia (in Japanese)* (pp. 8–17). Special Prize for Okochi Memorial Production.

Brown, M. S., & Goldstein, J. L. (1981). Lowering plasma cholesterol by raising LDL receptors. *New England Journal of Medicine, 305,* 515–517.

CS-514 Study Group. (1988a). Clinical efficacy of CS-514 (Pravastatin) for hyperlipidemia–multiclinic open study results-. *Clinician Pharmaceuticals, 4*(3), 409–437.

CS-514 Study Group. (1988b). Examination of the clinical usefulness of long-term administration of CS-514 (Pravastatin) for hyperlipidemia. *Clinician Pharmaceuticals, 4*(2), 201–227.

Editorial Board of Sankyo 100 Years. (2002). *Sankyo 100 years.* Sankyo Co., Ltd.

Furuta, K., Kimura, A., Miyadaka, S., & Ishikawa, Y. (2002). Secondary prevention of myocardial infarction with HMG-CoA reductase inhibitors. *Medical Journal of Kindai University, 27*(1), 17–26.

Goshima, Y., Yamamoto, A., Matsuzawa, Y., Nakaya, N., Hata, Y., Kita, T., et al. (1988). Evaluation of the clinical usefulness of pravastatin (CS-514) in hyperlipidemic patients a double-blind group comparative study using probucol as a single agent. *Medical History, 146*(13), 927–955.

Hamada, C. (2009). Role of statistics in healthcare (<Special Feature> increasing efficiency in healthcare). *Operations Research: Science of Management, 54*(7), 385–389.

Ikegawa, N. (2000). Research activity of Prof. Kyosuke Tsuda. *Journal of the Pharmaceutical Society of Japan, 120*(10), 817–824.

Ishigami, M., & Yamazoe, Y. (1998). Drug interactions of HMG-CoA reductase inhibitors involving cytochrome P450 (in Japanese). *Progress in Medicine, 18*(5), 972–980.

Kuroda, M. (1994). 107: Cholesterol-lowering agent, mevalotin. *Summary of Lectures of the Japanese Society for Biotechnology, 1994,* 8.

Li, J. J. (2009). *Triumph of the heart, the story of statins.* Oxford: Oxford University Press.

Mabuchi, H., Haba, T., Tatami, R., Miyamoto, S., Sakai, Y., Wakasugi, T., et al. (1981). Effects of an inhibitor of 3-hydroxy-3-methylglutaryl coenzyme a reductase on serum lipoproteins and ubiquinone-10 levels in patients with familial hypercholesterolemia. *New England Journal of Medicine, 305*(9), 478–482.

Naito, A. (2000). A half-century in the study of microbial transformation. *Journal of the Pharmaceutical Society of Japan, 120*(10), 839–848.

Nakamura, K. (2004). A unique cholesterol-lowering agent that no one had ever had before. *Atherosclerosis Supplements, 5,* 19–20.

Nakamura, Y. (2009). MEGA study (large-scale study of hyperlipidemia). *Journal of Geriatrics, 46*(1), 18–21.

Nakamura, H., Arakawa, K., Itakura, H., Kitabatake, A., Goto, Y., Toyota, T., et al. (2006) MEGA study group. Primary prevention of cardiovascular disease with pravastatin in Japan (MEGA Study): a prospective randomised controlled trial. *Lancet, 368*(9542), 1155–1163.

Nakaya, N., Homma, Y., Tamachi, H., Goto, Y., Shigematsu, H., & Hata, Y. (1988). Phase I clinical study of the antihyperlipidemic drug CS-514: Comparison of three doses by double-blind method in hyperlipidemic patients. *Clinician, 4*(2), 167–189. in Japanese.

NIH. (1984). *National Institutes of Health issued the statement on Lowering Blood Cholesterol to Prevent Heart Disease in Consensus Development Conference in December 1984.* https://consensus.nih.gov/1984/1984cholesterol047html.htm. Accessed: Jan 20 2014.

Okazaki, T., Enokita, H., Miyaoka, H., Otani, H., & Torinaga, A. (1989). *Annual Report of Sankyo Research Laboratories, 41,* 123–133.

Okazaki, T., & Naito, A. (1989). *Annual Report of Sankyo Research Laboratories, 38,* 80–89.

Okuda, S. (1991). Pravastatin R&D awarded by the Japan Society for drug Research. *Pharmacia, 27*(5), 459.

Shiomi, M., & Ito, T. (2009). The Watanabe heritable hyperlipidemic (WHHL) rabbit, its characteristics and history of development: A tribute to the late Dr. Yoshio Watanabe. *Atherosclerosis, 207*(1), 1–7.

Tanabe, K., Takeuchi, M., Ikezaki, T., Kitazawa, H., Toyomoto, T., Nakabayashi, T. (2008). Assessment of therapeutic equivalence of original and generic preparations of pravastatin sodium (Mevalotin vs. Mevan): A retrospective study. *Japanese Society of Pharmaceutical Health Care and Sciences, 34*(4), 347–354.

Tanaka, M. (2008). *Statin for the treatment of hyperlipidemia: Pravastatin and its science and R&D strategy. Case of drug discovery 20* (pp. 173–185). Maruzen Co., Ltd.

Tsujita, Y. (2000). Mevalotin for the treatment of hyperlipidemia (a drug developed in Japan) (in Japanese). *Cardiologists: Japanese Circulation Society Specialist Journal, 8*(1), 143–150.

Tsujita, Y. (2001). Development of Mevalotin antihyperlipidemic drug (drug discovery representing the 20th century) (<Special Feature> 21st century: Pharmaceutical Era) (in Japanese). *Farumashia, 37*(1), 20.

Vagelos, P. R. (1991). Are prescription drug prices high? *Science, 252,* 1080–1084.

Watanabe, Y., Ito, T., Shiomi, M., Tsujita, Y., Kuroda, M., Arai, M., et al. (1988). Preventive effect of pravastatin sodium, a potent inhibitor of 3-hydroxy-3-methylglutaryl coenzyme A reductase, on coronary atherosclerosis and xanthoma in WHHL rabbits. *Biochimica et Biophysica Acta, 960,* 294–302.

Chapter 4
Rosuvastatin (Crestor)

"Super Statin" That Became a Global Blockbuster Despite Its Late Entry

Yasushi Hara, Sadao Nagaoka and Koichi Genda

Abstract Rosuvastatin (Crestor) is a fully synthetic statin that was discovered by Shionogi and commonly referred to as a "super statin," because of its high potency, together with atorvastatin (Lipitor). Relative to the earlier statins, rosuvastatin was a very late entrant into the market. However, Crestor became a highly successful global blockbuster drug. In the global competition to discover synthetic statins with stronger efficacy than the statins from natural products, Shionogi succeeded by taking advantage of the company's wealth of experience and expertise in the field chemical synthesis, particularly with respect to sulfonamide additions. The superior characteristics of rosuvastatin are well illustrated by the ability of Crestor to expand and maintain its sales, despite the entry of generic drugs to the statin market following the expiration of patents for earlier statins. Crestor was launched only because AstraZeneca was able to resume its clinical development after Shionogi abandoned it. In doing so, AstraZeneca expanded the scale and diversity of clinical trials with the intention of a global launch of rosuvastatin. It also engaged in extensive comparative clinical trials with leading statins. This case illustrates the critical role of the complementary asset for turning a great drug invention into a successful innovation as well as the importance of combining capabilities across firms for such an objective.

Y. Hara (✉)
CEAFJP/EHESS, Paris, France
e-mail: yasushi.hara@r.hit-u.ac.jp

Faculty of Economics, Hitotsubashi University, Tokyo, Japan

S. Nagaoka
Tokyo Keizai University, Tokyo, Japan

K. Genda
Shionogi Seiyaku Kabushiki Kaisha, Osaka, Japan

© Springer Nature Singapore Pte Ltd. 2019
S. Nagaoka (ed.), *Drug Discovery in Japan*,
https://doi.org/10.1007/978-981-13-8906-1_4

4.1 Introduction

Rosuvastatin (Crestor) is a wholly synthetic statin and, together with Lipitor (ator-vastatin), is commonly referred to as a "super statin" because of its high potency. Indeed, the first-generation statins, pravastatin (Mevalotin), lovastatin (Mevacor),[1] and simvastatin (Lipovas), were derived from natural products and were not nearly as potent as rosuvastatin. Rosuvastatin has the following advantages when compared to atorvastatin (Lipitor), which was the best-in-class statin in the early 2000s: (1) an LDL (low-density lipoprotein) cholesterol-lowering effect that is equivalent to that of atorvastatin at one-quarter of the dose (see Table 4.1), (2) improved liver specificity, which promotes potent inhibition of HMG-CoA reductase in the liver and weak inhibition at other tissue sites, (3) it is more hydrophilic, and (4) decreased likelihood and severity of side effects because of drug combinations.

One of the key contributors to the success of the research and development (R&D) of rosuvastatin by Shionogi & Co., Ltd. (Shionogi) was that the company took advan-tage of its unique experience in chemical synthesis and applied it to the global search for a highly potent, synthetic statin that was more effective than statins derived from natural products. This allowed the drug to achieve peak sales levels above those of simvastatin and pravastatin, even though it was launched when there was already significant competition in the market, including that from generic drugs that were introduced to the market following the expiration of earlier statin patents. However, it is important to note that despite the excellent characteristics of the drug candidate, Shionogi suspended its development during the Phase II clinical trial; it was only launched with the collaboration with AstraZeneca.

4.2 Timeline of Rosuvastatin Synthesis and Development

The following are the major events that occurred between the discovery and launch of rosuvastatin:

Discovery and development of rosuvastatin by Shionogi
1990: Shionogi began the discovery program for a new statin.
1991: Successful synthesis of rosuvastatin, a patent application was filed on 1 July 1991, titled "Pyrimidine derivative."
1992–1994: Long-term toxicity studies were conducted in rats, dogs, and rabbits.
1993: Shionogi conducted Phase I clinical studies in Japan.
1995: Shionogi conducted Phase IIa clinical studies in Japan, and subsequently dis-continued development.

Rosuvastatin development by AstraZeneca
1998: AstraZeneca took over the development of rosuvastatin.

---

[1] Trade name in the USA.

Table 4.1 Properties of commercially available statins

Generic name	Product name	Origin	Octanol–water partition coefficient	Biological availability (%)	LDL-cholesterol reduction rate (%)				Absorption (%)	Half-life ($T_{1/2}$)	Time to peak plasma concentration (T_{max})
					10 mg	20 mg	40 mg	80 mg			
Lovastatin	Mevacor	Synthetic	1.70 lipophilic	<5		26	31	40	31	2.5–3.0	2.8
Pravastatin	Mevalotin/ Pravachol	Semi-synthesis (microorganism-derived)	−0.84 hydrophilic	17	20	24	30	–	37	0.8–3.0	0.9–1.6
Simvastatin	Lipovas/ Zocor	Semi-synthesis (microorganism-derived)	1.60 lipophilic	<5	26	35	39	46	65–85	1.9–3.0	1.3–2.4
Fluvastatin	Lescol	Synthetic	1.27 lipophilic	10–35		22	25	36	98	0.5–2.3	0.5–1.5
Cerivastatin	Baycol	Synthetic	1.69 lipophilic	60	25 (0.2 mg)	31 (0.3 mg)	35 (0.4 mg)	41 (0.8 mg)	>98	2–4	2
Atorvastatin	Lipitor	Synthetic	1.11 lipophilic	12	37	43	48	51	30	11–30	2.0–4.0
Pitavastatin	Livalo	Synthetic	1.49 lipophilic	>60	32.9 (1 mg)	39.7 (2 mg)	46.3 (4 mg)		50	11	0.5–0.8
Rosuvastatin	Crestor	Synthetic	−0.33 hydrophilic	20	46	52	55	–	50	20	3

Source Octanol–water partition coefficients (Yamamoto 2005) [lovastatin (Ohfuji et al. 2005), pitavastatin (medical interview form Livalo in Japan)]. LDL-cholesterol reduction rate (Brewer 2003) [pitavastatin (medical interview form Livalo in Japan)] obtained from in-house data at 8 weeks post-dose]. Origin, absorption rate, time to peak plasma concentration, bioavailability, and half-life obtained from Mukhtar and Reckless (2005)

November 1998: AstraZeneca began preclinical studies in rabbits and rats, which were completed in 1999.

January 1999: AstraZeneca began a late Phase II clinical trial comparing rosuvastatin with atorvastatin; this was completed in June 2000.

April 1999: A double-blind, late Phase II clinical trial was started by AstraZeneca and completed in December 1999.

June 1999: AstraZeneca began a late Phase II clinical trial comparing rosuvastatin to pravastatin and simvastatin; it was completed in April 2000.

April 2001: A Phase III clinical study of rosuvastatin was started by AstraZeneca.

2002: Shionogi submitted an "Application for Import Approval" in Japan.

November 2002: Rosuvastatin approved and launched in the Netherlands, which was the secretariat for the network of European Mutual Certification.

2003: FDA approval of rosuvastatin and launch in the United States.

2005: Shionogi obtained import approval and rosuvastatin was launched in Japan.

October 2009: Rosuvastatin approved in over 100 countries and sold in over 80 countries.

Prior to the R&D process for rosuvastatin, Shionogi had accumulated extensive experience in the development of drugs through the exploitation of pyrimidine and sulfonamide derivatives. Sulfonamide is a stable chemical moiety that is highly soluble in water, and has been incorporated into various drugs, including those developed by Shionogi, such as Bacta (a synthetic antimicrobial drug), Dimerin (a drug for diabetes), S-1452 (an asthma drug that was discontinued in Phase III clinical trials); and drugs licensed by Shionogi, such as acetazolamide (a diuretic agent), and furosemide (an antihypertensive agent). These drugs showed that the addition of sulfonamide could enhance the efficacy of pharmaceuticals. The experience gained through the development of these drugs significantly contributed to the synthesis of rosuvastatin.

The rosuvastatin research team consisted of the following five members, each of whom played a central role in the discovery. In particular, Dr. Kentaro Hirai, the lead researcher, had extensive experience in drug discovery through the use of pyrimidines.

- Kentaro Hirai (Principal Investigator)
- Teruyuki Ishiba
- Haruo Koike
- Masamichi Watanabe
- Shujiro Seo.

Shionogi began the discovery research that would lead to rosuvastatin in the early 1990s, immediately after the launch of natural statins such as pravastatin and lovastatin (Ohfuji et al. 2005); lovastatin had been launched in the United States in 1987 and was followed in 1989 by pravastatin (Mevalotin) in Japan. The efficacy and safety of statins, which competitively inhibit HMG-CoA reductase were widely accepted, and the global competition to discover synthetic statins had begun.

At the time, first-generation statins, which originated from natural products, were produced by fermentation methods. Therefore, if a new statin with the same mechanism of action could be produced by chemical synthesis, the statin's effects could be enhanced and the manufacturing costs drastically reduced. Thus, global pharmaceutical companies began the search for synthetic statins, although none had yet brought a drug to the market. With that said, patents had been filed and published for two synthetic statins; the patent for fluvastatin, from Sandoz, was filed in 1982 and published in 1984, while Warner-Lambert's atorvastatin patent was filed in 1986.

In Shionogi's search for rosuvastatin, fluvastatin, the first fully synthetic statin, was selected as the lead compound. Based on prior research that showed that hydrophilic statins, such as pravastatin, demonstrated better tissue specificity than lipophilic statins, such as lovastatin and fluvastatin (Roth et al. 1991), it was decided to make the new statin hydrophilic, which should correspond to more selective inhibition of cholesterol synthesis in the liver. Other groups had also published on the introduction of pyrimidines to increase inhibitory activity (Beck et al. 1990). Thus, the group opted for the introduction of pyrimidine to the statin backbone. However, rosuvastatin was ultimately created by the addition of sulfonamide, a process in which Shionogi had already completed with multiple drugs.

To protect the commercial value of the research, a substance patent was filed in Japan in July 1991, entitled "pyrimidine derivative." International patent applications, including for the United States, followed. In total, it took Shionogi approximately 2 years to synthesize rosuvastatin.

4.3   Discontinuation of the Clinical Trial by Shionogi

Shionogi began the clinical development of rosuvastatin in Japan in 1992 with long-term toxicity studies in rats, dogs, and rabbits, and these were completed by 1994. Early Phase II clinical trials for human use, which ran concurrently, were completed throughout 1993 and 1994.

At the same time, however, Takeda Pharmaceutical, which was also based in Japan, was in the process of developing the fully synthetic statin cerivastatin and was further along the development pipeline than Shionogi. Bayer discovered cerivastatin (Baycol), and a patent application for the drug was filed in 1990. As a result, Shionogi abandoned the development of rosuvastatin in 1994 because the company felt rosuvastatin was not sufficiently distinctive from other statins. This decision was made in light of Phase II clinical trial results that showed the cholesterol-lowering effect of rosuvastatin to be strong. Nevertheless, the Shionogi research group submitted a paper on the chemical synthesis and the pharmacological activity of rosuvastatin in July 1996, and it was published in the next year (Watanabe et al. 1997). Meanwhile, cerivastatin became commercially available in 1999, but was withdrawn because of severe side effects in 2001.

4.4 AstraZeneca Resumed Clinical Trials

In 1998, 4 years after the cessation of its clinical development but before the withdrawal of cerivastatin, AstraZeneca took over the development of rosuvastatin. This move was made in response to the company's need for new drug candidates after several AstraZeneca candidates were found to be failures. At AstraZeneca's request, Shionogi provided the results of preclinical studies, and AstraZeneca began a clinical trial in May 1998. Thus, global clinical development was advanced, with the extrapolation of overseas data to Japan in sight.

Preclinical studies in rabbits and rats were repeated in November 1998. Thereafter, in January 1999, a Phase II clinical trial was initiated comparing rosuvastatin with atorvastatin, and, in April 1999, a double-blind Phase II clinical trial was also initiated. In June of the same year, a Phase II clinical trial was started to compare rosuvastatin with pravastatin and simvastatin. These Phase II trials were completed by June 2000.

The results of last observation carried forward (LOCF) analysis of LDL-cholesterol reduction in 979 patients treated with 5 mg ($n = 243$) or 10 mg ($n = 231$) of rosuvastatin, 20 mg of simvastatin ($n = 250$), or 20 mg of pravastatin ($n = 255$) for 12 weeks are shown in Fig. 4.1. Rosuvastatin clearly reduced LDL-cholesterol by significantly lower doses than either simvastatin or pravastatin.

An open-label Phase III trial (STELLAR*TRIAL) then followed between April 2001 and March 2002 (Jones et al. 2003). This trial compared rosuvastatin with atorvastatin, simvastatin, and pravastatin. Cerivastatin was also initially included in the study; however, this comparative arm was terminated when cerivastatin was discontinued. In brief, the STELLAR*TRIAL was a 6-week comparative study of 2431 patients aged 18 years or older with LDL-cholesterol levels between 160 and

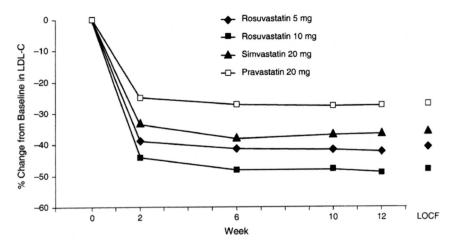

Fig. 4.1 Comparative LDL-cholesterol reduction by rosuvastatin, simvastatin, and pravastatin
Source Blasetto et al. (2003)

250 mg/dL. Study participants were randomly assigned one of the following treatment groups: 10–40 mg rosuvastatin ($n = 480$), 10–80 mg rosuvastatin ($n = 643$), 10–80 mg atorvastatin ($n = 641$), 10–80 mg simvastatin ($n = 655$), or 10–40 mg pravastatin ($n = 492$). The study demonstrated that rosuvastatin significantly reduced LDL-cholesterol and increased high-density lipoprotein (HDL)-cholesterol levels relative to the other statins.

Rosuvastatin was approved in November 2002, launching first in the Netherlands, which was the secretariat of the EU,[2] followed by the United Kingdom in March 2003. Approval was also granted by the FDA in 2003, and sales began that September in the United States. In Japan, Shionogi filed an application for import approval in 2002 and rosuvastatin launched in 2005. As of October 2009, rosuvastatin had been approved in over 100 countries and was marketed in over 80 countries.

4.5 Sources of Discovery: Fluvastatin as the Lead Compound

The discovery program for rosuvastatin began when the first-generation statins (pravastatin, lovastatin) were already in the market; thus, the R&D process involved the use of all major scientific advancements in the field and thorough study of the first-generation statins, including compactin. The Shionogi team credits the following three advancements by researchers outside the company as major contributors to their success. First, fluvastatin was selected as the lead compound for synthesizing a new compound. Fluvastatin was the first fully synthetic statin and was developed by Sandoz. Its patent application was already disclosed (see Fig. 2.1). The development of fluvastatin demonstrated that it was possible to chemically synthesize a drug that competitively inhibits HMG-CoA reductase. Second, Dr. Bruce Roth (Warner-Lambert) characterized the relationship between water solubility and organ selectivity of statins in a 1991 publication. This work led the Shionogi research team to consider the synthesis of hydrophilic statins, which would increase the drug's liver tissue specificity. Third, in 1990, researchers at Hoechst AG demonstrated the direction of chemical synthesis of a drug that could achieve high potency even at low concentration. This research prompted the introduction of pyrimidines and sulfonamides by the Shionogi team.

4.6 Crestor Sales: Peak Sales Exceed 7.1 Billion USD

Crestor sale trends between 2003 and 2017 are presented in Fig. 4.2 and include the sales of Crestor by AstraZeneca, the sales of Crestor by Shionogi, and the patent royalties earned by Shionogi from the AstraZeneca sales of Crestor. Of note, global

---

[2]Medical Interview Form: CRESTOR Tablets 2.5 mg/CRESTOR Tablets 5 mg (in Japanese).

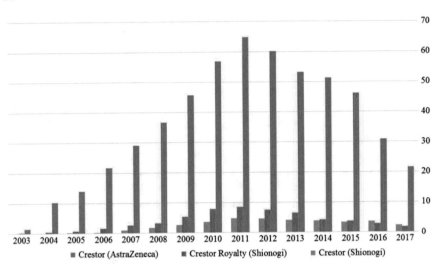

Fig. 4.2 Crestor sales, including sales by AstraZeneca, sales by Shionogi, and royalties earned by Shionogi (unit: 1M USD). *Sources* Cegedim Strategic Data K. K., Uto Brain Division "Pharma Future", Financial Annual Report by AstraZenca and Shionogi

net sales reached more than 7 billion USD in 2011. In addition, Shionogi received patent royalties that exceeded domestic market sales, which is significant because the patent royalty income directly contributed to operating profits. These figures also indicate that Crestor was widely accepted on the global market.

4.7 Crestor Versus Lipitor

As of 1990, when the search program for rosuvastatin (Crestor) was initiated, Sankyo had already completed the discovery of pravastatin (Mevalotin; Pravachol by Bristol-Myers Squibb), and Merck had launched both lovastatin (Mevacor) and simvastatin (Zocor). Similarly, fluvastatin (Lescol) had been discovered and was in clinical development at Novartis, and atorvastatin (Lipitor) had also been discovered at Warner-Lambert (Fig. 2.1). Furthermore, Bristol-Myers Squibb was exploring BMY-21950, which was discontinued during Phase II clinical trials, Hoechst had identified HOE-708, and Bayer had just applied for a patent for cerivastatin.

In this section, we compare the speed of the discovery and clinical development process of Pfizer's Lipitor with Shionogi and AstraZeneca's Crestor.

The discovery process of Lipitor began in 1982, with Waner-Lambert's R&D teams screening efforts to synthesize new HMG-CoA reductase inhibitors using publicly available compactin and lovastatin information as a reference. By 1985, they had successfully developed a method to synthesize a large amount of the candidate compound, PD-12358. However, the development program was abandoned after it was found that the compound was in conflict with a Sandoz patent (US4613610, "Cholesterol biosynthesis inhibiting pyrazole analogs of mevalonolactone and its derivatives") published in September 1986. Thereafter, the development continued using PD-123822, which had already been synthesized, as the lead compound. In late 1986, optical resolution of PD-123822 was performed to improve bioavailability, and it was found that the right-hand form had a longer half-life than conventional statins. This form was designated PD-134298-38A and would eventually be called Lipitor.

While lovastatin and pravastatin were being launched in the USA, Phase I clinical trials of Lipitor were initiated. In a placebo-controlled Phase II clinical study, in which administered doses were between 10 and 80 mg, Lipitor significantly decreased LDL-cholesterol (Nawrocki et al. 1995). Furthermore, a 2-year, long-term Phase III clinical trial comparing Lipitor with lovastatin, simvastatin, and pravastatin demonstrated efficacy and safety (Black et al. 1998). Lipitor was subsequently approved at the end of 1996 and launched in January 1997 (Shook 2007; Li 2009).

In comparing Lipitor and Crestor, the discovery period was much shorter for Crestor, requiring only 2 years, while it took 4 years from the start of the search to the patent application for Lipitor. Another 10 years passed before Lipitor applied for drug approval, followed by a further year to receive this approval; while it took 9 years for Crestor to reach the drug approval application stage and then a further year for the approval. Overall, Crestor had a 12-year discovery-to-approval period that was approximately 2 years shorter than Lipitor's 14-year period, but Crestor had a 4-year interruption that greatly delayed its entry into the market.

4.8   Properties of Commercially Available Statins

In Table 4.1, the major characteristics of the eight available statins are shown, including cerivastatin, which was later withdrawn. Comparing pravastatin and simvastatin, which were approved almost simultaneously in the US market, simvastatin has a longer time to maximal plasma concentration, higher absorption rate, and a higher LDL-cholesterol-lowering rate, which correspond to a longer duration of action. Similarly, super statins, such as atorvastatin and rosuvastatin, have a longer half-life and a higher LDL-cholesterol-lowering rate than first-generation statins.

High blood cholesterol concentrations cause various diseases, such as heart failure and cerebrovascular disease. These diseases are the second most common cause of death after cancer in Japan, and treatment with statins has been found to benefit many patients. Indeed, a large placebo-controlled epidemiological study comparing the changes in health status of patients before and after treatment with statins found that lowering the level of cholesterol in the blood of high-cholesterol patients significantly reduced mortality and was associated with minimal side effects.

4.9 Economic Value of Follow-on Innovation

Rosuvastatin is characterized by its ability to mediate a significant reduction in LDL-cholesterol and a concomitant increase in HDL-cholesterol levels at a low dose. It also has high hydrophilicity and low susceptibility to cytochrome-mediated metabolism (From Crestor interview-form). The high hydrophilicity of rosuvastatin correlates with its high liver specificity, and its low level of cytochrome-mediated metabolism reduces the risk of adverse events when administered with other drugs. From Table 4.1, it is clear that super statins, such as atorvastatin and rosuvastatin, have higher LDL-cholesterol lowering rates and longer half-lives than first-generation statins.

The value of rosuvastatin's product innovation is confirmed by the fact that its sales expanded to a high level despite the introduction of the generic statins. In general, the successful entry and sales growth of a new drug (hereafter referred to as "B") does not necessarily mean that there was a significant product innovation relative to the previous drug (hereafter referred to as "A"). Even if a drug is not associated with significant innovation, such as when the drug has the same mechanism of action and similar efficacy, if original drug A has a relatively high price because of its unique nature (monopolistic position), the introduction of an imitative drug could result in a duopoly. However, if new drug B remains significantly profitable, even with the entry of generic versions of drug A to the market, then there is a significant product innovation by drug B relative to drug A (see Box 4.1).

Box 4.1 Contribution of Follow-on Drugs to Innovation

One useful way to assess the extent to which follow-on drug B, which has the same mechanism of action as pioneer drug A, has made a new contribution in terms of efficacy and the reduction of side effects (contribution to innovation) is to assess the extent to which follow-on drug sales decline after the entry of generic versions of drug A. In this scenario, we assume that the government does not directly regulate the price of drugs, as is the case in the USA.

More specifically, if we denote the quality of drug A and drug B as q_A and q_B, respectively, and denote the price of drug B as P_B, then as a result of generic drug entry to the market, it is assumed that the price of drug A will fall to its production cost, c. For simplicity, the production cost of both drugs A and B are assumed to be equal to c. Drug B, which is a new drug, has market power because its patent is effective at that time. However, drug B's price is constrained by the competitive supply of drug A at a cost-equivalent level, meaning: $q_B - P_B = q_A - c$. Given that the consumer surplus from the purchase by a patient (or the insurance company) has to be the same, the price of drug B after entry of generic drug A should be:

$$P_B = (q_B - q_A) + c.$$

Drug B's price increases with the degree of product innovation (the difference in quality between two drugs = $(q_B - q_A)$ relative to Drug A). If there is no substantial product innovation, it will be established as $q_B - q_A = 0$, and the price of Drug B will also drop to its cost when the generic drug launch of drug A occurs. Therefore, the extent to which the sales of drug B decline after the entry of generic drug A indicates the magnitude of the product's innovation.

From this perspective, we analyzed long-term sales data for branded drugs whose main mechanism of action is inhibition of HMG-CoA reductase (Fig. 2.2). The fact that rosuvastatin made significant contributions to product innovation compared to earlier drugs is well illustrated by Crestor's expanding revenues, as shown in Fig. 4.2, despite the expiration of prior statin patents and the introduction of generic drugs for such statins.

Of note, rosuvastatin entered the market in 2003, but 2 years later, in 2005, the US patents for the two leading statins, Sankyo's pravastatin and Merck's simvastatin, expired. As can be seen in Fig. 2.2, the sales of these two products gradually declined. However, Crestor sales maintained their continuous growth and grew rapidly between 2005 and 2010. During this period, the sales of atorvastatin (Lipitor) also grew, albeit at a lower rate. Even though the introduction of first-generation statin generics lowered average statin prices, there was still strong market demand for super statins (rosuvastatin and atorvastatin), indicating that the advent of super statins was an important product innovation.

The US patent on atorvastatin (Lipitor) expired in September 2009, and subsequently, the sales of Lipitor declined substantially in 2012 with the introduction of generic drugs (see Fig. 2.1). Thus, the market entry of generic atorvastatin had a very strong effect, but the sales of rosuvastatin did not significantly decrease, indicating that the market appreciated the product innovation of rosuvastatin, even compared to atorvastatin.

Therefore, the high sales of rosuvastatin (Crestor), despite increased competition because of the expiration of other statin patents, indicates that the product innovation of rosuvastatin was important for patients.

4.10 AstraZeneca's Resumption of the Suspended Project

The previous sections showed that rosuvastatin made significant contributions to the statin market in terms of product innovation. This raises important questions on why Shionogi stopped the development of rosuvastatin and why AstraZeneca was able to successfully resume it.

We believe the following two differences in the clinical development strategies may answer these questions. First, there were significant differences in the way the two firms targeted markets. At least initially, Shionogi was focused only on the

Japanese market and was conducting clinical studies only for the Japanese market. In contrast, AstraZeneca developed the product for a global market from the beginning, a move that would cover Japan, the USA, and Europe. Global clinical trials are more diverse, and can clarify the efficacy of new drugs more fully. As an example, larger doses may be applied in the USA than in Japan. In the case of rosuvastatin, its strong effect for lowering LDL was well recognized in the Japanese clinical trial, but the trial might not have recognized the associated increase in HDL and reduced side effects. Furthermore, if clinical development is carried out considering global markets from the beginning, the cost can be lowered by utilizing bridge studies.

Second, AstraZeneca recognized that there were already commercially available statins, and conducted a comparative study with pravastatin, simvastatin, and atorvastatin as part of a clinical trial. This was likely to be a key step in establishing market assessments of Crestor. Such a clinical development strategy seems to have clearly established and conveyed the benefits of rosuvastatin to medical communities, which resulted in success that was mediated through the early establishment of market reputation.

4.11   Conclusion

Relative to the first generation of statins, rosuvastatin (Crestor) was a very late entrant into the market, because it was launched commercially just 2 years before the expirations of the US patents for the two pioneer blockbuster statins, Sankyo's pravastatin and Merck's simvastatin. However, Crestor became a highly successful global blockbuster drug despite such late entry. In the global R&D competition to discover synthetic statins with stronger efficacy than the statins from natural products, Shionogi succeeded by taking advantage of the company's wealth of experience and expertise in the field of chemical synthesis, particularly with respect to sulfonamide additions. Indeed, the superior characteristics of rosuvastatin are well illustrated by the ability of Crestor to expand and maintain its sales, despite the entry of generic drugs to the statin market, following the expirations of the patents for earlier statins.

Crestor was launched only because AstraZeneca was able to resume clinical development after it was abandoned by Shionogi. In doing so, AstraZeneca expanded the scale and diversity of clinical trials with the intention of globally launching rosuvastatin. It also engaged in extensive comparative clinical trials with leading statins. Such a clinical development strategy seems to have realized the advantages of rosuvastatin, which resulted in the early establishment of a favorable reputation and subsequent success. Thus, this case illustrates the critical role of the complementary asset for turning a great drug invention into a successful innovation as well as the importance of combining capabilities across firms for such an objective.

References

Beck, G., Kesselar, K., Baader, E., Bartmann, W., Bergmann, A., Granzer, E., et al. (1990). Synthesis and biological activity of new HMG-CoA reductase inhibitors 1. Lactones of pyridine- and pyrimide-substituted 3,5-dihydroxy-6-heptenoic(-heptanoic) Acids. *Journal of Medicinal Chemistry, 33*(1), 52–60.

Black, D. M., Bakker-Arkema, R. G., & Nawrocki, J. W. (1998). An overview of the clinical safety profile of atorvastatin (Lipitor), a new HMG-CoA reductase inhibitor. *Archives of Internal Medicine, 158*(6), 577–584.

Blasetto, J. W., Stein, E. A., Brown, W. V., Chitra, R., & Raza, A. (2003). Efficacy of rosuvastatin compared with other statins at selected starting doses in hypercholesterolemic patients and in special population groups. *The American Journal of Cardiology, 91*(5), 3–10.

Brewer, H. B. (2003). Benefit-risk assessment of rosuvastatin 10 to 40 milligrams. *The American Journal of Cardiology, 92*(4), 23–29.

Jones, P. H., Davidson, M. H., Stein, E. A., Bays, H. E., McKenney, J. M., Miller, E., et al. (2003). Comparison of the efficacy and safety of rosuvastatin versus atorvastatin, simvastatin, and pravastatin across doses (STELLAR* Trial). *The American Journal of Cardiology, 92*(2), 152–160.

Li, J. J. (2009). *Triumph of the heart: The story of statins*. Oxford University Press.

Mukhtar, R. Y. A., & Reckless, J. P. D. (2005). Pitavastatin. *International Journal of Clinical Practice, 59*(2), 239–252.

Nawrocki, J. W., Weiss, S. R., Davidson, M. H., Sprecher, D. L., Schwartz, S. L., Lupien, P. J., et al. (1995). Reduction of LDL cholesterol by 25% to 60% in patients with primary hypercholesterolemia by atorvastatin, a new HMG-CoA reductase inhibitor. *Arteriosclerosis, Thrombosis, and Vascular Biology, 15*(5), 678–682.

Ohfuji, K., Yano, S., Yamaguchi, M., Smith, G., Hirata, M., Shimada, H., et al. (2005). Pharmacological action and clinical effects of the new HMG-CoA reductase inhibitor rosuvastatin calcium (crestol). *Japanese Pharmacology Journal, 126*, 213–219.

Roth, B. D., Bocan, T. M. A., Blankley, C. J., Chuchlowski, A. W., Creger, P. L., Creswell, M. W., et al. (1991). Relationship between tissue selectivity and lipophilicity for inhibitors of HMG-CoA reductase. *Journal of Medicinal Chemistry, 34*(1), 463–466.

Shook, R. L. (2007). *Miracle medicines: Seven lifesaving drugs and the people who created them*. Portfolio Hardcover.

Watanabe, M., Koide, H., Ishiba, T., Okada, T., Seo, S., & Hirai, K. (1997). Synthesis and biological activity of methanesulfonamide primide- and N-methanesulfonyl pyrrole-substituted 3,5-dihydroxy-6-heptenoates, a novel series of HMG-CoA reductase inhibitors. *Bioorganic & Medicinal Chemistry, 5*(2), 437–444.

Yamamoto, A. (2005). Characteristics of new statin rosuvastatin. *Pharmacy, 56*(1), 149–159.

Chapter 5
Leuprorelin (Leuplin, Lupron, Viadur)

A Prostate Cancer Drug with Dual Innovations in Mechanism of Action and Drug Delivery System

Naoki Takada and Hideo Kawabe

Abstract Leuplin is a drug for the treatment of prostate cancer. It is a sustained-release injectable formulation of leuprorelin acetate, which is a derivative of luteinizing hormone-releasing hormone (LH-RH). Researchers at Takeda Pharmaceutical synthesized the active ingredient using the characterized primary structure of thyrotropin-releasing hormone (TRH) and LH-RH as a model, both discovered within academia. Their research was aided partly by luck and partly by researchers ready to take advantage of unexpected findings. Following the discovery of leuprorelin as active ingredient, the research team began the study of microcapsules using synthetic polymer compounds. This work led to the development of sustained-release leuprorelin, which is now an indispensable prostate cancer drug because it reduces the burden on the patient and on tertiary-care centers. A formulation of monolithic microcapsules containing leuprorelin was the first drug delivery system of its kind to be marketed globally. The phenomenon of downregulation was discovered during collaborative efforts between Takeda and Abbott Laboratories for developing leuprorelin. The creation of leuprorelin illustrates the significance of the link between pure science and in-house research and development. Leuplin combined multiple scientific advancements including recent discoveries in the fields of neuroscience, peptide synthesis, and microencapsulation technology.

N. Takada (✉)
Institute of Advanced Sciences,
Yokohama National University, Yokohama, Kanagawa, Japan
e-mail: takada-naoki-jx@ynu.ac.jp

H. Kawabe
Hiraki & Associates/Manufacturing Technology Association of Biologics,
Tokyo, Japan
e-mail: kawabe-h1604@hiraki-patent.co.jp

© Springer Nature Singapore Pte Ltd. 2019
S. Nagaoka (ed.), *Drug Discovery in Japan*,
https://doi.org/10.1007/978-981-13-8906-1_5

5.1 Introduction

Leuplin (leuprorelin) is a drug for the treatment of prostate cancer. It is a sustained-release injectable formulation of leuprorelin acetate, a derivative of luteinizing hormone-releasing hormone (LH-RH).

This drug was innovative for a number of reasons; in particular, the drug delivery system (DDS) was the first sustained-release injectable preparation to be globally marketed. Although the active ingredient leuprorelin is highly effective as a treatment for prostate cancer, conventional administration involves daily injections, which is practically inappropriate, considering the financial burden and the decreased quality of life of the patient. To solve this problem, Leuplin was created as a sustained-release preparation of leuprorelin. Specifically, the drug product is designed to microencapsulate the active ingredient leuprorelin so that the drug is retained in the patient's body for 3–4 weeks. Over that time, small amounts of the active ingredient are released into the blood at a constant rate (a 3-month sustained-release preparation is also available).

As outlined below, the active ingredient leuprorelin was discovered in 1973, and a self-administered daily preparation of Lupron was launched in the United States in 1985. However, daily administration and self-injection were considered inconvenient, and in Japan, self-injection by the patient was not permitted. Thus, Takeda Pharmaceutical began research into a leuprorelin DDS in the latter half of the 1970s. Leuplin, which utilized Takeda's new DDS technology was launched in the USA in 1989 and Japan in 1992. It is currently sold in 80 countries around the world. Leuplin has also been shown to have good clinical effects on endometriosis and uterine fibroids, as well as improvement in the symptoms of menorrhagia, lower abdominal pain, low back pain, anemia, premenopausal breast cancer, prostate cancer, and central precocious puberty (Takeda Pharmaceutical 2013).

5.2 Research and Development Timeline of Leuplin

The following is the timeline of the major events that took place over the 20 years from the initiation of research to the first market launch.

1971: Andrew Schally discovered the primary structure of LH-RH. Masahiko Fujino of Takeda Pharmaceutical began the synthesis of LH-RH derivatives.
1972: Joint development with Abbott Laboratories (Abbott) began.
1973: Discovery of leuprorelin and patent application.
1974: Successful synthesis of TAP-144 (leuprorelin). The clinical trial of leuprorelin as an anti-breast cancer agent began.
1977: Establishment of the Takeda–Abbott Product Partnership (TAP).
1980: Hiroaki Okada started to develop a sustained-release injectable formulation of leuprorelin.

December 1980: The target disease of TAP-144 was changed to prostate cancer, and new clinical trials began.

1983: A system was built for the large-scale synthesis of polymers without the use of metal catalysts.

1985: The launch of self-injectable Lupron injection in the USA, and clinical trials of sustained-injectable formulations started in the USA, and later in Japan and Europe.

June 1987: New drug applications submitted to the US Food and Drug Administration (FDA).

March 1989: A sustained-release injectable form of Lupron with a 1-month duration was launched in the USA.

1992: Manufacturing and marketing approval were granted for Leuplin in Japan. Leuplin was launched in Japan.

December 1995: A sustained-release injectable form of Lupron with a 3-month duration was launched in the USA.

5.3 Leuprorelin Mechanism of Action

Reproductive cancers, including uterine, breast, ovarian, and prostate cancer, are known as sex hormone-dependent cancers, and hormone therapy has traditionally been employed with the intention of suppressing sex hormone activity. Typical sex hormone targets include testosterone, progesterone, and estrogen. Hormone therapy involves the use of receptor antagonists against these hormones, as well as the use of the hormones themselves. For example, testosterone receptor antagonists and estrogens are frequently given in the treatment of prostate cancer. These sex hormones are secreted by the testes and ovaries in response to LH-RH in a multi-level process. In brief, LH-RH, which is released from the hypothalamus, acts on the pituitary gland to stimulate the release of gonadotropins, which induce the secretion of sex hormones from the testes or ovaries.

As noted above, leuprorelin, the active ingredient of Lupron/Leuplin, is a derivative of LH-RH, which implies that leuprorelin administration may result in increased sex hormone secretion and may promote the development of sex hormone-dependent cancers. However, leuprorelin, inhibits, rather than stimulates, the secretion of sex hormones. The key to this reversal is in the LH-RH receptor; these receptors mediate the activity of LH-RH on the pituitary. Compared to endogenous LH-RH, leuprorelin binds tightly to the receptor, ultimately inducing a decrease in the number of receptors because of receptor desensitization, and subsequent reduction in sex hormone secretion (Mann et al. 1984).

Although leuprorelin inhibits the progression of prostate cancer by the mechanism of action as described above, the molecule is a peptide and is ultimately degraded in vivo. Therefore, it should be given continuously to enhance efficacy. In this context, the necessity to increase the duration of action led to the development of leuprorelin microencapsulation, giving rise to Leuplin.

5.4   Academia Provides the Seed

Leuprorelin, the active ingredient of Leuplin, is a water-soluble peptide composed of nine amino acids. Its creation requires several peptide synthesis techniques. Coincidentally, Takeda Pharmaceutical had acquired peptide synthesis technology early on. According to the Takeda Pharmaceutical company history, peptide synthesis research became an in-house research theme in 1964, and the peptide synthesis research group was officially organized in 1967. To acquire the peptide synthesis technology necessary for their research, the company dispatched researchers to the Peptide Center of the Institute of Protein Research at Osaka University for experiential learning.[1]

At this time, research on peptide hypothalamic hormones was actively being carried out abroad. In particular, findings from the laboratories of Prof. Andrew Schally at Tulane University and Prof. Roger Guillemin at Baylor College of Medicine were particularly impressive, and the two groups, including their affiliates, were in a fierce competition to characterize hypothalamic peptide hormones, later called the "duel for the Nobel Prize." In 1969, the isolation of thyrotropin-releasing hormone (TRH) was reported almost simultaneously by both groups. Subsequently, the competition between the two groups to isolate LH-RH heated up, and in 1971, the structural determination of LH-RH was reported by Schally and Hisayuki Matsuo (Matsuo 1989).

During this struggle, Masahiko Fujino of Takeda Pharmaceutical studied under Darrell Ward (University of Texas), a colleague of Guillemin, which permitted him to learn of the progress in characterizing TRH and LH-RH directly. Upon his return to Takeda, Fujino started investigating the chemical synthesis of TRH. In 1969, when the chemical structure of TRH was elucidated, the industrial synthesis of peptides was rare. Fujino's research made significant advancements to the field and ultimately led to the successful synthesis of kilograms of TRH in laboratory facilities. This would not have been possible without the experience he cultivated during his studies abroad.

In 1971, Schally and colleagues successfully determined the structure of LH-RH, inspiring Fujino's group to begin research on the synthesis of LH-RH derivatives using the method developed for TRH synthesis. The basic idea in synthesizing a derivative of LH-RH is substituting one of the amino acids in LH-RH with another amino acid. Thus, although the synthesis of a single derivative does not require too much time, it is difficult to find a potent compound because millions of combinations exist.

5.5   Challenges and Luck in the Synthesis of Derivatives

Although the synthesis of LH-RH derivatives was challenging, some good luck and attention to detail allowed the research team to create an effective compound. The

---

[1]From an interview with Prof. Akira Okada.

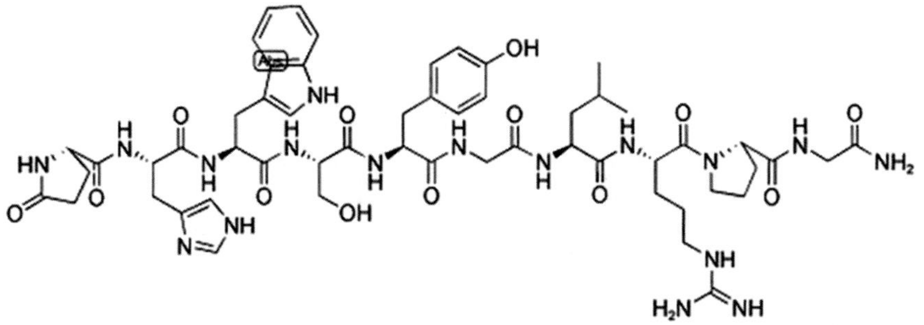

Fig. 5.1 The chemical structure of LH-RH. *Source* Chemical Book 9034 40-6

first instance of good luck was brought about by the failure of a young researcher. Fujino and Yasuaki Ogawa described the event in the following manner:

> Fujino ordered a young researcher to add CH_2CH_3 at the tenth position of $CONH_2$. Perhaps due to his inexperience, he [had] joined the company very recently, he mistakenly attached CH_2CH_3 at the 9th position of COHN, and when the activity of this compound was measured, it was surprisingly 6.7 times stronger than that of LH-RH. (Ogawa and Fujino 1994, pp. 176–177)

The structure of LH-RH is complicated, as shown in Fig. 5.1, and it is not surprising that a young researcher would make an error. However, careful investigation of this error, without discarding it, led to the discovery of a highly active derivative of LH-RH.

Moving forward, the research team investigated further activity enhancement by substituting other amino acids at this position. Initially, none of the newly synthesized substances demonstrated activity that was more than ten times the activity of LH-RH, and the synthetic work seemed to have once again been proved difficult. However, a second young researcher, Tsunehiko Fukuda, who had recently joined the company, enjoyed a second piece of luck in the drug's development when he synthesized a derivative with a substitution of the sixth amino acid. The resultant molecule demonstrated greater than 50 times the activity of LH-RH. Fukuda recalled that "it was considered to be unrelated to the activity until then, and I guessed that it would increase the affinity of the receptor if we replaced glycine-6 with a lipid-soluble substance" (Fukuda 2004, p. 177).

This finding promoted further synthesis research using the chemical compound synthesized by Fukuda as the lead compound. However, when Fukuda synthesized the same derivative again as per instructions from Fujino, the molecule did not demonstrate the same level of activity. For Fujino, it would have been a reasonable choice to treat the activity observed for the compound that Fukuda synthesized earlier as a measurement error and to start synthesizing the other substances immediately. However, Fujino carefully investigated Fukuda's synthesis protocol and eventually, they identified a compound that was 70 times more active than LH-RH. Ogawa and Fujino wrote the following:

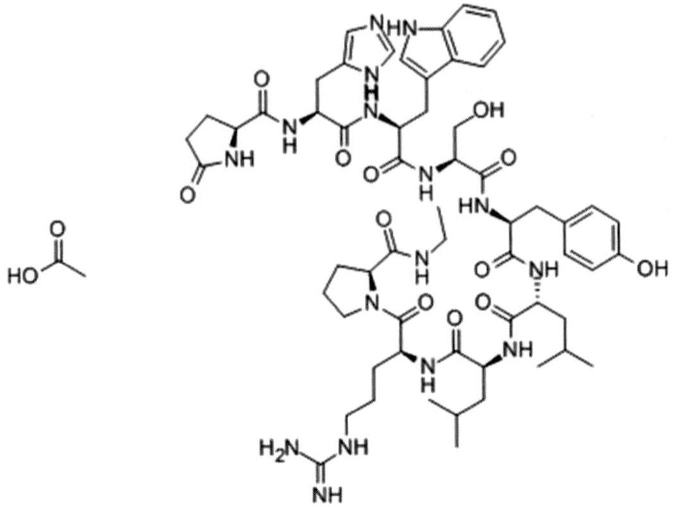

Fig. 5.2 The chemical structure of leuprorelin. *Source* Chemical Book 53714-56-0

While it would have been an end if the previous finding of high activity would have been simply dismissed as a mistake in the measurements, Fujino, driven by his intuition, thoroughly investigated the younger researcher's synthesis protocol. However, the activity did not increase even after they resynthesized the compound many times. After checking each of the raw materials used, it was found that one of the reactants was a racemate, a mixture of L and D isomers, instead of an L isomer, as planned. When the sixth amino acid was replaced by a D isomer, the activity became 70 times higher than that of LH-RH (Ogawa and Fujino 1994, p. 177).

Several derivatives of this substance were synthesized, and in 1973, leuprorelin (Fig. 5.2), which was 80 times more potent than LH-RH, was discovered. Because the compound was the 144th synthesized derivative, it was coded TAP-144 and proceeded to the next stage. Considering that the LH-RH receptor had not been fully characterized at the time, the discovery of leuprorelin was incredibly novel. It is important to note that many rival companies were also trying to synthesize LH-RH derivatives at this time. Fujino's group acquired a substantial lead over the other companies because they successfully synthesized highly active derivatives within 2 years of the structural determination of LH-RH (Fujino et al. 1974).

5.6 The Start of Microencapsulation

Leuprorelin was highly effective against prostate cancer, as discussed below, but the necessity for daily injections imposed an enormous burden on the patients. Furthermore, self-injections were not permitted in Japan. Then, Hiroaki Okada and Takatsuka Yashiki began studying the production of synthetic polymer compounds in the spring of 1980. In 1983, Yasuaki Ogawa and Masaki Yamamoto joined the research group. The original research goal was to develop polymers that degrade in vivo for the production of capsules in which leuprorelin is uniformly dispersed.

First, biodegradable polymers, which are the raw materials of capsules, were selected. Okada focused on PLA (polylactic acid) and PLGA (lactic acid/glycolic acid) copolymers, which were used as dissolving sutures. According to Okada, the company was going to purchase PLA and PLGA from the USA, but the compounds were very expensive; thus, Okada chose to synthesize the compounds in-house. As a result, Okada clarified the physical properties and the biodegradability of the polymers through his own polymer synthesis experience, which led to the early selection of PLGA for Leuplin (Okada 1993).

Following the selection of PLGA, polymer synthesis research began in cooperation with Wako Pure Chemical Industries, an affiliated company, for the development of the production process. Given that Takeda Pharmaceutical's department that was responsible for polymer synthesis was busy with other projects, and the high cost of raw materials, the need to conduct joint research was evident (Ogawa and Fujino 1994). At the time, it was common to use tin, a heavy metal, as a catalyst in the synthesis of polymers. However, given that drugs, especially injectables, are expected to contain no heavy metals, they devised a method for synthesizing polymers of lactic acid and glycolic acid without the use of a catalyst. An Eli Lilly patent (US4273920A) was used as a reference, and in 1983, the Takeda–Wako alliance succeeded in synthesizing polymers without the use of metallic catalysts (Nevin 1981).

Once the most suitable polymer was obtained for the DDS, a method of producing monolithic microcapsules that contained leuprorelin was investigated. Two encapsulation methods were known at the time. These were phase separation and in-liquid drying (solvent removal), and trials indicated that the phase separation method gave the highest encapsulation rate. Okada chose the in-liquid drying method because water was readily available as a high-quality solvent for injection. Following many experimental trials, Okada devised a submerged drying method using a water/oil/water (W/O/W) emulsion that succeeded in producing microcapsules with a high drug encapsulation rate. The sustained release of the produced microcapsules was confirmed by subcutaneous or intramuscular administration to rats and dogs, followed by measurement of the drug concentration in the blood. From this point forward, the optimization of preparation conditions and industrialization research was carried out mainly by Masaki Yamamoto, and a method for reliably obtaining an encapsulation rate of 95% or more was established.[2]

---

[2]From an interview with Prof. Akira Okada.

5.7 Discovery of Downregulation and a Change in the Target Disease

Given that the primary function of LH-RH is to maintain the levels of sex hormones via stimulation of the synthesis and secretion of sex hormones, research on the effects of LH-RH derivatives and preclinical studies were conducted with the intent to accelerate ovulation. To begin, researchers at Takeda Pharmaceutical evaluated the abilities of drugs to induce ovulation, and conducted safety tests. However, in 1972, Abbott initiated a joint research program that assessed the pharmacological properties of LH-RH derivatives (Ogawa 1995). The collaborative research coordinator, Dr. Katsura Morita, decided that the duration of the collaborative research would be 3 years and that both parties would retain equal rights to substances discovered through the collaboration for a decade (Morita 2000).

The onset of this collaboration resulted in significant changes to the preclinical research objective, which had initially been to elucidate the potential of leuprorelin as an ovulation accelerator. However, Abbott pharmacologists reported that leuprorelin treatment inhibited the production of sex hormones; specifically, leuprorelin down-regulated sex hormones. The discovery of this unexpected effect led to a complete change in the target disease. Nevertheless, the discovery of downregulation was fortunate because it created opportunities for leuprorelin in the treatment of reproductive system cancers, which was a large market.

Furthermore, the discovery of LH-RH-mediated sex hormone downregulation forced the drug development groups to address two challenges. One was to understand the mechanism of this downregulation, which was elucidated in the late 1970s and into the 1980s. The other was the need to reconsider the area of application by focusing on a disease that could be treated by suppression of gonadal function rather than accelerating ovulation. The companies were thus presented with three potential directions with which to test the drug: (a) contraceptive, (b) ovulation promoter, and (c) therapeutic for sex hormone concentration-dependent diseases. Of these, development as a contraceptive was abandoned early because it was too time-consuming, while ovulation acceleration was not seriously considered because sales representatives from Takeda and Abbott were interested in this type of drug. Therefore, the selection was to target the sex-hormone dependency. They first examined the effect in the treatment of breast cancer, confirming that the drug was effective against drug-induced breast cancer. Therein, TAP-144, an anti-breast cancer drug, was investigated in clinical trials, followed by clinical development in the USA in collaboration with Abbott. In 1980, TAP-144 was designated by the US FDA for fast track review. Although it was the first Japanese-made drug to receive such a designation, it was not effective in all types of patients, prompting the change in direction to the development of a prostate cancer drug.

A key factor in the decision to switch disease targets was the work of Charles Huggins. Huggins' research showed that removing the testicles prevented the growth and progression of cancer, suggesting pharmaceutical treatments for prostate cancer that did not require surgical intervention were possible (Huggins and Hodges 1941).

The possibility of such therapy was confirmed during clinical research in the USA, which was a deciding factor in the switch to developing Leuprorelin as a treatment for prostate cancer. Clinical trials of TAP-144 as a prostate cancer drug began in December 1980, and approval was granted in the USA in 1985. It was subsequently sold in the USA under the trade name Lupron Injection. However, as mentioned above, leuprorelin is degraded in the body, and its duration of action is short; patients, therefore, required daily injections. Consequently, Lupron was sold as an injectable solution in the US, where patients can self-administer injections. Conversely, it was not possible to sell Lupron in Japan because self-injection was not permitted.

In addition to the joint research with Abbott, the drug formulation laboratory of Takeda Pharmaceutical began to research mucosal formulations, which could be self-administered via the vaginal or nasal mucosa, in 1976. These formulation and anticancer effect studies were completed by Hiroyuki Mima and Hiroaki Okada in collaboration with Iwao Yamazaki and Hisanori Kawaji of the biological laboratory at Takeda Pharmaceutical. More specifically, levels of the drugs, luteinizing hormone (LH), and follicle-stimulating hormone (FSH) in the blood were measured, the gonadal organs were evaluated, and anticancer effects in rats with breast cancer were studied (Okada et al. 1983). Okada also assisted Yamazaki in evaluating the blood levels of the drug by preparing alternatives for radio immune assays (RIA) in rabbits. The rat LH and FSH reference standards required for animal experiments were supplied by the US National Institutes of Health. Mucosal formulations studied in this way had a contraceptive effect (Shimamoto 1980); however, mucosal formulations have not been marketed, as this paled in comparison to the importance of its development as a prostate cancer drug. Nevertheless, the methods developed for the study of a mucosal preparation, including quantitative RIA technology, led to the establishment of sustained-release injectables in a short period.

5.8 Commercialization Challenges

As the research moved from preclinical tests to clinical trials, new challenges were encountered. Among these, the introduction of quality assessment methods and the development of manufacturing facilities for sustained-release injectable products were particularly pressing issues. Because of FDA regulations, it was necessary to establish the evaluation and test methods necessary for industrialization before conducting the clinical trials. However, the innovative formulation of Leuplin meant that existing production and quality control evaluation methods could not be used. Therefore, in 1983, the development of manufacturing processes began, and in 1985, Hajime Toguchi, the leader of the Lupron Project, began to build a scaled-up manufacturing process.

For the industrial manufacture of Leuplin, the conventional safety requirement of heat sterilization of injectables presented a significant problem. For ordinary injectables, heat sterilization can be performed in the last process of production. In contrast,

Leuplin could not be heat sterilized because the encapsulation material is temperature sensitive. Therefore, it was necessary to construct a new manufacturing facility that could carry out all the processes under aseptic conditions. The sterilization process involved dissolving all the raw materials in water and organic solvents and filter sterilizing the solution. Once sterilized, the solution was emulsified, liquid dried, washed, prepared, freeze-dried, and the packaging filled—all of these steps were performed under aseptic conditions. In particular, significant time and effort were required to ensure that the post-sterilization emulsification process was carried out under aseptic conditions. While the manufacturing cost was not significantly higher than traditional parenteral injection manufacturing, this was the first opportunity to examine the industrialization of a sustained-release preparation. Therefore, it was necessary to independently develop many technologies, such as fine particle control and micro-filling of the dry powder. Finally, ten aseptic facilities were created, the aseptic processes were simulated using culture medium, and a system was designed for continuous monitoring. In May 1988, the FDA inspected the new facilities (Kawamura 2006).

The first clinical trial took place in the United States. Once the method for producing sustained-release injectable products was established, TAP took the initiative in clinical development and negotiations with the FDA; the drug's clinical development was promoted to the extent that Takeda Pharmaceutical alone could not have realized. The FDA also positively addressed the unprecedented problem posed by the evaluation of long-term sustained-release formulations, and Takeda Pharmaceutical was able to obtain useful advice from the FDA. Following the favorable results of the US clinical trial, clinical trials were quickly carried out in Japan and Europe. In 1989, TAP launched Leuplin in the USA as the 1-month sustained-release form, it was marketed in Japan in 1992.

5.9 Scientific Theory Behind Leuplin: Application of Nobel Prize-Winning Compounds

In 1948, Jeffrey Harris of the University of Cambridge proposed that the secretion of protein hormones, such as LH and FSH, secreted by the anterior pituitary gland, may be controlled by substances present in the hypothalamus (Harris 1948). To prove this hypothesis, it was necessary to isolate and determine the structure of substances that are involved in the closed circuitry of the hypothalamic–pituitary–endocrine system.

This hypothesis initiated the competition among groups of researchers to identify this hormone in the 1960s. The competition between Andrew Schally of Tulane University and Roger Guillemin of the Baylor College of Medicine to determine the structure of hypothalamic hormones was particularly fierce. At the time, hypothalamic hormones were considered to be very difficult to investigate, and indeed the two researchers had considerable difficulty in isolating hypothalamic hormones.

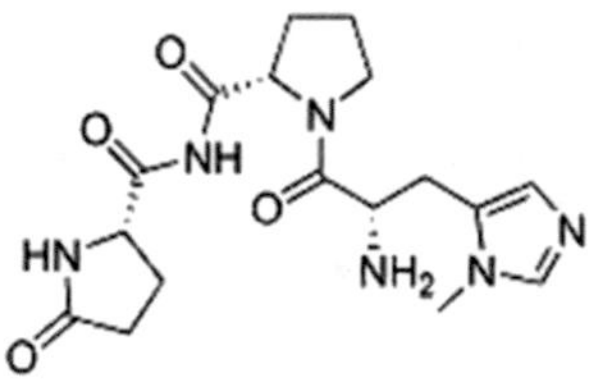

Fig. 5.3 The chemical structure of TRH. *Source* Chemical Book 34367-54-9

Both Schally and Guillemin chose TRH as their first research subject. To study TRH, Schally was provided with the hypothalamus of 300,000 pigs by Oscar Mayer Company and Guillemin purchased the hypothalamus of 500,000 sheep. Overall, the structure of TRH was reported by both groups in 1969, respectively. Furthermore, both groups identified the same structure for TRH, as shown in Fig. 5.3, regardless of the species of the source (Bøler et al. 1969; Burgus et al. 1969).

Notably, the structure of TRH was more straightforward than the researchers had assumed, and thus the isolation of LH-RH became realistic. The two groups then worked toward the structural determination of LH-RH. At that time, there was a Japanese scientist, Akira Arimura, who had been studying hypothalamic hormones with Schally since 1965 and was in charge of testing LH-RH activity. In the period between 1960 and 1970, Schally and Arimura isolated and purified 250 μg of LH-RH from the hypothalamus of 165,000 pigs, but the identification was left to Yoshihiko Baba who joined the group in 1968. However, Baba reached an impasse.

Then, in 1970, Arimura invited Hisayuki Matsuo, who was affiliated with Osaka University's Institute of Protein Research at the time, to determine the structure of LH-RH. The group reached out to Matsuo for assistance because he had developed the tritium labeling method in 1967. This method was advantageous, because it was possible to use a very small amount of sample for analysis, while other peptide structure determination methods at the time required a large amount of sample. Matsuo's tritium labeling method succeeded in identifying a peptide consisting of ten amino acids as LH-RH in April 1971 (Kitamura 2011).

5.10 DDS: Optimizing Drug Therapy

Through the study of sustained-release formulations, which began in the 1960s, PLA and PLGA were identified as materials that fulfill the requirements for sustained-release formulations. Using this information, the microcapsule preparation method using PLA and PLGA was also advanced in the 1960s. In 1968, Alejandro Zaffaroni, of ALZA Corporation, proposed the concept of a DDS as a new formulation design theory for the optimization of pharmacotherapy (Mizushima 2009).

Drugs designed with a DDS were first administered to the human body in the USA in 1974, with the glaucoma drug, pilocarpine hydrochloride (Ocusart), being the first. To administer this drug, ethylene–vinyl acetate copolymer containing pilocarpine hydrochloride, the active ingredient, was applied as a contact lens, and the drug effects lasted for approximately 1 week. Multiple other sustained-release formulations followed, such as scopolamine transdermal therapeutic system, Aftach, and Rhinocort, before the launch of Leuplin (Japan Health Sciences Foundation 2008). The speed with which DDSs were launched suggests that knowledge of DDSs had accumulated prior to the beginning of the 1980s.

5.11   Scientific Foundation of Takeda Pharmaceutical

Takeda Pharmaceutical had already begun to acquire peptide synthesis technology, which was pioneering technology at the time, prior to the start of the leuprorelin research and development (R&D) process. Specifically, the ability to synthesize peptides in large quantities was established by dispatching researchers to the Institute of Protein Research at Osaka University during the latter half of the 1960s, and by employing graduate students who had trained at the institute. The establishment of in-house state-of-the-art peptide synthesis technology would be one of the factors that led to Takeda's success in synthesizing the first TRH and LH-RH derivatives. The flexibility demonstrated by Takeda's R&D activities were superior because the company had a policy of open innovation, which is best demonstrated by the joint research with Abbott, an overseas company. Such a move was rare for a Japanese company at the time.

In 1981, Takeda reorganized its pharmaceutical research laboratories according to their specialized research field. However, Okada's microcapsule research group could conduct inter-laboratory research using a wide range of technologies to achieve their goals, such as synthesis of the bases for encapsulation and the evaluation of absorption. This cross-disciplinary collaboration was unique to Takeda Pharmaceutical, as sectionalism by department was a strong theme within pharmaceutical companies at the time.

Leuplin, thus, was developed by the collaborative efforts of multiple Takeda departments. This structure was possible under the management of Hiroyuki Mima, the director of the drug formulation laboratory at that time. Mima was impressed by ALZA Corporation, the world's first DDS start-up company, and created the first DDS research group in Japan. This environment not only gave Okada the opportunity to study under Prof. Takeru Higuchi at the University of Kansas, but also to develop Leuplin on his return. Similarly, Yamamoto studied at the University of Texas. Thus, Mima, who was a visionary for DDS research, encouraged collaboration and provided researchers with an environment that challenged new possibilities. This was a major impetus for research and development at Takeda.

The new production technology, which was necessary for commercialization of Leuplin, was constructed thanks to the organizational capability of Takeda. Given that Leuplin could not be supplied as an injectable solution because of the microcapsule's biodegradable polymers, solid injectables were prepared that could be dissolved prior to administration with a syringe and were easy to distribute to medical professionals. Drugs with such characteristics need to be easy to prepare in clinical practice, and the high quality of the reconstituted injectable solutions must be maintained, requiring the close cooperation of relevant company departments for production. Leuplin was a new drug that required the construction of new production facilities, and this was possible with the strong financial capability of Takeda. The manufacture of this new facility was necessary because drugs must be sterilized to be supplied as an injection. Leuplin is a solid, which is not as easy to sterilize as a liquid, and the polymer of the microcapsule degrades with heat, thus discounting heat sterilization as an option. To overcome these issues, a manufacturing facility was constructed that allowed all manufacturing processes subsequent to the filter sterilization of the pharmaceutical intermediate to be carried out aseptically. To achieve this, Takeda required not only cooperation between the factory production and quality assurance divisions, but also a good understanding of the future of pharmaceuticals and the financial capability to build the facilities.

5.12   Scientific Contributions to Leuplin R&D

Cooperation between Takeda and the Institute of Protein Research at Osaka University was a major contributor to the science of the discovery process. Notably, Takeda Pharmaceutical began peptide synthesis research in 1964 and introduced its researchers to synthesis technology through training programs at the institute's Peptide Center. Takeda officially established a peptide synthesis research group in 1967. The Institute of Protein Research was established as a center for use by researchers across the country and was primarily an academic institution that was managed under the concept of open innovation. Furthermore, academic research in the Kansai area, such as at Osaka University and Kyoto University, boasted the highest level of peptide synthesis in the world at the time. Moreover, the protein research laboratory at Osaka University supplied raw peptide material to various organizations. This example highlights the importance of academia's initiative in the dissemination of cutting-edge technologies to industry.

The experience gained by dispatching researchers to foreign universities also greatly affected the research. Masahiko Fujino, who played a central role in leuprorelin R&D, attended the University of Texas for additional training at a time when the competition for structural determination of LH-RH was taking place. His training provided him with firsthand experience in the study of TRH and LH-RH, which led to his motivation to create LH-RH derivatives. Furthermore, considering that Hisayuki Matsuo completed the structure determination of LH-RH at Tulane University, Fujino was in a position to quickly obtain state-of-the-art scientific information on the structure of peptide hormones.

Leuplin was a commercial success partly because of its development as a treatment for prostate cancer, a change in the target disease that was based on Abbott's discovery of downregulation. Abbott's description of downregulation was a major scientific contribution to the field. One of the reasons that Abbott initiated the collaborative research was that the vice-president of Abbott was an acquaintance of Sueo

Tateoka, the Director of Takeda's Central Laboratory. Both researchers had studied at the University of Illinois. In this regard, the formation of human networks and the absorption of science through international study programs contributed to the drug discovery.

In terms of studying abroad, Okada, who played a central role in the development of sustained-release formulations, studied under Prof. Takeru Higuchi at the University of Kansas, where he accumulated knowledge and hands-on DDS experience. Notably, Prof. Higuchi was one of the founders of ALZA Corporation, the world's first DDS start-up. Similarly, Masaki Yamamoto, who joined the research group in 1983, studied the production of biodegradable polylactic acid microspheres under Prof. James McGinity at the University of Texas in 1982. Thus, Okada learned the science of DDS and Yamamoto learned polylactic acid technology while in the USA; the combination of these skills created the underlying technology of Leuplin (Yamamoto 1984).

The discovery that leuprorelin acts through sex hormone downregulation was another factor, and largely enabled the development of the TAP clinical program for the treatment of prostate cancer. The concept of downregulation as a pharmacological action was gaining traction in American medical circles at the time, and the early incorporation of this scientific knowledge into collaborative corporate research enabled clinical trials to be carried out for the disease in which the largest market demand was expected. In addition, an American joint venture company conducted clinical studies in the USA in consultation with the FDA, which enabled the smooth and prompt completion of these clinical trials. Completion of the clinical trials in the USA was also a major contributor to the success of the clinical program.

Importantly, clinical trials and commercialization of Leuplin were dependent on a stable supply of the drug and its components. To supply Japanese patients with leuprorelin, a method of controlled drug release was necessary, which led to the development of the microcapsule. However, injection of a microencapsulated compound had not yet been applied to clinical use. From this perspective, it can be implied that the development of excellent production technology was necessary to realize the practical applications of a microencapsulated, controlled-release injection in the clinic. However, such technology does not typically originate from academic research, but rather from applied research, which is central to the pharmaceutical sciences. Thus, the technology was developed in a vertical collaboration between firms by referring to the technology disclosed in prior patent documents. Notably, the synthesis of pharmaceutical-grade polymer bases, a raw material of the microcapsule, was carried out in cooperation with Wako Pure Chemical Industries, an affiliated company. Leuplin was supplied in a syringe-injectable form, and a new customized design of the syringe was carried out jointly with the syringe producer. In the pharmaceutical field, where quality standards and anti-pollution measures are very stringent, the shape and operability of the syringe were important because the syringe is filled with Leuplin in the powder state. The technology and experience of the syringe manufacturer also contributed to the finalization of the product.

5.13 Market Success of Leuplin

Leuplin with a 1-month duration was first marketed in the USA in 1989, and in Japan in 1992, as an injection kit. Takeda Pharmaceutical marketed the product in Europe and some Asian countries including Japan, while TAP sold into the USA and Abbott sold into the remaining markets. Leuplin SR, a 3-month formulation, was launched in 2002. The global sales of Leuplin between 2000 and 2011 are shown in Fig. 5.4. Of note, the total sales from all leuprorelin series, including SR types, was approximately 2.4 billion USD in 2011.

Leuplin suppresses cancer progression with high response rates, even for advanced prostate cancer. The benefit to the patient is greater than that of surgery because the therapeutic effect is comparable to that of traditional orchiectomy without the painful procedure. Other indications include premenopausal breast cancer, endometriosis, central precocious puberty, and uterine fibroids. When compared with other chemotherapeutic agents, it has fewer side effects and less resistance to the dose, mediating a sharp decrease in the number of visits to hospital. This is economically advantageous in that it prevents treatment-related declines in work productivity of patients and increases the quality of life.

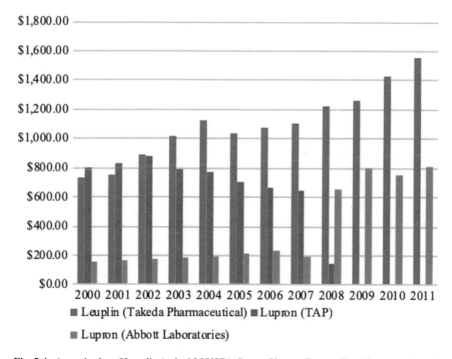

Fig. 5.4 Annual sales of Leuplin (unit: 1 M USD). *Source* Pharma Future (Cegedim strategic data)

5.14   R&D Competition

In 1973, Takeda Pharmaceutical succeeded in the creation of leuprorelin as an effective LH-RH-based drug ingredient. During 1974 and 1975, however, other companies and Andrew Schally himself filed patent applications for LH-RH derivatives (Table 5.1). This suggests that pharmaceutical companies around the world were actively engaged in R&D competition, although they may not have been aware of each other's projects, for the synthesis of LH-RH derivatives, following the structural determination of LH-RH by Andrew Schally in 1971. Takeda's success in this context may be attributed to its early acquisition of peptide synthesis technology and understanding of the scientific advances in LH-RH.

In comparison, R&D competition was apparent during the development stage of the sustained-release microencapsulation preparation. The patent application for the microencapsulation of Leuplin was filed on 4 November 1983, but two companies had filed patent applications for the microencapsulation of LH-RH derivatives prior to Takeda. The details of these applications are given in Table 5.2.

Of these applications, Syntex advanced to the FDA-filing stage, but it failed to resolve the sterilization issue, which allowed Leuplin to go beyond FDA approval. However, the content of the patent filed by Syntex was very similar to that for Leuplin. In particular, Syntex adopted the W/O emulsion method for the preparation of microcapsules, which was a prototype of the W/O/W method used in the preparation of Leuplin microcapsules. Takeda Pharmaceutical, therefore, signed an agreement with Syntex for the use of sustained-release formulation technologies between 1990 and 2005 to avoid patent infringement.

Imperial Chemical Industries (now AstraZeneca) discovered goserelin acetate which is an LH-RH agonist, and successfully developed a PLGA implanted slow-release formulation of goserelin, which was approved in the UK in 1986 for the treatment of prostate cancer. The sustained-release formulations are marketed under the name Zoladex (Kissei Pharmaceutical 2011).

Table 5.1 Patent applications for LH-RH derivatives prior to 1977

Application year	Applicant
September-1973	Takeda Pharmaceutical
April-1974	Sankyo Company Limited
May-1974	American Home Products Corporation
March-1975	American Home Products Corporation
June-1975	Andrew V. Schally
January-1976	Andrew V. Schally
August-1975	Hoechst Aktiengesellschaft
May-1976	Imperial Chemical Industries (ICI)

Source Thomson Innovation

Table 5.2 Patent applications for microcapsule products

Publication number and date	Applicant	Invention name	Priority date	Application number and date
S57-118512, 1982/7/23	Syntex	Capsulation of soluble polypeptide	1980/11/18 (US)	S56-184342, 1981/11/17
S57-150609, 1982/9/17	Imperial chemical industries	Uniform co-polymer and its production method of pharmaceutical compositions, lactic acid, and glycolic acid	1981/2/16 (England)	S57-23497, 1982/2/16

Source Thomson Innovation

5.15 Competition After the Launch

Leuplin is an innovative sustained-release formulation, but it is not the only commercially available prostate cancer therapy. Table 5.3 lists drugs (excluding generics) currently on the market for prostate cancer. Although these drugs are intended for the treatment of prostate cancer, they do not necessarily have the same mechanism of action as Leuplin.

In Table 5.3, only two drugs have the same mechanism of action as Leuplin; these are, goserelin acetate (Zoladex) and triptorelin acetate (Decapeptyl and Trelstar). These drugs also contain LH-RH agonists that mediate a reduction in LH-RH receptors through downregulation and are sustained-release preparations of PLGA and PLA. However, other prostate cancer drugs utilize a different mechanism of

Table 5.3 Pharmaceuticals targeted at prostate cancer

Generic name	Product name	Origin
Leuprorelin	Leuplin, Lupron, enantone	Takeda Pharmaceutical
Goserelin	Zoladex	AstraZeneca
Bicalutamide	Casodex	AstraZeneca
Triptorelin	Decapeptyl	Ipsen
	Trelstar	Watson Pharmaceuticals (now Actavis)
Cabazitaxel	Jevtana	Sanofi-aventis
Abiraterone	Zytiga	Johnson & Johnson
Estramustine	Estracyt	Leo Läkemedel AB
Cisplatin	Briplatin, Randa	Bristol-Myers Squibb
Chlormadinone	Prostal	ASKA Pharmaceutical
Flutamide	Ulexine, Odyne	Schering-Plough Corporation

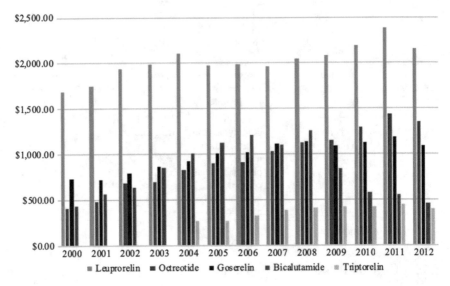

Fig. 5.5 Annual sales of prostate cancer drugs (unit: 1 M USD). *Source* Pharma Future (Cegedim strategic data)

action and thus have different characteristics from Leuplin and Zoladex. Therefore, it is reasonable to consider competition in the product market not only for LH-RH agonist preparations but also for other drugs with prostate cancer indications.

Figure 5.5 shows the sales of a selection of the drugs from Table 5.3, with aggregated sales presented by the generic name of the drug, because many are sold by multiple companies through licensing agreements. Notably, Fig. 5.5 excludes cabazitaxel, abiraterone, estramustine, cisplatin, chlormadinone, and flutamide, because each of these had sales of less than 0.3 billion USD in 2012.

Sales trends indicate that leuprorelin has had the largest sales among the drugs indicated for prostate cancer. This may be attributable not only to the high efficacy of leuprorelin but also to the dramatic improvement in the quality of life of prostate cancer patients, which is a significant benefit of the sustained-release formulation because of the decreased physical burden on patients.

5.16 Conclusion

The creation of Leuplin began with significant scientific advancements in academia. The discovery of LH-RH by Schally and Matsuo was a great achievement that emerged from the field of neuroendocrinology. Fujino and other Takeda Pharmaceutical researchers harnessed this information and with some good luck synthesized leuprorelin, launching it in the USA as Lupron. In addition, the availability of peptide synthesis technology at Takeda Pharmaceutical, which had been acquired by training Takeda researchers at Osaka University's Institute for Protein Research, was a major factor in the company's success. Leuplin has also succeeded in introducing the science of interfacial chemistry to pharmaceutical technology. Okada and other research groups began their research on sustained-release formulations based on a thorough understanding of the significance of new technologies, such as DDS and microencapsulation.

The creation of Leuplin also highlights the organizational capability of Takeda Pharmaceutical, which was fully exploited in the Leuplin R&D process. Traditional drug development strategies divide the labor across R&D units as standard practice for drug discovery efficiency. However, in the discovery of Leuplin, Okada insisted that all essential tasks, such as drug base synthesis, and pharmacological and bio-pharmaceutic testing, be completed by one research group. The scope of this project expanded further as manufacturing started, and the manufacturing facilities suitable for sustained-release formulations were constructed by the team. This became possible through close communication between the engineers, who were aware of the problems to be solved for mass production of innovative DDS products, and management, who recognized the technical value of innovative DDS products and believed that it was a worthwhile investment.

The creation of Leuplin illustrates the significance of the link between basic science and in-house R&D. Leuplin was based on new scientific advancements, including discoveries in the fields of neuroscience, peptide synthesis technology, and microencapsulation technology. At the same time, Leuplin emerged only because a new seed from science was nurtured by the knowledge and experience of Takeda's in-house researchers. Recently, pure science has attracted attention as a source of innovation in the context of academic–industrial alliances; however, companies still play a major role in realizing such potential.

References

Bøler, J., Enzmann, F., Folkers, K., Bowers, C. Y., & Schally, A. V. (1969). The identity of chemical and hormonal properties of the thyrotropin releasing hormone and pyroglutamyl-histidyl-proline amide. *Biochemical and Biophysical Research Communications, 37*(4), 705–710.

Burgus, R., Dunn, T. F., Desiderio, D., & Guillemin, R. (1969). Molecular structure of the hypothalamic hypophysiotropic TRF factor of ovine origin: mass spectrometry demonstration of the PCA-His-Pro-NH2 sequence. *Comptes Rendus hebdomadaires des Séances de l'Académie des Sciences, 269*(19), 1870–1873.

Chemical Book. Leuprorelin |53714 56-0. http://www.chemicalbook.com/ ProductChemicalPropertiesCB8398455_JP.htm. Accessed in December 30, 2015.

Chemical Book. LHRH |9034 40-6. http://www.chemicalbook.com/ChemicalProductProperty_JP_ CB5706343.htm. Accessed in December 30, 2015.

Chemical Book. TRH |34367 54-9. http://www.chemicalbook.com/ChemicalProductProperty_JP_ CB8408947.htm. Accessed in December 30, 2015.

Fukuda, T. (2004). From a little old lab to young people. *Peptide Newsletter Japan, 52*, 1–3.

Fujino, M., Fukuda, T., Shinagawa, S., Kobayashi, S., Yamazaki, I., Nakayama, R., et al. (1974). Synthetic analogs of luteinizing hormone releasing hormone (LH-RH) substituted in position 6 and 10. *Biochemical and Biophysical Research Communications, 60*(1), 406–413.

Harris, G. W. (1948). Neural control of the pituitary gland. *Physiological Reviews, 28,* 139–179.

Huggins, C., & Hodges, C. V. (1941). Studies on prostatic cancer. *Cancer Research, 1,* 293–297.

Japan Health Sciences Foundation. (2008). *Research report post-genome drug development and new development of DDS technology.* HS Report No. 63.

Kawamura, K. (2006). Development of Japanese GMPs and drugs: Results and expectations for tomorrow. *PDA Journal of GMP Validation Japan, 8*(1), 2–17.

Kissei Pharmaceutical. (2011). *Pharmaceutical interview Form/Zoladex 1.8 mg depot* (7th ed.).

Kitamura, K. (2011). People who supported cardiovascular disease research. *Cardiac Practice, 22*(4), 74–75.

Mann, D. R., Gould, K. G., & Collins, D. C. (1984). Influence of continuous gonadotropin-releasing hormone (GnRH) agonist treatment on luteinizing hormone and testosterone secretion, the response to GnRH, and the testicular response to human chorionic gonadotropin in male rhesus monkeys. *Journal of Clinical Endocrinology and Metabolism, 58*(2), 262–267.

Matsuo, H. (1989). Road to Stockholm. In K. Maruyama (Ed.), *Game of the Nobel Prize: Myths and reals of scientific discovery* (171–220).

Mizushima, T. (2009). *What is DDS?* Pharmaceutical Daily website. http://www.yakuji.co.jp/entry8822.html. Accessed in November 11, 2015.

Morita, K. (2000). *"New Drug" created in this way-a secret story on development that will be revealed by the president of the company.* Nihon Keizai Shimbun-sya.

Nevin, S. R. (1981). Copolymer of lactic and glycolit acids using acid ion exchange resin catalyst. US4273920A.

Ogawa, Y. (1995). Development of Leuplin®, a microcapsule that slowly releases a drug over a month. *Membrane, 20*(2), 149–153.

Ogawa, Y., & Fujino, M. (1994). Luprin: The drug to cure prostate cancer. In *Japan Society of agricultural chemistry, medicine in this topic: Background and therapeutic effects of development* (pp. 173–190).

Okada, H. (1993). Development of Lupron depo: DDS is the second drug discovery. *Pharmacia, 29*(11), 1253–1255.

Okada, H., Sakura, Y., Kawaji, H., Yashiki, T., & Mima, H. (1983). Regression of rat mammary tumors by a potent luteinizing hormone-releasing hormone analogue (leuprolide) administered vaginally. *Cancer Research, 43*(4), 1869–1874.

Shimamoto, T. (1980). Contraceptive effects of intranasal administration of gonadotropin-releasing hormone agonists. *Pharmacia, 16*(1), 48–49.

Takeda Pharmaceutical. (2013). *Pharmaceutical interview form/LH-RH derivative microcapsule extended release preparation for Leuplin injection 1.88/3.75 Kit.*

Yamamoto, M. (1984). Learning from the University of Texas DDI. *Journal of Powder Engineering, 21*(5), 304–307.

Chapter 6
Ofloxacin and Levofloxacin (Tarivid/Cravit)

Best-in-Class Antimicrobial Agents

Yasushi Hara and Yuji Honjo

Abstract Both ofloxacin (Tarivid), a second-generation fluoroquinolone, and lev-ofloxacin (Cravit), a third-generation fluoroquinolone, were developed by Daiichi Pharmaceutical. Both are excellent antimicrobials with wide antibacterial spectrums, and levofloxacin is the world's first optically active fluoroquinolone. Daiichi Pharmaceutical conducted research and development (R&D) over approximately 20 years to discover and develop these two drugs. The R&D process for ofloxacin was based on the notion that the physical properties of drugs affect not only blood levels but also metabolic and water-soluble pharmacokinetics, which determine overall efficacy. The project was initially implemented as unauthorized research, suggesting that researchers' individual initiatives are important to pursue diverse ideas for innovation. Levofloxacin's R&D process involved the difficult task of optical resolution of ofloxacin and resulted in two-times the antibacterial activity and a drastic reduction in side effects. Nevertheless, Daiichi Pharmaceutical was just 2 days ahead of its competitor in filing the patent application for levofloxacin, demonstrating the importance of time-based competition in the discovery of drugs with known mechanisms of action. A new asymmetric synthesis method was introduced for the mass production of levofloxacin. This enabled Daiichi Pharmaceutical to achieve its original goal of reducing the production costs of levofloxacin to no more than twice that of ofloxacin.

Y. Hara (✉)
CEAFJP/EHESS, Paris, France
e-mail: yasushi.hara@r.hit-u.ac.jp

Faculty of Economics, Hitotsubashi University, Tokyo, Japan

Y. Honjo
Chuo University, Tokyo, Japan

© Springer Nature Singapore Pte Ltd. 2019
S. Nagaoka (ed.), *Drug Discovery in Japan*,
https://doi.org/10.1007/978-981-13-8906-1_6

6.1 Introduction

Ofloxacin (Tarivid), a second-generation fluoroquinolone, was developed by Daiichi Pharmaceutical Co., Ltd. (Daiichi-Sankyo) and was launched in 1985. Shortly thereafter, Daiichi Pharmaceutical developed levofloxacin (Cravit), a third-generation fluoroquinolone, as a successor to ofloxacin. Levofloxacin was launched in 1993 and was the world's first optically active fluoroquinolone. Quinolone antibiotics are a group of compounds with broad-spectrum bactericidal effects; fluoroquinolones are a subset of quinolones that contain a fluorine atom.

Sulfonamides were used as antimicrobials before the Second World War; however, resistance to sulfonamides quickly emerged, which led to the introduction of synthetic quinolone antimicrobials in the 1960s. Unfortunately, conventional synthetic quinolones did not show strong antibacterial activity against Gram-positive bacteria, so these antimicrobials were only used to treat limited diseases, such as urinary tract and enteric infections, which are primarily caused by Gram-negatives (Hayakawa 2010).

Conversely, ofloxacin demonstrated a broad spectrum of antibacterial effects, including efficacy against both Gram-negative and Gram-positive bacteria. It was evaluated as an oral antibacterial agent with a wide range of indications, including respiratory infections. However, ofloxacin was associated with some side effects, including mild insomnia. Thus, the company synthesized levofloxacin in an effort to reduce side effects and increase antibacterial activity. In addition, levofloxacin demonstrated strong activity against penicillin-resistant *Streptococcus pneumoniae* (PRSP), which became a serious issue in the 1990s, and was highly valued in clinical practice (Hayakawa et al. 2010). The bacterial species targeted by levofloxacin are given in Table 6.1.

The main features of levofloxacin are: (1) strong antibacterial activity (stronger than ofloxacin) and (2) and lower toxicity than ofloxacin (racemic mixture) by optical resolution to the *S*-isomer.

This chapter discusses the research and development (R&D) history of ofloxacin, the first new quinolone synthetic antibacterial agent launched by Daiichi Pharmaceutical and the source of levofloxacin. The subsequent development of levofloxacin is also described.

6.2 Mechanism of Action of Fluoroquinolones

Fluoroquinolones exhibit antibacterial activity by acting on the Type II topoisomerases, DNA gyrase (Gyr) and topoisomerase IV (TopoIV), which are the enzymes involved in bacterial DNA replication, to inhibit DNA replication (Totsuka et al. 2009).

DNA gyrase is a tetramer that consists of two A subunits (GyrA) and two B subunits (GyrB). It plays a key role in DNA replication, transcription, recombination,

Table 6.1 Bacterial targets of levofloxacin

Gram-positive bacteria	1. *Staphylococcus* sp.
	2. *Streptococcus* sp.
	3. *Pneumococcus* sp.
	4. *Enterococcus* sp.
	5. *Bacillus anthracis*
	6. *Peptostreptococcus* sp.
	7. *Propionibacterium acnes*
Gram-negative bacteria	1. *Neisseria gonorrhoeae*
	2. *Moraxella catarrhalis*
	3. *Escherichia coli*
	4. *Shigella* sp.
	5. *Salmonella* sp.
	6. *Citrobacter* sp.
	7. *Klebsiella* sp.
	8. *Enterobacter* sp.
	9. *Serratia* sp.
	10. *Proteus* sp.
	11. *Morganella morganii*
	12. *Providencia* sp.
	13. *Yersinia pestis*
	14. *Vibrio cholerae*
	15. *Haemophilus influenzae*
	16. *Pseudomonas aeruginosa*
	17. *Acinetobacter* sp.
	18. *Brucella* sp.
	19. *Francisella tularensis*
	20. *Campylobacter* sp.
	21. *Coxiella burnetii*
	22. *Chlamydia trachomatis*
	23. *Legionella* sp.

Source Hayakawa et al. (2010)

and repair. DNA gyrase binds to double-stranded DNA, makes a double-stranded break at the binding site, and rejoins the DNA. In this process, the DNA passes through the enzyme to induce the formation of a negative super helical structure in the double-stranded DNA (Fig. 6.1). The quinolones inhibit DNA synthesis by fitting into the DNA loops at the breakpoints created by GyrA and prevent reassociation (Fig. 6.2). GyrB also exhibits ATPase activity and is involved in the conformational change of GyrA during the hydrolysis process.

Model for the DNA transport reaction of type II topoisomerases

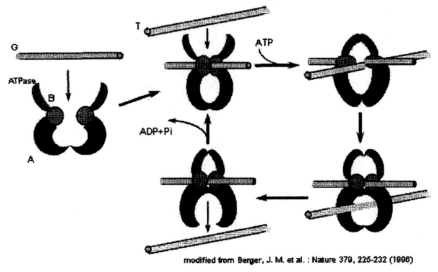

modified from Berger, J. M. et al. : Nature 379, 225-232 (1996)

Fig. 6.1 Functional model of DNA gyrase. *Source* Sato (2006)

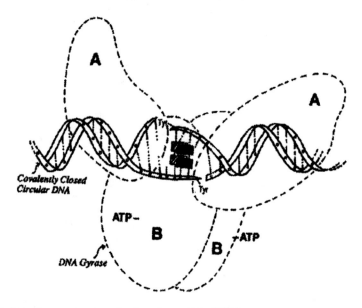

Fig. 6.2 Quinolone mechanism of action. *Source* Sato (2006)

Similarly, TopoIV catalyzes the decatenation of replicated DNA to daughter cells and is a tetramer of two molecules each of ParC and ParE. Notably, the structures of ParA and ParB are similar to those of Gyr but are characterized by a decatenation function that cleaves entangled dimers during late DNA replication.

Fluoroquinolones exhibit their antimicrobial properties by acting on bacterial Gyr and/or TopoIV, although the inhibitory effect differs between species. In particular, Gram-negative bacteria are more susceptible, wherein Gyr is the primary target and TopoIV is the secondary target. Conversely, in Gram-positive bacteria, TopoIV is the primary target and Gyr is the secondary target (Sato 2006). Thus, fluoroquinolones are characterized by their potent bactericidal activity against a broad spectrum of bacterial species, including Gram-positive and Gram-negative bacteria that grow aerobically and anaerobically (Takahashi et al. 2003).

6.3   Ofloxacin R&D

The history of R&D for ofloxacin is shown below and covers the period from the introduction of nalidixic acid, a synthetic fluoroquinolone, to the market up until the market launch of ofloxacin by Daiichi Pharmaceutical.

1962: Nalidixic acid (Wintomylon) was discovered and commercialized by Winthrop Laboratories in the USA under the leadership of George Lescher.
1964: Daiichi Pharmaceutical began importing and selling nalidixic acid in Japan. Daiichi Pharmaceutical began in-house discovery research of weakly acidic quinolones, using nalidixic acid as a lead compound.
1971: The clinical study of candidate DB-2563 began; its development was later discontinued.
1971: Isao Hayakawa joined the quinolone-related research group.
1978: The development of candidate DJ-6783, synthesized by Hayakawa, was abandoned.
1979: Hayakawa invented ofloxacin based on his research of ampholytic quinolones, which was inspired by the patent disclosure of norfloxacin by Kyorin Pharmaceutical.
1980: Preclinical studies of ofloxacin were initiated.
1981: Clinical trials of ofloxacin began.
June 1985: Ofloxacin was launched in West Germany.
September 1985: Ofloxacin was launched in Japan.
December 1985: Ofloxacin was launched in Malaysia and Hong Kong.
December 1990: Ofloxacin was launched in the United States.

Before and after the Second World War, Daiichi Pharmaceutical developed and marketed sulfa chemotherapy fungicides including the first-stage sulfamin Terapol in 1937, the second-stage sulfamine Terrapol in 1939, the sulfapyridine Terebilidine in 1939, the acetosulfamine Neotelapol in 1940, the sulfadiazine Teradiacin in 1948, and the sulfamonomethoxine Dimethone in 1962 (Daiichi Seiyaku 2007).

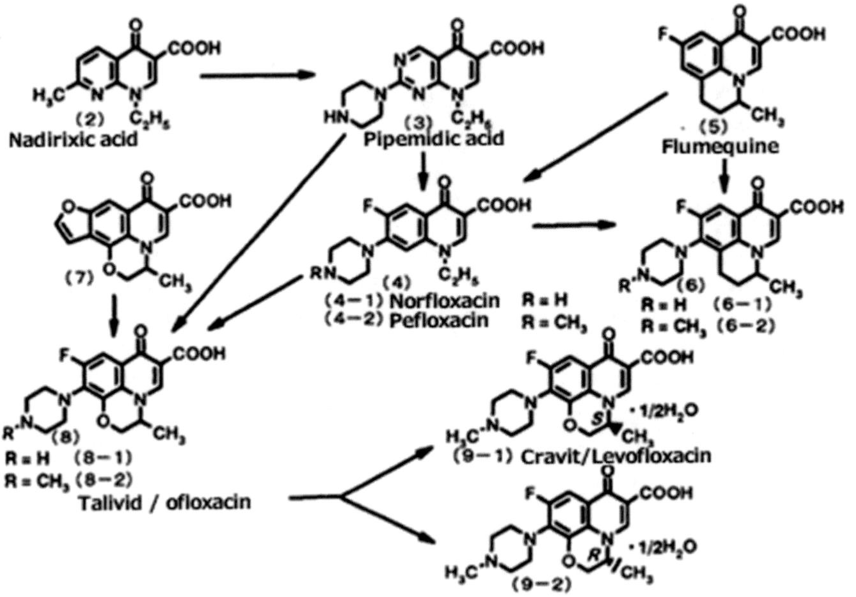

Fig. 6.3 Modifications to the chemical structure of nalidixic acid to produce ofloxacin and levofloxacin. *Note* (7) denotes an old quinolone with strong antimicrobial activity across a broad spectrum and low acute toxicity, obtained by Hayakawa et al. (6-1) and (6-2) denote compounds with antimicrobial profiles similar to that of ofloxacin, but (6-1) is associated with high acute toxicity, and (6-2) is associated with side effects such as insomnia because of its high lipophilicity. *Source* Adapted from Hayakawa (1999)

The move by Daiichi Pharmaceutical to start developing synthetic quinolones was triggered by experience in the import and sales of nalidixic acid (Fig. 6.3). Indeed, Daiichi's licensing of nalidixic acid in 1964 led to the development of ofloxacin and levofloxacin.

When first introduced to the market, nalidixic acid was characterized by a new mechanism of action, good oral absorption, a low frequency of antibiotic resistance, and no cross-resistance to sulphonamides or beta-lactams (acquired resistance to other drugs as well as resistance to other types of drugs). Nalidixic acid was indicated for urinary tract and intestinal infections because it had an effective antibacterial spectrum against Gram-negative bacteria. However, nalidixic acid was not approved for respiratory infection, because it was not effective against Gram-positives. Therefore, the development of antibacterial agents that were effective for a broader spectrum of pathogens and capable of reaching of the whole body, including the respiratory tract, were required. Because nalidixic acid was improved to expand the antibacterial spectrum against Gram-positive bacteria and the level of the overall antibacterial activity was increased, quinolones that were effective for various infectious diseases, including respiratory infections, could be developed. Indeed, several companies began searching for the next generation of fluoroquinolones.

Daiichi Pharmaceutical began importing and marketing nalidixic acid as a drug for urinary tract infections in 1964; they also began the R&D process for a new compound using nalidixic acid as a lead compound (Daiichi Sankyo 2012a). With the intention of launching the new drug as soon as possible, Daiichi Pharmaceutical allocated over 20 full-time staff to the R&D of ofloxacin (Daiichi Seiyaku 1997). Thioxacin (DJ-1611) was selected as a candidate compound in the synthesis process from more than 1000 derivatives. Although thioxacin was found to show excellent antibacterial activity, its development was abandoned because of poor absorption in the intestines (Daiichi Seiyaku 2007). In 1971, another compound (DJ-2563) entered clinical trials, but its development was discontinued for the same reason (Daiichi Sankyo 2012a).

Continued R&D led to the discovery of DJ-6783, which was an acidic quinolone compound with strong activity against Gram-positive and Gram-negative bacteria, including *Pseudomonas aeruginosa*. The drug also reached high concentrations in the blood and was selected as a drug candidate in 1978; clinical trials began in 1979 (Hayakawa et al. 2010). The chemical synthesis of this compound was undertaken by Isao Hayakawa, who would later lead the levofloxacin R&D team (Hayakawa 1991). However, DJ-6783 was metabolized in the human body and excreted in urine as a glucuronide conjugate with almost no antibacterial activity; thus, it had no effect on urinary tract infections in early clinical trials, which resulted in its discontinuation (Daiichi Seiyaku 2007).

Nonetheless, Hayakawa encountered a new fluoroquinolone chemical structure (norfloxacin) that had been developed by Kyorin Pharmaceuticals. Despite low blood concentrations, high rates of tissue penetration and high concentrations of norfloxacin were found in tissue, suggesting the fluorine and adjacent basic sites enhanced both the activity and physical properties of the compound (Hayakawa et al. 2010). Given this new information, Hayakawa began an unauthorized secret research project ("*yami* research") in early 1979 to synthesize amphoteric quinolones with carboxylic acid and basic moieties along a tricyclic backbone (pyridobenzoxazine) with an oxazine ring. This led to the synthesis of DJ-6779 (Fig. 6.3, compound 7), which showed strong and broad antibacterial activity, but its development was discontinued because of poor oral absorption and low blood concentrations (Research Industry Association of the Japan Machinery Federation 1995).

Six months later, Daiichi Pharmaceutical synthesized DL-8112, which was similar to the new quinolones of other companies. DL-8112, which had a piperazinyl group, was highly active but acutely toxic in mice, and was poorly absorbed when administered orally in rats (Hayakawa 2010). Ofloxacin (DL-8280) was synthesized in 1979; preclinical studies demonstrated a broad-spectrum drug with strong antimicrobial activity and good pharmacokinetics (Hayakawa et al. 1984). The chemical structure of ofloxacin featured not only an *N*-methylpiperazine that granted high lipophilicity and efficacy against Gram-positive bacteria, but also an oxazine ring that contributed low lipophilicity, high water solubility, and safety. Ofloxacin was also weakly toxic in mice. These features resulted in excellent intestinal absorption and tissue penetration (Sato et al. 1982).

6.4   Efficient Mass Production of Ofloxacin

Researchers faced two major problems in developing a mass-production process for ofloxacin that would meet the needs of market launch: how to efficiently isolate high-purity closed-ring compounds and how to avoid regioisomer by-products in the last substitution reaction with cyclic amines (Hayakawa et al. 1996).

Initially, N-methylpiperazine was reacted with the intermediate compound in the synthesis process of ofloxacin. However, this process required a reaction temperature of 100–140 °C for approximately 3 h, which was not suitable for mass production given the unavoidable generation of regioisomer by-products (Joumuro and Tusurumi 1983). Fortunately, new technology was developed for obtaining chelated BF2, by which the instantaneous ring closure reaction was advanced by dropwise addition of BF3 and tetrahydrofuran to the intermediate compound dissolved in acetic anhydride. The resulting chelate was reacted with N-methylpiperazine at room temperature or below to obtain pure ofloxacin (Hayakawa et al. 1996). In following this protocol, it became possible to obtain ofloxacin at comparatively low temperatures with high yield and purity relative to the conventional method.

6.5   Ofloxacin Clinical Research

In 1981, Daiichi Pharmaceutical began a clinical trial of ofloxacin; this Phase I clinical trial was carried out in 40 adult males (Ichihara et al. 1984) and was followed by a double-blind study comparing ofloxacin and amoxicillin for respiratory infections from December 1982 to July 1983 (Kobayashi et al. 1984). Similarly, a double-blind comparative study with cefaclor as a comparator was conducted between March and September in 1983 for patients with superficial suppurative diseases (Fujita et al. 1984). In the Phase III clinical trial, eight double-blind comparative studies were conducted, covering various disease fields. Ofloxacin showed non-inferiority in all fields; indications for the treatment in seven fields, including the respiratory tract, were selected (Hayakawa 2010).

Ofloxacin was launched as Tarivid in West Germany in June 1985 and then in Japan in September 1985. By 1988, ofloxacin had been launched in 48 countries. In the United States, Daiichi Pharmaceutical gained marketing approval from the US Food and Drug Administration (FDA) in December 1990; ofloxacin was then launched as Phloxine in February 1991 (Daiichi Seiyaku 2007).

6.6   Timeline of Levofloxacin R&D

The R&D timeline for levofloxacin from the optical resolution of ofloxacin to the launch of levofloxacin is shown as follows:

April 1985: Successful optical resolution of ofloxacin.

June 20, 1985: Patent application filed for levofloxacin.
September 1985: Ofloxacin launched in Japan.
October 11, 1985: The patent application for an intermediate of levofloxacin was filed.
1986: Patent application for the production of levofloxacin by the tosylproline method was filed.
1988: Phase II clinical trials began.
1989–1991: Phase III clinical trials began.
October 1993: Levofloxacin was approved by Pharmaceuticals and Medical Devices Agency, PMDA
December 1993: Levofloxacin launched in Japan.
2000: Levofloxacin acquired approval for Typhoid and Paratyphoid in Japan.
2002: Levofloxacin acquired additional indications of Cravit for Plague, Tulariseck, Brucellosis, and Q Fever in Japan.
2006: Levofloxacin was approved for the treatment of *Legionella* sp. in Japan.
2011: Levofloxacin acquired approval for the treatment of *Chlamydia pneumoniae* and *Mycoplasma pneumoniae*.

6.7   Increased Competition for Ofloxacin and the Discovery of Levofloxacin

After the launch of ofloxacin, Daiichi Pharmaceutical assumed that drugs capable of outperforming ofloxacin would not be commercially available for some time.[1] Accordingly, the synthetic antibacterial agents research laboratory of Daiichi Pharmaceutical was reduced to two persons. However, Bayer's ciprofloxacin attracted clinical attention as a highly effective new antimicrobial agent, which prompted more than 20 companies to begin the development of synthetic quinolones with high potency. As a result of the competition among these companies, drugs with antimicrobial activity that outperformed ofloxacin were quickly developed and put on the market (Hayakawa 2007).

In response, Daiichi Pharmaceutical increased the number of R&D researchers to focus on synthesizing the next fluoroquinolones. The research team examined the structure–activity relationships of chemical compounds that were similar to ofloxacin, while avoiding the scope of other companies' patents, as well as to responding to input from the sales department. However, the research team was unable to synthesize any chemical compounds with higher activity than ofloxacin (Hayakawa et al. 1996).

---

[1] After the market launch of ofloxacin, Daiichi believed that "there was a fortunate coincidence in the derivation of ofloxacin. Competitors of a superior antimicrobial agent will not immediately reach the market. Thus, it must be possible to monopolize the market with ofloxacin" for at least 10 years (Research Industry Association of the Japan Machinery Federation 1995).

Ofloxacin is a racemic mixture that has equal amounts of the left-handed and right-handed enantiomers of its asymmetric carbon atom. During the development of ofloxacin, the activities of the 3S- and 3R-isomers were considered to be almost equal and to be sufficiently effective as a racemic mixture; thus, it was commercialized in that form. Therefore, it was not important to examine physical properties of the enantiomers, partly because separating them was considered to be unnecessarily difficult. Indeed, researchers failed to optically resolve the mixture using a diastereomeric derivatization method in which the functional group of ofloxacin was derivatized and separated by liquid chromatography. It was later recognized in the literature that the diastereomer derivatization method could not separate and discriminate enantiomers of ofloxacin because of its chemical structure (Ohrui et al. 2003).

Consequently, three new methods of synthesis of optically active substances, namely, chiral column separation, oxygen, and rate sil-propylin methods, were considered as methods to obtain optical resolution of ofloxacin. A formal project team was set up with Shogo Arako (Research Director) and Shoichi Sakano, who was studying asymmetric hydrolysis of lipases in the Natural Products Fermentation Research Group.

Accordingly, Arako and Sakano identified that it may be possible to separate the enantiomers of ofloxacin's intermediate directly using a chiral column prior to the asymmetric hydrolysis process (Hayakawa 2010). To achieve this, they asked Sumitomo Chemical to produce a preparative Sunipax OA-4200 column. Using the preparative column, they obtained 2 mg of optically active intermediate four times a day, and, in total, managed to obtain 250 mg over a 1-month period. Moreover, they identified the optimum conditions by carrying out a modeling experiment for each step and ultimately succeeded by obtaining about 10 mg of each ofloxacin enantiomer from each optically active intermediate. These experiments were conducted by Shuichi Yokoyama, and the method was later patented.

This optical resolution process of ofloxacin generated DR-3355, DR-3354, and the S- and R-isomers, respectively (Daiichi Seiyaku 2007). Subsequently, the assessment of activity in vitro and in vivo using mouse models revealed that the 3S-isoform was twice as active as ofloxacin and it was less acutely toxic in mice than the 3R-isoform (Hayakawa et al. 1986, 2010). Therefore, it was expected that the insomnia associated with ofloxacin would be relieved. The patent application for levofloxacin was filed on 20 June 1985, while Bayer almost simultaneously submitted a patent for the same compound (22 June 1985), indicating that intense competition occurred for optical resolution.

Numerous synthetic compounds were obtained via N-tosylprolylamide as a diastereomer method, yet none of them became development candidates because of safety and pharmacokinetic issues (Hayakawa et al. 1996). In addition, other proposals were made to evaluate highly lipophilic compounds with different substituents. However, the results from other companies showed that highly lipophilic compounds posed serious risks of side effects involving the central nervous system, and the results of animal experiments revealed that DR-3354 had a strong arousal

effect and an effect on central nervous system. Therefore, clinical development of the
S-isoform, DR-3355, began in the autumn of 1986; it would become levofloxacin.

6.8   Development of the Mass Production Process of Levofloxacin

In the process of establishing the production system for levofloxacin, Daiichi Pharmaceutical aimed at reducing the production costs of levofloxacin to no more than
twice that of ofloxacin, because the antibacterial activity of levofloxacin was twice
that of ofloxacin (Hayakawa et al. 1996). A number of methods were tested, including the N-tosylprolylamide production method (Fujiwara and Yokota 1989), optical
resolution with diastereomeric salts (Fujiwara and Satoh 1989), and racemization
through imines (Fujiwara and Tsurumi 1989). Eventually, the asymmetric synthetic
elements method was chosen.

However, many competitors, which had already developed new fluoroquinolones,
such as Bayer (Silliver and Growe 1987), Kyorin Pharmaceutical, Dainippon Pharmaceutical, and Abbott, had also filed patents for synthesis methods using L-aralinol
as an asymmetric synthetic element. For this reason, Daiichi Pharmaceutical sought
to develop an original synthetic method.

While developing their method, Daiichi Pharmaceutical referenced (R)-1,2-
propanediol derivatives derived from methyl D-lactate, which were stimulated by
the tricyclic quinolone synthesis and oxazine ring closure; this fact was disclosed
by Pfizer's scientific papers and patents (Parikh et al. 1988; Gilligan et al. 1984).
(R)-1,2-Propanediol derivative sodium salt was used to obtain the intermediate compound after a series of chemical processes (Hayakawa et al. 1996). By transforming
the intermediate compounds into levofloxacin in a manner similar to that used for
ofloxacin, the production method for levofloxacin was established (Fig. 6.4) (Fujiwara and Ebata 1989, 1990).

6.9   Levofloxacin Clinical Development

During the preclinical development of levofloxacin, Hayakawa temporarily moved
to another department to lead a new R&D project (Hayakawa 2010). However, he
continued to coordinate the preclinical tests, including Good Laboratory Practice
tests, sample procurement for different groups, supplying the compound for tests,
and monitoring the progress of manufacturing tests.

After preclinical testing, Hayakawa transferred the clinical development to a clinical research group. As a result of the clinical trial, levofloxacin was recognized by
medical doctors to be applicable for a wide range of infectious diseases because it
demonstrated strong activity against Gram-negative and Gram-positive bacteria. In

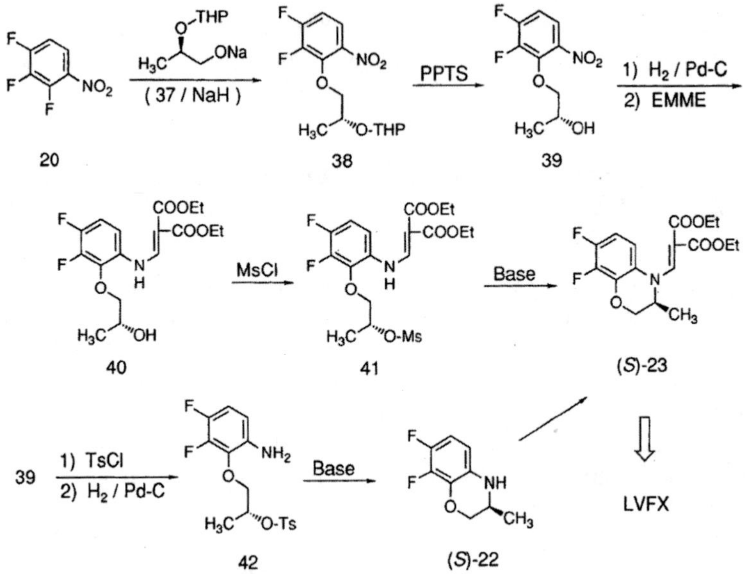

Fig. 6.4 Levofloxacin synthesis using (*R*)-1,2-propanediol derivatives. *Source* Hayakawa et al. (1996)

addition, the slight insomnia associated with ofloxacin was suppressed; hence, its evaluation improved. Furthermore, levofloxacin was equally effective as ofloxacin at half of the oral dose, 600 mg per day, which reduced the severity of side effects.

The Phase I trial was conducted in adult males (Shiba et al. 1992; Nakashima et al. 1992), and the results were presented in the 28th Interscience Conference on Antimicrobial Agents and Chemotherapy (Nakashima et al. 1988). Next, a dose-controlled clinical study (Phase II), a double-blind comparative study (Phase III), and an open clinical study were conducted. The target diseases and clinical study durations are presented in Tables 6.2, 6.3, and 6.4.

Phase III clinical studies of levofloxacin were conducted overseas, similar to the clinical studies in Japan. The overseas clinical trials involved more than twenty-two university institutions and 313 patients aged 18 years or older were enrolled (Preston et al. 1998). Levofloxacin was approved in October 1993 in Japan, listed in the National Health Insurance drug price standard in November 1993, and launched in December 1993 (Daiichi Seiyaku 2007).

Subsequently, an increase in pneumococcal resistance to fluoroquinolones in elderly persons was reported, likely because new fluoroquinolones, including ofloxacin and levofloxacin, were widely used. As a result, a request for cooperation on the development of proper use guidelines for quinolone antimicrobials was submitted to the Minister of Health, Labour and Welfare by the Japanese Society of Chemotherapy in July 2005. This led to a review of the twice-daily or thrice-daily administration of 100 mg levofloxacin at the time of approval and the thrice-daily

Table 6.2 Dose-controlled clinical study (Phase II) of levofloxacin

Targeted disease	Participant number	Duration	Reporting paper
Chronic bronchitis, acute exacerbation of diffuse panbronchiolitis, bronchiectasis with infection, bronchial asthma, emphysema, pulmonary fibrosis, old pulmonary tuberculosis	76 cases (52 presumptive cases of phlogogenic bacteria; 41 single bacterial infections)	April 1988 to June 1989	Saito et al. (1992a)
Complicated urinary tract infections	201 patients (162 evaluable)	May 1988 to December 1988	Kawada et al. (1992)

Table 6.3 Double-blind comparative study (Phase III) of levofloxacin

Targeted disease	Number	Duration	Reporting paper
Acute and chronic exacerbations of otitis media or otitis externa	201 patients (180 were evaluable for clinical response)	September 1989 to August 1990	Ishii et al. (1992)
Lower than moderate chronic lower respiratory tract infection	165 patients (148 were evaluable for clinical response)	October 1989 to August 1991	Soejima et al. (1992)
Bacterial pneumonia	159 patients (140 were included in the evaluation of clinical response)	November 1989 to August 1991	Soejima et al. (1992)
Peridontitis and urinary tract infections	327 patients (261 patients evaluable for clinical response)	August 1989 to March 1990	Kawada et al. (1992)

administration of 200 mg. Furthermore, a clinical study was conducted to introduce once daily administration of 500 mg as in the United States and Europe. In a Phase I study, a single dose study of levofloxacin 250, 500, 750, and 1000 mg in healthy Japanese adult males, and a once-daily repeated dose study of levofloxacin 500 mg in adult males and elderly males were conducted to investigate the pharmacokinetics and safety of levofloxacin (Shiba et al. 2009). Subsequently, a Phase III clinical trial in patients with respiratory and urinary tract infections was conducted (Totsuka et al. 2009). Thereafter, marketing authorization was obtained for levofloxacin 250, 500 mg, and fine granules (10%) in April 2009.[2]

---

[2] Drug Interview Form: Levofloxacin Tablets Cravit 250, 500 mg/Cravit Fine Granules 10%.

Table 6.4 Open clinical study of levofloxacin

Targeted disease	Number	Duration	Reporting paper
Respiratory tract infections (pneumonia, chronic lower respiratory tract infections, pharyngolaryngitis/tonsillitis, acute bronchitis)	463 patients (420 patients included in the evaluation of clinical response)	September 1989 to September 1991	Saito et al. (1992b)
Infectious enteritis (e.g., shigellosis, Salmonella enteritis, Campylobacter enteritis, cholera)	266 patients (259 patients included in the assessment of adverse reactions; 216 patients included in the assessment of laboratory values or better)	July 1989 to March 1991	Murata et al. (1992)
Urogenital infection	539 patients (517 evaluable)	December 1989 to May 1991	Kawada et al. (1992)
Surgical infections (superficial suppurative diseases, mastitis, perianal abscess, cholecystitis/cholangitis)	292 patients (of whom 267 were evaluable for clinical response)	October 1989 to September 1990	Yura et al. (1992)
Dermatological infection	40 patients (clinical response evaluable in 39 patients)	November 1989 to May 1990	Takahashi et al. (1992)
Obstetrical and gynecological infections (intrauterine infection, uterine adnexitis, bartholinitis/abscess, cervicitis, and mastitis)	290 patients (197 included in the evaluation of clinical response)	August 1989 to October 1990	Matsuda et al. (1992)
Otitis media and otitis externa	201 patients (180 were evaluable for clinical response)	September 1989 to August 1990	Ishii et al. (1992b)
Tonsillitis, peritonsillar tumors, pharyngolaryngitis, and suppurative sialadenitis	110 patients (88 patients evaluable for clinical response)	September 1989 to August 1990	Ohyama et al. (1992)

(continued)

Table 6.4 (continued)

Targeted disease	Number	Duration	Reporting paper
Sinusitis	74 patients (68 patients evaluable for clinical response)	September 1989 to August 1990	Baba et al. (1992)
Infections in the field of dentistry and oral surgery (periodontitis, pericoronitis, jaw inflammation)	223 patients (203 were evaluable for clinical response)	September 1989 to September 1990	Sasaki et al. (1992)

Box 6.1: Optical activity and drug discovery (by Hideo Kawabe)

Many pharmaceuticals are small organic compounds, and there are many compounds in which carbon is bonded as a chain to a basic skeleton. A carbon atom can have four covalent bonds, but a carbon atom that features a different covalent partner on each bond is called an asymmetric center. A compound with an asymmetric center has two stereo structures, or mirror-image compounds, that would exist if a mirror were placed in the symmetry plane. Because these mirror-image compounds exhibit different properties with respect to light, dextrorotatory compounds are called D- or R-isomers, levorotatory compounds are called L- or S-isomers, and a mixture of them is called a racemic mixture.

Many drugs exert their actions by specifically binding to target receptors and enzymes in vivo. Pharmaceuticals are also degraded and excreted in the body by drug-metabolizing enzymes. The binding of a drug to a receptor or enzyme is exemplified by a lock-and-key relationship, but the strength or weakness of the drug binding is influenced by factors such as steric structure, charge, and hydrophobicity. Depending on the nature of the receptor or enzyme, some drugs bind only a small region of their conformation (highly specific) while others accept a wide range of drugs with different conformations (nonspecific). Receptors and enzymes, which are particularly structurally specific, are significantly affected by changes in conformation because of the optical activity of the drug. Therefore, depending on the nature of the drug's target, drugs with the same structural formula may differ markedly because of differences in optical activity such as the D- and L-isomers, or the R- and S-isomers.

Thalidomide was a key compound that made the medical community recognize the importance of optical activity and drug action. Thalidomide, a non-barbital hypnotic agent, became rapidly popular as a drug that removed acute respiratory depression, a side effect of barbital, in 1957, but it was found to have teratogenic side effects in the 1960s and its manufacturing approval was withdrawn. However, it was confirmed in 1979 that the teratogenicity of thalidomide was caused by the S-isomer and the hypnotic effect was caused by the R-isomer. Although many drugs are supplied as racemic intermediates, the

FDA has since established new guidelines based on examples, such as thalidomide, to provide a closer look at the safety of optically active compounds and racemic intermediates.

Since then, much has been learned about the relationship between optical activity and pharmacological activity. Many new drugs have been discovered with the optical activity of the chemical compound in mind. As discussed in this chapter, the research on levofloxacin began with exploratory studies of the optically active forms of ofloxacin, which was on the market, and is a racemic mixture. Optical resolution of ofloxacin showed that the S-$(-)$ form was twice as active as the racemic form and had fewer side effects. Although there were many difficulties to be overcome in the commercialization process, such as the construction of commercialization technology for the production of optically active substances consisting only of S-$(-)$ compounds, levofloxacin was marketed as an antibacterial agent with higher potency and less side effects than ofloxacin. Advances in optical resolution techniques made this possible.

Furthermore, the success of levofloxacin prompted the industry effort to develop a drug consisting only of optically active compounds with new and superior properties to overcome the disadvantages of existing racemic drugs racemic drug. For example, cetirizine (Zyrtec), which was marketed as an antiallergic agent, was a racemic mixture consisting of the R-isomer (levocetirizine) and the S-isomer (dextrocetirizine). The inhibitory effect on histamine receptors was 30 times stronger in the R-isomer than in the S-isomer, and the duration of inhibition of histamine receptors by the R-isomer was 108 min longer. Furthermore, the histamine receptor binding rate in the brain was weaker in the R-isomer, causing less drowsiness. Thus, R-only levocetirizine was developed and commercialized as Xyzal, which is now well received because it requires only half the dose of Zyrtec and has a long duration of action. Similarly, omeprazole (Omepral) was an anti-ulcer agent that inhibited gastric acid secretion by inhibiting gastric proton-potassium ATPase, but the efficacy varied among patients. Omeprazole is also a racemic mixture. Examination of the optically active isomers revealed that the S-isomer was not significantly affected by drug-metabolizing enzymes, whereas the R-isomer was rapidly degraded. The R-isomer was also found to be ineffective in patients with strong drug-metabolizing enzymes. Therefore, a new drug esomeprazole (Nexime), which consists only of the S-form, was developed and marketed as an anti-ulcer drug with few individual differences.

Notably, there are some cases in which successful delivery of optically active substances as a new drug occurred at the time of drug discovery. Tamsulosin hydrochloride, the active ingredient in Harnal, as discussed in Chap. 7, was found to bind α_1-adrenoceptors in the R-form about 600 times more tightly than the S-form to the α-adrenoceptor in the basic research stage of drug discovery. Comparing the selectivity of α_1-adrenoceptors and α_2-adrenoceptors in the R- and S- forms, the former was 9550-fold stronger, whereas the latter was only

6.6-fold. In the case of tamsulosin, it was evident that the *R*-isomer of the compound was more accurate and effective as a drug, but at the same time, in a limitation of the drug, the development of advanced production technology was necessary to produce a large amount of an optically active compound at low cost and to supply it as an industrial product. Astellas Pharma succeeded in constructing a technology for mass production of the *R*-isomer, and, at present, high-purity optically active tamsulosin hydrochloride with *S*-isomer content of 0.02% or less is used as an active ingredient in Harnal.

6.10   Technological Advances Since Sulfa-Based Antimicrobials

The development of antimicrobial chemotherapeutic agents started with the sulfa drugs in the 1930s. Penicillin had been discovered by Alexander Fleming in 1929, and the clinical development of benzylpenicillin was conducted in the 1940s. Later, macrolide and aminoglycoside antimicrobial agents were developed and put on the market.

The quinolone compounds that led to ofloxacin and levofloxacin were first synthesized in 1949. A degradation product obtained during the manufacturing process of melicopin was isolated from citrus plants at ICH, and became known as the first quinolone compound. Although this compound had antimicrobial activity, it was not developed as an antimicrobial agent for safety reasons. At the time, researchers at Sterling Winthrop demonstrated the antibacterial activity of 7-chloroquinolone derivatives, which were produced as by-products of the production of the antimalarial chloroquine. In 1962, Sterling Winthrop synthesized nalidixic acid for conversion of the quinolone ring into the basic skeleton of 1,8-naphthyridine, which did not conflict with ICH's patents, by adding chemical modifications (Takahashi et al. 2003). Subsequently, competition emerged in the development of fluoroquinolones based on nalidixic acid.

6.11   Ofloxacin and Levofloxacin R&D Team

The levofloxacin R&D team members were as follows:

1. Isao Hayakawa: Principal inventor of ofloxacin and levofloxacin

   Joined Daiichi Pharmaceutical in April 1969.
   Joined the quinolone development group in 1971.
   Became the head of the first Discovery Research Institute in June 1993.

Became the director of the First Research Institute for Drug Discovery and Development in October 1995.

Became the president of the Drug Discovery and Development, and Chemistry Institute in October 2000.

Appointed as a director in June 2001.

Head of drug discovery research in April 2004.

Became the Fellow of R&D Headquarters, Daiichi-Sankyo.

2. Shogo Arako, Shuichi Yokoyama, Masazumi Imamura,[3] and Shouichi Sakano

    These names were included in the basic patent.

3. Toshihiro Fujiwara and Tsutomo Ebata

    Inventors of the process for preparing condensed oxazines and propoxybenzene derivatives
    In 1995, Fujiwara was the director of the Institute of Production and Technology, and Ehata was a chief engineer of the Institute of Production Technology.

In the development of levofloxacin, Hayakawa was appointed as the team leader, and he tackled the issues of optical resolution. Hayakawa also engaged in the synthesis of compounds, and had overseen the synthesis of fluoroquinolones since his second year at Daiichi Pharmaceutical. Prior to the discovery of ofloxacin, the discontinuation of DJ-6783 in 1978 led Hayakawa to acquire a broader knowledge of drug discovery and synthesis methodologies. He also developed a database that included all the experimental and clinical trial information for ofloxacin and levofloxacin (Daiichi Sankyo 2012a). In addition, Hayakawa actively sought opportunities to interact with other departments, specifically, by asking other departments to make large quantities of compounds.

When Hayawaka was assembling a research team to develop levofloxacin, he recruited researchers from other laboratories who studied the optical resolution method. The development team consisted of five persons, and, although there were disagreements within the group regarding the development of compounds by optical resolution, Hayakawa was convinced of the potential of the project; his resolve led to the development of levofloxacin.

During clinical development of levofloxacin, Hayakawa was appointed as an inhouse "coordinator," and he was temporarily transferred to the Clinical Development Laboratory, as mentioned above. At that time, the institute introduced a system that integrated all information on a single development candidate within one coordinator. Hayakawa became the first coordinator for levofloxacin (Daiichi Sankyo 2012b). With regard to this move, the director of the Institute of Clinical Development said "there's no precedent, but I'll take responsibility if it's unsuccessful" and gave Hayakawa most of the decision-making authority (Tsukazaki 2013).

Ofloxacin was the highest-selling drug developed by Daiichi Pharmaceutical to that time, and its development was based on company-wide support, including the

---

[3] After the development of levofloxacin, Imamura acquired a patent attorney qualification and left the company in 1988.

use of labor-saving and automated manufacturing facilities (Daiichi Seiyaku 1997). Similarly, many research departments were involved in the development stage of levofloxacin, such as the First R&D, Development, and Production Technology Research Laboratories, and the First Pharmaceutical Development Department (Daiichi Seiyaku 1997).

6.12   Scientific Contributions to the Development of Ofloxacin and Levofloxacin

The main scientific sources that played direct roles in the R&D process for ofloxacin and levofloxacin are as follows:

(1)   The lead compound, nalidixic acid, introduced by Daiichi Pharmaceutical.
(2)   Kyorin Pharmaceutical's R&D of its new fluoroquinolone, norfloxacin.
(3)   Adoption of asymmetric synthesis in the mass production of ofloxacin and levofloxacin

The first two scientific contributions originated from the R&D of other companies. Universities also contributed to the development of the asymmetric synthesis, including the contribution of the general principles of asymmetric synthesis described by Prof. Ryoji Noyori of Nagoya University.

The first scientific source was the use of nalidixic acid as a lead compound. After importing the in-license product in 1962, Daiichi Pharmaceutical promoted the R&D of fluoroquinolones using nalidixic acid as a lead compound. It took approximately 20 years to derive ofloxacin and levofloxacin, and one of the early papers on nalidixic acid was cited in a paper regarding levofloxacin (Lesher et al. 1962).

The second scientific source was the development of fluoroquinolones by other companies, particularly norfloxacin by Kyorin Pharmaceutical. In the development of ofloxacin, an in-house survey of existing chemical structures revealed that the properties of norfloxacin, such as high tissue penetration and high concentration of free bodies, was likely caused by the structure of the compound, which led the team towards amphoteric quinolones. The basic substance patent for norfloxacin was widely cited in the basic patent for the first-generation fluoroquinolones sold in early 1980. Ofloxacin was created by combining these findings with an acidic quinolone that had been obtained during discovery research and had demonstrated strong antibacterial activity.

The third scientific source was the adoption of an asymmetric synthesis method for the mass production process of levofloxacin. The asymmetric synthesis was chosen to decrease the manufacturing costs of levofloxacin to less than half the production cost of ofloxacin. However, many competitors used conventional synthesis using L-aralinol, which motivated Daiichi Pharmaceutical to develop an original synthesis method. The production process of intermediates using (R)-1,2-propanediol derivatives was established after combining experience from the synthesis of tricyclic quinolones, as reported by the patent and the scientific papers by Pfizer and its researchers (see Sect. 6.9), and with knowledge internally accumulated by Daiichi Pharmaceutical.

The general principles of asymmetric synthesis found by Prof. Ryoji Noyori were utilized in the manufacturing process of levofloxacin (Yomiuri 2010). Finally, the BINAP ruthenium asymmetric hydrogenation method developed in the late 1980s was used to produce intermediates for levofloxacin (Noyori 2008).

6.13 Sales of Levofloxacin: Approximately 3 Billion USD at Peak

The global sales of levofloxacin are shown in Fig. 6.5. The sales peaked at 2.99 billion USD (300 billion yen) in fiscal 2007. Among the basic patents of levofloxacin, US 5053407 extended its protection period by 810 days, from 1 October 2008 to 10 December 2010. Thereafter, the sales of levofloxacin declined, following the expiration of the patent protection period in 2011.[4]

At the time, patients with chronic respiratory disease were hospitalized for acute aggravation, and medical treatment, including injections, was required. However, ambulatory treatment became possible with the emergence of ofloxacin; the cost

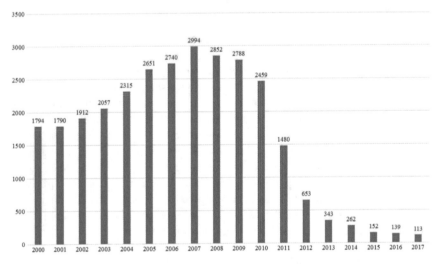

Fig. 6.5 Global sales of levofloxacin between 2000 and 2017 (unit: 1 M USD). Sales are represented as the sum of levofloxacin sales, including Cravit by Daiichi Pharmaceutical, Tavanic by Sanofi-Aventis, and Levaquin by Johnson & Johnson. *Source* Pharma Future Sezidem Strategic Data Uto Brain Division and financial reports

[4]Patent Terms Extended Under 35 USC §156. http://www.uspto.gov/patent/laws-and-regulations/patent-term-extension/patent-terms-extended-under-35-usc-156. Accessed: June 30, 2015.

reduction in medical treatment was estimated to be 279 billion yen per year in 1988 (Hayakawa et al. 2010).

6.14 Fluoroquinolone Development Competition

Following the approval of nalidixic acid in 1962, acid quinolones, such as pyromic acid (Dainippon Pharmaceutical), cinoxacin (Eli Lilly), and the first amphoteric quinolone pipemidic acid (Dainippon Pharmaceutical) were developed and marketed prior to the discovery of ofloxacin by Daiichi Pharmaceutical. In 1978, the research group at Kyorin Pharmaceutical discovered norfloxacin. Norfloxacin, unlike the conventional acidic quinolones, had excellent metabolic stability and tissue penetration, despite low blood levels. Norfloxacin also had high free body concentrations in diseased tissues. In addition, it showed strong antibacterial activity against *Pseudomonas aeruginosa*. Because of these characteristics, norfloxacin was launched in 1984 as the world's first fluoroquinolone.

After the launch of norfloxacin, competition occurred among pharmaceutical companies to develop a drug with better efficacy, and many new fluoroquinolones, including Daiichi's ofloxacin and levofloxacin, were commercialized. While many fluoroquinolones, such as norfloxacin, are bicyclic compounds with two six-membered rings, two drugs, ofloxacin and levofloxacin, are tricyclic compounds with three six-membered rings.

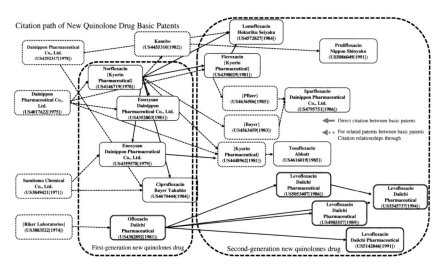

Fig. 6.6 Basic patent citation relationships between fluoroquinolones. *Source* Prepared by the author

Figure 6.6 shows the citation relationships between the basic fluoroquinolone patents. The basic patents for these fluoroquinolones all have either primary or secondary citations for the basic patent of norfloxacin (US4146719) issued by Kyorin. This suggests that the knowledge spillover effect of Kyorin's invention was significant. In contrast, the basic patent of ofloxacin (US4382892) issued by Daiichi Pharmaceutical is cited only by the basic patent of levofloxacin (US5053407; US4985557; US5142046; US5545737). The levofloxacin basic patent does not have either primary or secondary backward citations related to other fluoroquinolones, unlike the drugs of other firms. These results may indicate that ofloxacin and levofloxacin have more unique chemical structures than the fluoroquinolones synthesized by other companies.

6.15 Levofloxacin Gains Half the Fluoroquinolone Market Share

Figures 6.7 and 6.8 show trends in the sales and market share of the new, major fluoroquinolones from 2000 to 2012, while Fig. 6.5 summarizes the total share of Cravit, Tavanic, and Levaquin. The total market share of levofloxacin increased in the 2000s, and it maintained approximately half of the total share of fluoroquinolones. In addition to the share of ofloxacin, the total share accounts for approximately 80% in the latter half of the 2000s.

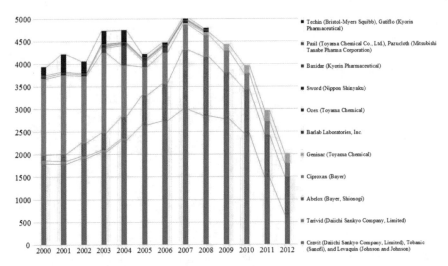

Fig. 6.7 Trends in the sales of major fluoroquinolones (unit: 1 M USD) *Source* Pharma future, Uto Brain Division, Sezidem strategic data

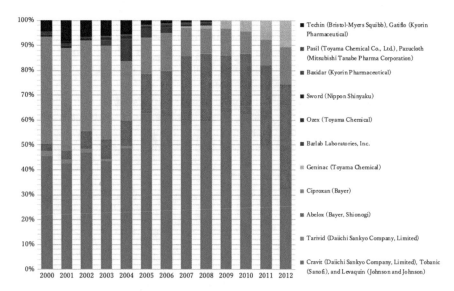

Fig. 6.8 Trends in the market share of fluoroquinolones. *Source* Pharma future, Uto Brain Division, Sezidem strategic data

6.16   Conclusion

Daiichi Pharmaceutical conducted R&D over approximately 20 years, using nalidixic acid (in-licensed product) as a lead compound, and eventually launched ofloxacin and levofloxacin, which were excellent antimicrobials with broad-spectrum antibacterial effects. Both ofloxacin and levofloxacin, which were innovative drugs with broad efficacy and limited side effects, achieved high market shares and sales over time.

The R&D of ofloxacin was based on the notion that the physical properties of drugs affect not only blood levels but also metabolic and water-soluble pharmacokinetics, which would determine its overall efficacy. Taking advantage of the firm's experience in chemical synthesis, as well as suggestions from the chemical structure of Kyorin Pharmaceutical's norfloxacin, ofloxacin, which was an amphoteric quinolone with high activity and a broad antibacterial spectrum was successfully synthesized. The combination of testing in-house development candidates with suggestions from the external patents was key to the innovation process. The project was initially implemented as an unauthorized research project, which suggests that researchers' initiatives are important to pursue diverse ideas for innovation.

In the R&D of levofloxacin, the difficult task of optical resolution of racemic ofloxacin resulted in a twofold increase in antibacterial activity, as well as a drastic reduction in side effects. In addition, the method of splitting the racemic ofloxacin at the intermediate, rather than the final product, was discovered to be optimal, and a non-conventional method (chiral column method) was chosen. Furthermore, a new asymmetric synthesis method was introduced to reduce production costs, and a new production process to obtain intermediates using (R)-1,2-propanediol derivatives was established by combining the learning from the synthesis method developed by Pfizer and the knowledge accumulated internally. This enabled Daiichi Pharmaceutical to achieve its goal of reducing the production cost of levofloxacin to no more than twice that of ofloxacin, despite of its twofold increase of efficacy.

Intense research competition for fluoroquinolones also occurred during the search for ofloxacin and levofloxacin. Indeed, Bayer filed a patent application for levofloxacin just 2 days after Daiichi Pharmaceutical. This demonstrates that the discovery of drugs with known mechanisms of action can become a time-based competition and a patent race.

References

Baba, S., et al. (1992). Open clinical study on sinusitis of Levofloxacin. *Chemotherapy, 40*(S3), 365–378.

Daichi Sankyo. (2012a). *Cravit(R) Development story, born from the solitary flight, Daichi Sankyo R&D Information.* http://daiichisankyo.co.jp/rd/newdrug_story/cravit/01/index.html. Accessed April 22, 2012. (in Japanese).

Daichi Sankyo. (2012b). *Cravit(R) development story, fight with myself, battle with regime, Daichi Sankyo R&D Information.* http://daiichisankyo.co.jp/rd/newdrug_story/cravit/02/index. html. Accessed April 22, 2012. (in Japanese).

Daiichi Seiyaku. (1997). Daiichi Seiyaku Hachiju-nenshi (in Japanese).

Daiichi Seiyaku. (2007). Daiichi Seiyaku Kuju-nenshi (in Japanese).

Fujita, K., et al. (1984). Double-blind comparative study results of DL-8280 and Cefaclor for superficial suppurative diseases. *Journal of Infectious Diseases, 58*(9), 793–819.

Fujiwara, T., & Ebata, T. (1989). H1–250369: Process for manufacturing fused Oxazines. Japanese Patent Office.

Fujiwara, T., & Ebata, T. (1990). H2-732: Propoxybenzene derivatives and methods for preparation. Japanese Patent Office.

Fujiwara, T., & Satoh, Y. (1989). H1-180867: Optical resolution. Japanese Patent Office.

Fujiwara, T., & Tsurumi, H. (1989). H1-131168: Manufacturing method of Benzoxazine derivatives. Japanese Patent Office.

Fujiwara, T., & Yokota, T. (1989). H1-175975: (S)-Benzoxazine derivatives method. Japanese Patent Office.

Gilligan, P. J., McGuirk, P. R., & Witty, M. J. (1984). US4636506: 7-heterocyclic-1,4-dihydroquinolones. USPTO.

Hayakawa, I. (1991). When fluoride was added, the drug was produced. *Pharmacia, 27*(10), 1009–1011.

Hayakawa, I. (1999). Development history of Cravit/Levofloxacin (New Quinolone synthetic antibiotics. *Chemistry and Education, 47*(11), 740–743.

Hayakawa, I. (2007). Recall of Levofloxacin ("Cravit") developments. Daiichi Seiyaku Kuju-nenshi (in Japanese), 255–256.

Hayakawa, I. (2010). Quinolone synthetic antimicrobials: Drug discovery language of ofloxacin, levofloxacin, and citafloxacin. *Pharmacia, 46*(6), 526–531.

Hayakawa, I., Atarashi, S., Yokohama, S., Imamura, M., Sakano, K.-I., & Fukuyama, M. (1986). Synthesis and antibacterial activities of optically active Ofloxacin. *Antimicrobial Agents and Chemotherapy, 29*(1), 163–164.

Hayakawa, I., Fujiwara, T., & Ebata, T. (1996). Development and commercialization of the quinolone synthetic antibacterial agent ofloxacin (Tarivid(R)) and levofloxacin (Cravit(R)). *Journal of the Society of Synthetic Organic Chemistry, 54*(1), 62–67.

Hayakawa, I., Handa, H., Furuhama, K., Sai, R., & Satoh, K. (2010). R&D of levofloxacin, a controversial antimicrobial agent for a wide range of trees. Okouhi Memorial Prize, 43–58.

Hayakawa, I., Hiramitsu, T., & Tanaka, Y. (1984). Synthesis and antibacterial activities of substituted 7-Oxo-2,3-dihydro-7 h-pyrido[1,2,3-de][1,4]benzoxazine-6-carboxylic acids. *Chemical & Pharmaceutical Bulletin, 32*(12), 4907–4913.

Ichihara, H., Takezawa, H., Tsuruma, M., Sai, R., & Sato, K. (1984). Phase I trial of DL-8280. *Chemotherapy, 32*(S1), 118–149.

Ishii, T., et al. (1992). Phase III open labeled clinical trials for levofloxacin otitis media and otitis externa. *Chemotherapy, 40*(S3), 334–351.

Joumuro, S., & Tsurumi, H. (1983). S58-29789: Peridobenzoxazine derivatives and their manufacturing methods. Japanese Patent Office.

Kawada, S., et al. (1992). Examination of the clinical dosage of levofloxacin for complicated urinary tract infection. *Chemotherapy, 40*(S3), 210–229.

Kobayashi, H., et al. (1984). Comparative efficacy study of DL-8280 and Amoxicillin for respiratory infections. *Journal of the Japanese Association for Infectious Diseases, 58*(6), 525–555.

Lesher, G. Y., Froelich, E. J., Gruett, M. D., Bailey, J. H., & Brundage, R. P. (1962). 1,8-Naphthyridine derivatives. A new class of chemotherapeutic agents. *Journal of Medical Pharmaceutical Chemistry, 5,* 1063–1065.

Matsuda, S., et al. (1992). Clinical study of Levofloxacin (LVFX) for gynaecological infectious diseases. *Chemotherapy, 40*(S3), 311–325.

Murata, M., et al. (1992). Clinical studies of Levofloxacin (DR-3355) for infectious enterocolitis and examination of fecal drug levels and intestinal flora in patients with infectious enterocolitis. *Chemotherapy, 40*(S3), 170–187.

Nakashima, M., Uematsu, T., Kanamaru, M., Okazaki, O., & Hakusui, H. (1992). Phase I study of Levofloxacin, (S)-(-)-Ofloxacin. *Japan Journal Clinical Pharmacology Therapeutics, 23*(2), 515–520.

Nakashima, M., Uematsu, T., Kanamaru, M., Okazaki, O., Hashimoto, S., & Tachiawa, H. (1988). Pharmacokinetics of DR-3355, a new quinolone, in healthy volunteers. In *Program Abstract of 28th Interscience Conference on Antimicrobial Agents and Chemotherapy* (p. 278). Washington, D.C. Abstr. 951.

Noyori, R. (2008). Watashi No Rirekisho. Nikkei Shimbun, September 23, 2008.

Ohrui, Y., Akasaka, K., & Imaizumi, K. (2003). Development of ultrasensitive distant asymmetric discrimination method overcoming the problem of the geostereomer method and its application to absolute positioning of biologically active natural products. *Science and Biology, 41*(10), 691–698.

Ohyama, K., et al. (1992). Examination of the usefulness of Levofloxacin for tonsillitis, pharyngolaryngitis and sialadenitis. *Chemotherapy, 40*(S3), 352–364.

Parikh, V. D., Fray, A. H., & Kleinman, E. F. (1988). Synthesis of 8,9-Difluoro-2-Methyl-6-Oxo-1,2-Dihydropyrrolo[3,2,1-Ij]Quinoline-5-Carboxylic acid. *Journal of Heterocyclic Chemistry, 25*(5), 1567–1569.

Preston, S., et al. (1998). Pharmacodynamics of Levofloxacin A new paradigm for early clinical trials. *JAMA, 279*(2), 125–129.

Research Industry Association of the Japan Machinery Federation. (1995). *Development of antimicrobials "Tarivid/Cravit". A hundred spirit of engineers: How the product was invented* (Vol. 1, pp. 98–105).

Saito, A., et al. (1992a). Dose-comparative study of Levofloxacin (LVFX) for chronic lower respiratory tract infections. *Chemotherapy, 40*(S3), 75–96.

Saito, A., et al. (1992b). Clinical evaluations of Levofloxacin for medical infections. *Chemotherapy, 40*(S3), 147–169.

Sasaki, J., et al. (1992). Levofloxacin clinical studies on infections in dental and oral surgery. *Chemotherapy, 40*(S3), 379–391.

Sato, K. (2006). Research and development of quinolone synthetic antimicrobials, report of mycotics research center. *Chiba University, 10,* 69–73.

Sato, K., Matuura, Y., Inoue, M., Une, T., Osada, Y., Ogawa, H., et al. (1982). In vitro and in vivo activity of DL-8280, a new oxazine derivative. *Antimicrobial Agents and Chemotherapy, 22*(4), 548–553.

Shiba, K., Fukase, H., & Sugiyama, A. (2009). Phase I study of Levofloxacin 250–1000 mg in healthy adult and elderly men. *Journal of Japanese Society of Chemotherapy, 57*(S2), 1–11.

Shiba, K., Sakai, O., Shimada, J., Okazaki, O., Aoki, H., & Hakusui, H. (1992). Effects of antacids, ferrous sulfate and ranitidine on absorption of DR-3355 in humans. *Antimicrobial Agents and Chemotherapy, 36*(10), 2270–2274.

Silliver, M., & Growe, K. (1987). S62-145088: Enantiomerically pure 1,8-bridged 4-quinolone-3-carboxylic acid. Japanese Patent Office.

Soejima, R., et al. (1992). Double-blind comparative study of Levofloxacin and Ofloxacin for chronic lower respiratory tract infections. *Chemotherapy, 40*(S3), 97–120.

Takahashi, S., Hayakawa, I., & Akimoto, K. (2003). Development and transition of quinolone system synthetic antibacterial agent. *Journal of Pharmaceutical History, 38*(2), 161–179.

Takahashi, K., et al. (1992). Clinical investigation of Levofloxacin dermatological infections. *Chemotherapy, 40*(S3), 286–305.

Totsuka, K., Kono, S., Matsumoto, T., Sunagawa, K., & Shiba, K. (2009). Levofloxacin 500 mg 1-credibility and dosage-once a day. *Journal of the Japanese Society of Chemotherapy, 57*(5), 441–442.

Tsukazaki, A. (2013). Cravit (Levofloxacin). Japanese scientists taking on new drugs. Kodansha, 34–45.

Yomiuri, S. (2010). *A forum for Nobel Prize Laureates. messages to the next generation.* http://info.yomiuri.co.jp/yri/n-forum/nf2010923.htm. Accessed May 25, 2012.

Yura, J., et al. (1992). Levofloxacin, a quinolone oral antibacterial agent, for treatment of surgical infections. *Chemotherapy, 40*(S3), 270–285.

Chapter 7
Tamsulosin (Harnal, Flomax, OMNIC)

Breakthrough Drug that Drastically Changed the Treatment of Prostatic Hyperplasia

Yasushi Hara, Akira Nagumo and Hajime Oda

Abstract Tamsulosin hydrochloride (Harnal) is a drug for dysuria associated with prostatic hyperplasia that was developed by Yamanouchi Pharmaceutical Co., Ltd. It works by blocking adrenergic α_1-adrenoceptors. Dysuria covers a variety of urogenital issues, including prostatic hyperplasia, polyuria, and urinary incontinence. Among these conditions, tamsulosin is effective at reducing dysuria caused by hypertrophy of the prostatic gland. Tamsulosin is prescribed as a once-daily dose and has transformed prostate hyperplasia therapy. Indeed, tamsulosin is both a first-in-class and best-in-class drug that has dominated market share. The research and development (R&D) process for tamsulosin was distinctive in two ways. First, R&D into tamsulosin did not originally begin with the intent of creating a drug to treat dysuria. In fact, a new application of an already-known compound originally developed as an antihypertensive drug was discovered as a result of unauthorized research. Second, continued R&D into tamsulosin led to progress in the scientific understanding of the mechanism of action in the prostate, which was clarified by an academia–industry collaboration.

Y. Hara (✉)
CEAFJP/EHESS, Paris, France
e-mail: yasushi.hara@r.hit-u.ac.jp

Faculty of Economics, Hitotsubashi University, Tokyo, Japan

A. Nagumo
Medical Affairs, Medical Science Liaison, MSD K.K., Tokyo, Japan
e-mail: akira.nagumo@merck.com

H. Oda
Faculty of Business Administration, Tohoku Gakuin University, Miyagi, Japan

© Springer Nature Singapore Pte Ltd. 2019

111

S. Nagaoka (ed.), *Drug Discovery in Japan*,
https://doi.org/10.1007/978-981-13-8906-1_7

7.1 Introduction

Tamsulosin hydrochloride (Harnal) was developed by Yamanouchi Pharmaceutical Co., Ltd. (Astellas Pharma Inc.) as a drug for dysuria associated with prostatic hyperplasia. Its mechanism of action is to block adrenergic α_1-adrenoceptors.[1]

Dysuria is a symptom of various urogenital issues including, polyuria, urinary incontinence, and prostate hyperplasia. Among these conditions, tamsulosin effectively treats dysuria caused by prostatic hypertrophy. The prostate gland is an organ that surrounds the urethra in men and tends to enlarge with age. Because the hypertrophied prostate gland presses on the urethra, urination becomes difficult and may cause urinary disturbances, such as urinary frequency, urinary urgency, and dysuria. Prostatic hypertrophy is estimated to affect 85.47 men per 10,000 in the United States (Sarma et al. 2005).

Since the discovery of adrenaline as a sympathomimetic substance in 1901, it became clear in 1948 that two subtypes of sympathetic receptors exist, α and β. In 1968, Edvardsen and Cetecreb reported the presence of α receptors in the bladder neck of rabbits and cats, suggesting the application of α-blockers to prostatic hyperplasia may ease symptoms (Edvardsen and Setekleiv 1968).

Tamsulosin acts on α_1-adrenoceptors distributed in the smooth muscle of the prostate and inhibits the binding of adrenergic receptors to their endogenous ligands, such as noradrenaline, thereby reducing the pressure on the urethra caused by the enlarged prostate. When an endogenous ligand binds to an α_1-adrenoceptor, calcium ions are released from the endoplasmic reticulum, the site of intracellular calcium storage, and voltage-dependent calcium channels open in the plasma membrane, which leads to an influx of extracellular calcium ions. This increase in intracellular calcium ions induces the contraction of smooth muscle and attendant pressure on the urethra. Tamsulosin relieves smooth muscle contraction caused by stimulation of α_1-adrenoceptors, relaxing the smooth muscle and facilitating urination.

The treatment concept of tamsulosin was different from that of the existing dysuria therapies for prostatic hyperplasia, which were targeted at reducing the hyperplasia. Therefore, treatments such as surgical removal of the prostate or antiandrogen therapy to shrink the prostate, were common at the time. However, patients continued to report dysuria following both surgical operations and hormone therapy. Furthermore, hormone therapy took months to take effect and was associated with some side effects, such as sexual function degeneracy (Takenaka et al. 1995).

Importantly, tamsulosin does not shrink the prostate itself; it is a symptomatic treatment that makes it easier to empty the urine by inhibiting smooth muscle contraction. Moreover, the effect of tamsulosin appears within 2 weeks, and side effects such as dizziness and light-headedness are less than those of other adrenergic blockers, making it a transformative drug in the therapy of prostatic hypertrophy dysuria.

The research and development (R&D) process for tamsulosin was distinctive in two ways. First, R&D toward tamsulosin was not originally designed to create a dysuria drug. Instead, the original project was a new application of an already-known

---

[1] Astellas Inc., Drug Interview Form: Harnal D Tablet.

compound, which was explored and developed as an antihypertensive drug. After the potential of the compound to act as a drug for dysuria was recognized, improvement and commercialization in that direction were made. Second, advances in Tamsulosin R&D led to progress in scientific understanding of the mechanism of action in the prostate. Increased understanding of disease states became not only a foundation of R&D, but the process of developing new drugs generated new scientific questions and served as a starting point for advancing scientific knowledge. Indeed, scientific knowledge and drug discovery influenced each other and have proceeded together.

7.2   Tamsulosin R&D Timeline

1964: Completion of Yamanouchi Central Research Laboratory (Phase 1).

1965: Yamanouchi began structure–activity relationship studies of sympatholytic drugs (Takenaka et al. 1995).

1972: Indenolol (Prusan), a β-blocker, was invented (Takenaka et al. 1995).

1974: Langer et al. classified α-adrenoceptors into α_1- and α_2-adrenoceptors (Barnes et al. 1974).

1976: Discovered amosulalol (Logan), which blocks both α- and β-adrenoceptors, a drug for hypertension (Takenaka et al. 1995).

1979: Examination of the optical isomers of amosulalol revealed the S-isomer primarily blocked α-adrenoceptors, while the R-isomer primarily blocked β-adrenoceptors. Conducted structural modifications and invented desoxy derivative of amosulalol (YM-11133), an α_1-blocking agent.

1980: Optical resolution of the racemate YM-12617, which was produced by structural modifications of YM-11133, showed that the α_1-blocking activity of the $R(-)$-isomer was 600 times stronger than that of the $S(+)$-isomer. Researchers began development of the $R(-)$-isomer, YM617, which would become tamsulosin (Takenaka et al. 1995).

1986: The clinical trial in Japan began with the purpose of reducing dysuria in patients with prostatic hypertrophy. The development of extended-release formulations avoided the side effects of orthostatic disturbance because of the rapid elevation of blood concentration.

1988: Minneman described the presence of distinct subtypes of α_1-adrenoceptors (Takenaka et al. 1995; Minneman 1988).

1993 (July): Tamsulosin was launched in Japan as Harnal.

1997: Tamsulosin was launched in the United States as Flomax, as OMNIC in Italy, Ukraine, and Greece, and in Austria as Alna Retardkapseln.

1998: Tamsulosin was launched in Iceland as OMNIC.

(Available in 97 countries as of 2012).[2]

---

[2] Astellas Inc., Drug Interview Form: Harnal D Tablet.

7.3 Development History of Tamsulosin

Toichi Takenaka, the lead developer of tamsulosin, joined Yamanouchi in 1964 after majoring in veterinary medicine at Gifu University. Upon joining the company, he was granted study leave to Kumamoto University and the University of Tokyo, where he studied experimental pharmacology techniques. Structure–activity relationship studies of sympatholytic drugs began at Yamanouchi in 1965. Takenaka played a key role in these studies wherein he focused on the effects of suspected sympatholytic drugs in dogs and rats.[3] In brief, the compound was administered to the animal and vasoconstriction was monitored in the search for new antihypertensive drugs. It was through these studies that Yamanouchi developed the β-blocker indenolol (Prusan) in 1968 and formoterol (Atock) for bronchial asthma in 1972; both of these drugs were successfully marketed (Takenaka et al. 1995).

In 1968, a selective α_1-blocker, prazosin, a quinazoline derivative, was developed by Pfizer and early clinical trials showed promise for the treatment of hypertension and heart failure. Thereafter, antihypertensive drug candidates with different mechanisms of action were studied one after another. Antihypertensive drugs, which would apply to a large patient population, thus became an important aspect of Yamanouchi's drug development plan (Takenaka 2015). When the antihypertensive effect of a dihydropyridine derivative, a calcium antagonist, was confirmed by Yamanouchi in 1968, structural modifications of the lead compound were conducted in an effort to synthesize a compound that was more water-soluble and more light-stable. As a result of these studies, Takenaka discovered nicardipine in 1972 (Yamanouchi Pharmaceutical 1975). Nicardipine was launched in 1981 under the product name Perdipine and was referred to as a calcium antagonist with a novel mechanism of action.

Interestingly, nicardipine was discovered during a random drug screen, which contrasts with subsequent discovery research conducted by Takenaka wherein drug synthesis research was conducted with clear drug targets. For example, Takenaka's antihypertensive drug research was targeted toward blockage of the adrenoceptors associated with hypertension. Consequently, in 1976, Yamanouchi discovered amosulalol (Logan), an α- and β-blocker, that the company developed as an antihypertensive drug (Takenaka et al. 1995).

7.4 Basic Research: Amosulalol R&D and the Creation of Tamsulosin

The lead compound of tamsulosin was amosulalol (Logan), a racemate capable of blocking both α-and β-adrenoceptors, which was developed as an antihypertensive agent (Fig. 7.1). Amosulalol, like other β-blockers, has an asymmetric carbon atom; two isomers exist around this carbon, which creates a racemate with the actions

---

[3]Interview with Dr. Toichi Takenaka, in May 18, 2012.

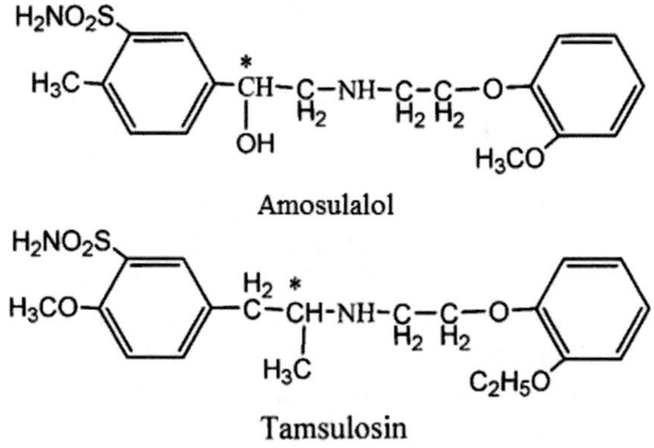

Fig. 7.1 Structure of amosulalol and tamsulosin. *Source* Honda (2006)

Amosulalol

Tamsulosin

of both isomers, resulting in an α- and β-adrenoceptor dual blocker (Honda 2006). Of these, the $S(+)$-isomer blocked α-adrenoceptors, and the $R(-)$-isomer blocked β-adrenoceptors. Given that the OH group of the $S(+)$-isomer was presumed to be placed opposite the receptor, the desoxy derivative of amosulalol (YM-11133), a compound without the OH group, was synthesized. Assessment of the pharmacological activity of deoxyamosulalol demonstrated that it was also an α-blocker.

Based on these results, R&D was carried out to create an α_1-blocker that was more potent than prazosin, which was the most potent antihypertensive agent at the time. In examining the structure–activity relationship of the carbon side chains in YM-11133 (Fig. 7.2), it was found that the introduction of a CH_3 group at R_3 increased the activity tenfold and that optimal groups at R_2 and R_4 were H and C_2H_5O, respectively. For the R_1 group, CH_3O gave the best pharmacological selectivity, metabolic stability, and overall physical properties when compared with CH_3 or OH analogs. As such, YM-12617, a racemate, was obtained. Among the optical isomers of YM-12617, it was found that the α_1-blocking action of the $R(-)$-isoform was 600 times stronger than that of the $S(+)$-isoform. The purified $R(-)$-isoform, YM-617, was later given the name tamsulosin and was marketed as Harnal.

However, the antihypertensive effect was not sufficiently potent, and tamsulosin was positioned as a backup compound to be prepared if the development of amosulalol was discontinued (Honda 2006). Therefore, the development of tamsulosin as an antihypertensive agent was abandoned in 1979, and its application to urological disease, which was not an in-house research focus of Yamanouchi, started as *Yami research* (unauthorized research) (Asano and Takenaka 2010).

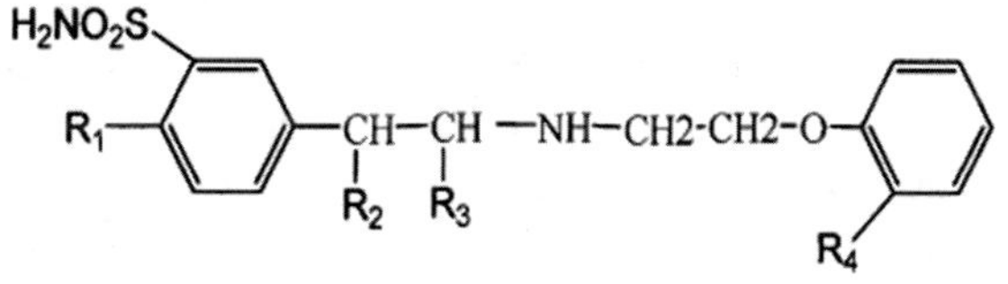

Fig. 7.2 Structure–activity relationship of carbon sidechains in YM-11133. *Source* Honda (2006)

7.5   Application to Prostatic Hyperplasia

In a study meeting at Yamanouchi, Takenaka and his colleagues learned that Caine, an Israeli urologist, had demonstrated the effectiveness of an α-blocker (phenoxybenzamine) in alleviating dysuria in patients with prostatic hyperplasia in an investigator-initiated clinical trial (Caine et al. 1975). Furthermore, the researchers were aware of results published by Langer in 1974 that characterized the presence of α-adrenoceptors in the urinary tract. According to the authors, α_1-adrenoceptors were located in the smooth muscle and α_2-adrenoceptors were located in the preganglionic fibers of the nervous system (Langer and Salomón 1974). However, the phenoxybenzamine used by Caine acted on α_1- and α_2-adrenoceptors, resulting in relaxation of the prostate, but also caused dizziness, lightheadedness upon standing, and hypotension; there were too many side effects to utilize it for dysuria.

Further information came from a Yamanouchi medical representative in Nagoya City, who explained that phenoxybenzamine, which was used to treat urethral obstruction, was no longer available because of suspected carcinogenicity, and a urologist at Nagoya University who had read the aforementioned Caine paper was looking for a new α-blocker (Tsukazaki 2014). This event played an important role in recognizing the needs of tamsulosin as a therapeutic drug for dysuria (Honda 2006).

During in-house experiments testing the action of YM-12617 and related compounds in rabbit urethras, it was confirmed that the subtype of the receptor involved in contraction was the α_1-adrenoceptor (Honda et al. 1985). Results were then confirmed in human tissue by carrying out collaborative research with Kazuki Kawabe of the University of Tokyo Department of Urology (Takenaka 2011; Kunisawa et al. 1985). The receptor subtype was confirmed to be α_1 in both rabbit and human prostate; thus, it was possible to use the results from rabbit experiments as preclinical data. Similarly, YM-12617 was also shown to be the most potent α_1-blocker in both rabbit and human lower urinary tract smooth muscle (Honda and Nakagawa 1986).

Based on these investigations, the development of a new drug that could reduce dysuria, not by reducing the size of the prostate, but by inhibiting the smooth muscle contraction of the prostate using an α_1-blocker, was indicated. Takenaka and his colleagues, however, reported that they were studying α-blockers and proceeded with their research without official reporting. They conducted the early-stage experiments for urinary drugs in this manner because the company's strategy was focused on the three discovery fields of gastrointestinal, cardiovascular, and respiratory disease (Asano and Takenaka 2010).[4] They also sought to obtain internal approval to implement a clinical trial for the treatment of dysuria associated with neurogenic bladder, but were unsuccessful because of the small size of the neurogenic patient population and limited sales volume (Honda 2006).

It then became clear that the needs of patients with benign prostatic hyperplasia were high for a dysuria drug. To confirm the needs of this patient population, investigators conducted interviews with urologists. These interviews described the approach to treat benign prostatic hyperplasia at the time, which was to surgically

---

[4]Interview with Dr. Toichi Takenaka, 18 May 2012.

remove part of the prostate to make it smaller, or to eliminate the cause of the enlarged prostate by using drugs to suppress the action of androgens. However, to chemically suppress the action of androgens it was necessary to administer the drug for a minimum of 6 months, even though the drug did not adequately reduce the symptoms of dysuria in all cases.

It was clear to the investigators that conventional therapy did not effectively treat the symptoms of prostate hyperplasia, and there was room for the entry of a new drug, particularly one with a novel mechanism of action. The research team then demonstrated that tamsulosin selectively lowered urethral pressure without affecting heart rate and blood pressure relative to other α_1-blockers when it was tested using anesthetized dogs (Sudo et al. 1990). In a collaborative study with Shizuo Yamada of the University of Shizuoka using human prostate and vascular membrane samples, tamsulosin was found to be a prostate-selective α_1-blocker with 30 times greater affinity for the prostate than blood vessels (Yamada et al. 1994). Thus, it became possible to obtain internal approval for the clinical development of tamsulosin as a dysuria drug for prostate hyperplasia, because the selectivity of tamsulosin for prostate tissue had been demonstrated.

7.6 Clinical Trial Program: Overcoming Side Effects with Extended-Release Formulations

Clinical trials began in 1986 and were initially conducted with the racemic mixture. However, the difficulty in obtaining approval for the racemic mixture in the USA led to full-scale clinical development only after preclinical studies, such as toxicological studies, had been completed with the $R(-)$-isoform of the compound, which eventually became tamsulosin.

The results of the Phase I clinical trial in healthy adult male volunteers showed orthostatic hypotension in the subjects, which was not dose-dependent and often became known as the "first dose phenomenon," a response that occurred transiently after the initial administration (Hoffman and Lefkowitz 1990). Orthostatic disturbances were related to the initial absorption rate of the drug and could be partially prevented by titrating the drug gradually from a lower dose (Shintani et al. 1978). The concept of gradually increasing the dose led to the development of the extended release formulation, which, even if administered as a full-dose the first time, was not associated with orthostatic disturbances (Sako 2006; Nishiura and Mizumoto 2008). Furthermore, in the aforementioned interviews with urologists, it was recommended that the drug be developed as a once-daily formulation, which supported the development of an extended-release formulation (Takenaka 2015).

The Phase II clinical trial, and all subsequent trials, were conducted in a placebo-controlled, double-blind manner. To determine the optimal dose, a double-blind, once-daily, 4-week oral dose-finding study was conducted with placebo and tamsulosin (Kawabe et al. 1990a, 1990b). The use of a placebo-controlled, double-blind study was adopted because the improvement of subjective symptoms was the most important factor in the evaluation of efficacy during the dose-finding stage, and placebo was essential to set the dose with high accuracy. Takenaka adopted a double-blind approach by persuading the clinical investigators and used a scientifically robust experimental protocol (Asano and Takenaka 2010).[5] Subjective symptoms, such as irritation and obstructive symptoms, were assessed by patient diaries and scored according to US Food and Drug Administration guidelines. Objective signs were assessed by urodynamic studies, including urinary flow rate and residual urine volume (Takenaka et al. 1995). In the dose-finding study, tamsulosin did not cause orthostatic hypotension or affect blood pressure, and the optimal dose was set as 0.2 mg/day. In a subsequent Phase III clinical trial, the double-blind comparative study was conducted with 0.2 mg/day tamsulosin and placebo (Kawai et al. 1991).

A one-year long-term efficacy and safety study was also conducted between June 1988 and May 1990 (Yoshida et al. 1991). Throughout the study, (1) no orthostatic hypotension or ejaculatory disturbances were observed, and (2) the improvement rate of subjective symptoms, as measured by the impression of the patient and attending physician increased with the duration of treatment up until 16–20 weeks, at which point they remained constant. Finally, the improvement rate of the objective signs was maintained at a constant value after 4 weeks; these findings confirmed tamsulosin was suitable for long-term administration.

In parallel with these clinical trials, basic research was being conducted to confirm the mechanism of action. The molecular explanation as to why tamsulosin acted preferentially on the prostate and not on blood vessels, as indicated by the lack of a hypotensive effect, was not clear. The collaborative research related to the α_1-adrenoceptor increase in prostatic hypertrophy was carried out with the University of Tokyo Department of Urology, and research with Essen University Medical School (now University of Duisburg-Essen) characterized the selectivity of tamsulosin for α_1-adrenoceptors of the lower urinary tract (Taguchi et al. 1997).

Afterward, in the 1990s, gene cloning technology of adrenoceptor subtypes advanced, and it soon became clear that there were three subtypes of the α_1-adrenoceptor: α_{1A}, α_{1B}, and α_{1D}. Furthermore, concentrations of the α_1-adrenoceptor subtypes in the human prostate and blood vessels were significantly different (Takenaka 2010). In normal blood vessels, the three subtypes were equally distributed; however, the prostate gland was rich in α_{1A} and α_{1D}, but poor in α_{1B}-adrenoceptors. The research team proved that differences in subtype distribution affected the efficacy and specificity of the drug. Given that tamsulosin acted strongly on α_{1A}- and α_{1D}-adrenoceptors, the drug's activity was specific to the prostate gland and hypotension was not observed as a side effect. However, it would be more than 10 years after the discovery of tamsulosin before this mechanism was fully understood (Takenaka 2010; Suto and Takenaka 1993).

---

[5]Interview with Dr. Toichi Takenaka, 18 May 2012.

Tamsulosin was launched in Japan in July 1993. Initially, Eli Lilly had been conducting the clinical trials in the United States; however, they withdrew the development of tamsulosin during Phase II studies because of the perceived small market of prostate hypertrophy patients. Yamanouchi then continued the development process (Tsukazaki 2014). Subsequently, tamsulosin was launched in 1997 in the United States as Flomax; in Italy, the Ukraine, and Greece as OMNIC, and in Austria as Alna Retardkapseln. As of 2012, tamsulosin was available in 97 countries.[6]

7.7 Scientific Basis for R&D of Tamsulosin

While the causes of benign prostatic hyperplasia remain unclear, lifestyle and social factors have been implicated in disease development, and the number of patients has been consistently increasing in recent years (Fig. 7.3). The prostate gland also enlarges because of abnormal modulation of steroid hormones (androgens), with onset often occurring in the 50 s or older. As mentioned above, surgical resection and hormone therapy were not sufficient for the relief of associated dysuria.

The discovery of adrenaline and its receptor subtype served as a basis for R&D of tamsulosin, which provides a symptomatic treatment with very few side effects. Specifically, tamsulosin treats dysuria by relaxing smooth muscle contraction. Adrenaline was discovered as a sympathomimetic substance in 1901, and the existence of two kinds of sympathetic nerve receptor subtypes (α, β) was indicated in 1948. Their presence was considered common knowledge when Toichi Takenaka was a student, and the R&D of tamsulosin was promoted on the basis of such knowledge. Furthermore, research published by Langer in 1974, describing subtypes of α-adrenoceptors, namely α_1- and α_2-adrenoceptors, distributed throughout the smooth muscle and nervous system, provided an opportunity for the application of a new drug.

Fig. 7.3 Estimated number of patients with prostatic hyperplasia in Japan, and the total number of patients in Japan (in thousands). *Source* Patient Survey 2011, Statistics and Information Department, Minister's Secretariat, MHLW (Classification of Diseases and Injury)

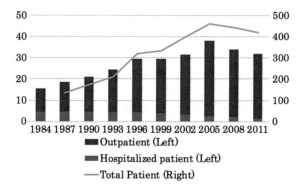

■ Outpatient (Left)

■ Hospitalized patient (Left)

— Total Patient (Right)

---

[6]Astellas Inc., Drug Interview Form: Harnal D Tablet.

7.8 Yamanouchi R&D Structure

The Yamanouchi R&D system was established in the 1960s, and, in 1962, Masuo Murakami of the Osaka University Institute of Industrial Science was named as director of the Yamanouchi Central Research Institute (Yamanouchi Pharmaceutical 1975). In terms of facilities, first-phase construction of the Central Research Institute was completed in 1964; subsequent expansions, the second phase in 1968 and the third phase in 1971, were also implemented, enhancing the research quality and facilities. It is presumed that before the start of tamsulosin R&D in the 1970s, the research organization of the Central Research Institute had been established (Yamanouchi Pharmaceutical 1975). As the R&D leader, Murakami focused on the three research areas of cardiovascular, digestive, and respiratory diseases. He also established a manufacturing method that bypassed existing manufacturing patents for the development of so-called zolo drugs (*zoroshin*; equivalent to me-too drug). At the same time, Takenaka and colleagues were granted study leave to universities to acquire expertise in experimental methods that laid the foundation for the development of an original drug.[7]

During the development of tamsulosin, Takenaka was transferred to the clinical development department to initiate clinical development. Given the low priority of urological development projects, Takenaka was responsible for promoting the clinical development of tamsulosin, which was aided by his experience in developing amosulalol and completing the tamsulosin basic research (Tsukazaki 2014).

7.9 Scientific Contributions to Basic Research of Tamsulosin

The contribution of science to the basic research of tamsulosin consisted of the following three findings. First, advances in science had a direct impact on identifying a use for the discovered compound, notably the symptomatic treatment of prostate hyperplasia. Takenaka obtained information from clinical research by Caine that showed the effectiveness of an α-blocker against dysuria associated with prostatic hyperplasia. This data prompted the concept of applying tamsulosin, an α_1-blocker, to the urinary system (Caine et al. 1975). Next, the need for the drug was confirmed by conducting interviews with clinicians and grasping the treatment trends.

Second, an evolving scientific understanding of the prostate-specific adrenergic receptor subtypes was acquired from the scientific literature. In the course of a literature review, Edvardsen and Setekleiv (1968) reported the presence of α-receptors in the bladder neck of rabbits, cats, and guinea pigs. Langer also described the presence of α-receptor subtypes in humans, α_1- and α_2-adrenoceptors, and indicated that α_1-adrenoceptors were located in smooth muscle tissues, while α_2-adrenoceptors were

---

[7]Interview with Dr. Toichi Takenaka, 18 May 2012.

located in nerves (Barnes et al. 1974). With improved understanding these receptors, researchers were able to predict that tamsulosin would be an effective therapeutic agent for prostate hyperplasia by selectively acting on α_1-adrenoceptors, with minimal side effects. This prediction was proved through animal experiments. Basic research continued in parallel with clinical trials to clarify the tamsulosin mechanism of action; these experiments were conducted in collaboration with academia.

Third, education of the company's researchers contributed to the basic research of tamsulosin. Takenaka learned of the presence and subtypes of adrenergic receptors through his university studies. Upon joining Yamanouchi, he was granted study leave to a university, where he learned the experimental methods and techniques used in drug efficacy evaluations.

7.10 Scientific Contributions to Clinical Research of Tamsulosin

Scientific contributions to the clinical research of tamsulosin consisted of the following two points. First, the Phase II clinical trial was designed by the R&D team on scientifically strict evaluation criteria, namely the use of double-blinding, which had not yet been institutionalized in Japan. It was a pioneering effort in Japan where open-label clinical trials were common, and the focus was on basic research, while clinical research was viewed as unimportant. The results of these studies were published in a paper entitled "Use of an alpha 1 blocker, YM617, in the treatment of benign prostatic hypertrophy" in the *Journal of Urology*, which was the leading academic journal in urology (Kawabe et al. 1990a).

Second, collaborative research with the University of Tokyo and Essen University were conducted to clarify the mechanism of action of tamsulosin (Kunisawa et al. 1985). In collaborative research with the University of Tokyo, it was clarified that the α_1-adrenoceptor mediated urethral contraction in human tissue. The lack of readily available human samples also encouraged collaborative research, because it was easier for universities to obtain this material. Meanwhile, collaboration with the University of Essen allowed researchers to characterize the specificity of tamsulosin's effects on the lower urinary tract.

7.11 Sales of Tamsulosin

Tamsulosin was launched in Japan in 1993, and as of 2014, Harnal and other drugs with the same active ingredients were available in more than 60 countries worldwide.[8] Peak sales reached 3.2 billion USD in 2009, as can be seen in Fig. 7.4, which shows the sales of Harnal (Astellas) and Flomax (Boehringer). Notably, the sales of brand

---

[8]Astellas Inc., Drug Interview Form: Harnal D Tablet.

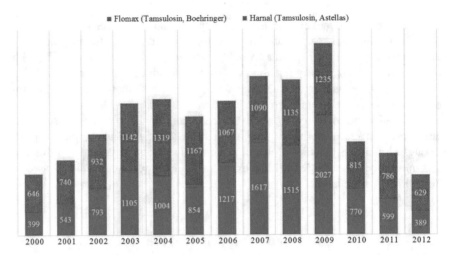

Fig. 7.4 Global sales of Harnal and Flomax between 2000 and 2012 (unit: 1 M USD). *Source* Pharma Future Sezidem Strategic Data, Ut Brain Division

name tamsulosin have declined since 2009, although the prescription of tamsulosin has not decreased. This effect is likely related to the increasing use of generic drugs after the patent expiration for tamsulosin.

Previously, surgical resection of hypertrophied prostate was a conventional treatment for dysuria associated with prostate hyperplasia before the launch of tamsulosin. The frequency of this surgery has been significantly decreased by the introduction of tamsulosin. Antiandrogen therapy was also a common treatment, but patients had to take the drug for longer than 6 months to notice an effect and negative side effects, such as sexual dysfunction, were observed.

In comparison, the effect of tamsulosin appeared within 2 weeks, and the associated side effect of orthostatic disturbance was limited. Given the high selectivity of the drug, it transformed prostate hyperplasia therapy, particularly as it was a prescription for a once-daily dose of an extended-release formulation.

7.12 Competition with Other Companies: First α_1-Blocker to Treat Prostate Hypertrophy

Tamsulosin was the first-in-class α_1-blocker with prostate selectivity in the world. In the preclinical stage, it was compared with existing α_1-blockers, such as prazosin and phentolamine, and it was confirmed that the effect of tamsulosin on blood pressure was very limited, especially in comparison to the other α_1-blockers (Takenaka et al. 1995; Asano and Takenaka 2010). Following the launch of tamsulosin, many α_{1A}- and α_{1D}-blockers were developed, including Asahi Kasei Pharma's naftopidil (Flivas),

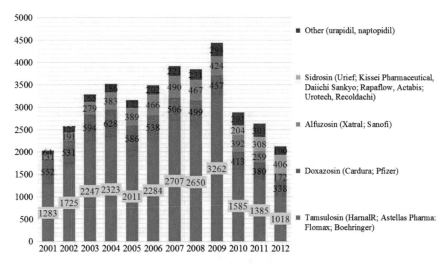

Fig. 7.5 Trends in global α_1-blocker sales by active ingredient between 2001 and 2012 (unit: 1 M USD). *Source* Pharma Future Sezidem Strategic Data, Ut Brain Division

Daiichi Sankyo's silodosin (Ureif), and others.[9] Recently, prescriptions increasingly involve the use of a combination of an α_1-blocker and a 5α-reductase inhibitor, which weakens the action of male hormones (Heisei Chozai Yakkyoku 2013).

Figure 7.5 shows the sales of α_1-blockers for benign prostatic hyperplasia by active ingredient. Tamsulosin is not only the first-in-class but also the best-in-class drug, as indicated by its large share of the market.

7.13 Conclusion

Tamsulosin was the world's first α_1-blocker with prostate selectivity, and therefore was associated with very few side effects such as orthostatic disturbance. Through the development of an extended-release formulation, it could be prescribed as a once-daily dose. These properties allowed tamsulosin to transform prostate hyperplasia therapy. It was not only the first drug in its class, but also remains the best drug in its class, as indicated by its dominant market share.

At the start of the R&D process for tamsulosin, the project's primary objective was the creation of an antihypertensive drug that strongly blocked the α_1-adrenoceptor. After the drug did not demonstrate antihypertensive effects, the drug was only brought to the market following the discovery of a new application. It was based on the results of clinical research showing that the use of an α-blocker was effective for dysuria associated with prostate hyperplasia. This revised target then allowed for the continued development of tamsulosin. Furthermore, the scientific discoveries

---

[9]The Handbook for Healthcare Drugs.

of α-adrenoceptor subtypes and their different anatomical locations indicated that tamsulosin was selective for the prostate and had very few side effects.

Given that the newly discovered antidysuria drug did not fit with the Yamanouchi priority research areas, the R&D of tamsulosin was carried out as unauthorized research (*yami research*). Through receiving information from medical representatives and conducting interviews with urologists, it was possible to obtain in-house approval for the clinical development of tamsulosin. This was aided by the need for an antidysuria drug in the market and verification of the effectiveness and safety of tamsulosin in animal tests.

Furthermore, the mechanism of action was clarified in collaborative research between academia and industry, which in turn confirmed the effectiveness of tamsulosin in humans. In the process of confirming the effectiveness of tamsulosin for dysuria in humans, the human sample testing was carried out by collaborators at the University of Tokyo. Collaborative research with Essen University also played an important role in the process of identifying the mechanism of action of tamsulosin. Finally, Yamanouchi was a pioneer in Japan by carrying out the placebo-controlled, double-blind clinical trial for Phase II and after.

References

Asano, M., & Takenaka, T. (2010). Drugs for the treatment of urinary disorders associated with prostate hypertrophy-harnal. *Laboratory Medicine, 28*(18), 3035–3040.

Barnes, E. M., Langer, S. Z., & Weiner, N. (1974). Release of norepinephrine and dopamine-β-hydroxylase by nerve stimulation. I. Role of neuronal and extraneuronal uptake and of alpha presynaptic receptors. *Journal of Pharmacology and Experimental Therapeutics, 190*(3), 431–450.

Caine, M., Raz, S., & Zeigler, M. (1975). Adrenergic and cholinergic receptors in the human prostate, prostatic capsule and bladder neck. *British Journal of Urology, 47*(2), 193–202.

Edvardsen, P., & Setekleiv, J. (1968). Distribution of adrenergic receptors in the urinary bladder of cats, rabbits and guinea-pigs. *Acta Pharmacologica et Toxicologica, 26*(5), 437–445.

Heisei Chozai Yakkyoku. (2013). Use of major α1-blockers and 5α-reductase inhibitors in benign prostatic hypertrophy. http://www.heisei-ph.com/pdf/H25.2.21_y.pdf. Accessed March 14, 2016.

Hoffman, B. B., Lefkowitz, R. J. (1990) *Goodman and Gilman's the pharmacological basis of therapeutics* (8th ed., p. 221). Pergamon Press.

Honda, K., Miyata-Osawa, A., & Takenaka, T. (1985). alpha 1-Adrenoceptor subtype mediating contraction of the smooth muscle in the lower urinary tract and prostate of rabbits. *Naunyn-Schmiedeberg's Archives of Pharmacology, 330*(1), 16–21.

Honda, K., & Nakagawa, C. (1986). Alpha-1 adrenoceptor antagonist effects of the optical isomers of ym-12617 in rabbit lower urinary-tract and prostate. *Journal of Pharmacology and Experimental Therapeutics, 239*(2), 512–516.

Honda, Kazuo. (2006). Tamsulosin hydrochloride inventory. *Chemistry and Education, 54*(3), 134–137.

Kawabe, K., Ueno, A., Takimoto, Y., Aso, Y., & Kato, H. (1990a). Use of an α1-blocker, YM617, in the treatment of benign prostatic hypertrophy. *Journal of Urology, 144*(4), 908–911.

Kawabe, K., Seishi, U., Naitsu, T., Kato, A., & Kitagawa, R. (1990b). Optimal dose setting test of YM617 for urination disorder associated with prostatic hypertrophy. *Urology Surgery., 3,* 1247–1259.

Kawai, K., Ueno, S., Nakamoto, T., Kato, J., Kitagawa, R., Imamura, K., Oshima, H., et al. (1991). Clinical evaluation of YM617 for dysuria associated with benign prostatic hyperplasia (in Japanese). *Urological Surgery, 4*, 231–242.

Kunisawa, Y., Kawabe, K., Niijima, T., Honda, K., & Takenaka, T. (1985). A pharmacological study of alpha adrenergic receptor subtypes in smooth muscle of human urinary bladder base and prostatic urethra. *The Journal of Urology, 134*(2), 396–398.

Langer, Salomón Z. (1974). Presynaptic regulation of catecholamine release. *Biochemical Pharmacology, 23*(13), 1793–1800.

Minneman, Kenneth P. (1988). Alpha 1-adrenergic receptor subtypes, inositol phosphates, and sources of cell Ca2+. *Pharmacological Reviews, 40*(2), 87–119.

Nishiura, M., Mizumoto, T. (2008). Modified-release fast-disintegrating tablet (Harnal D) containing fine, modified-release particles. *Drug Delivery System, 23*(1), 77–80.

Sarma, A. V., Jacobson, D. J., McGree, M. E., Roberts, R. O., Lieber, M. M., & Jacobsen, S. J. (2005). A population based study of incidence and treatment of benign prostatic hyperplasia among residents of Olmsted County, Minnesota: 1987 to 1997. *The Journal of Urology, 173*(6), 2048–2053.

Sako, K. (2006). Current status and issues in drug discovery: Pharmaceutical products. In *Symposium of Banyu Life Science Foundation, Public Interest Incorporated Foundation*. https://www.msd-life-science-foundation.or.jp/banyu_oldsite/symp/about/symposium_2006/seizai/sako.pdf. Accessed March 14, 2016.

Shintani, Fujio, Hayashi, Masahiro, & Hanano, Manabu. (1978). Transition of plasma concentration of prazosin hydrochloride and its efficacy and subjective symptoms (in Japanese). *Clinical and Research, 55*, 4037–4044.

Sudo, K., Inagaki, O., Asano, M. (1990). *The Japanese Journal of Pharmacology, 52*, 131.

Suto, Katsumi, & Takenaka, Toichi. (1993). How to proceed to medical care for the elderly-concerning the α-receptor subtype for benign prostatic hypertrophy. *New Drug and Treatment, 43*(4), 15.

Taguchi, K., Saitoh, M., Sato, S., Asano, M., & Michel, M. C. (1997). Effects of tamsulosin metabolites at alpha-1 adrenoceptor subtypes. *Journal of Pharmacology and Experimental Therapeutics, 280*(1), 1–5.

Takenaka, T. (2010). Legends in urology. *The Canadian Journal of Urology, 17*(5), 5345.

Takenaka, T. (2015). Drug discovery and medical history of tamsulosin (Harnal) (in Japanese). *Journal of Clinical and Experimental Medicine, 252*(4), 323–325.

Takenaka, Toichi. (2011). From drug discovery to management. *Pharmacia\ 47*(4), 291–293.

Takenaka, T., Fujikura, S., Honda, K., Asano, M., & Niigata, K. (1995). R&D of a novel α1-receptor blocker, tamsulosin hydrochloride. *Pharmaceutical Journal, 115*(10), 773–789.

Tsukazaki, A. (2014). Drugs by samurai: Part II, aiming at drug discovery from Japan Part II, Tamsulosin hydrochloride (in Japanese). *Medical Asahi, 74*–77.

Yamada, S., Suzumi, M., Tanaka, C., Mori, R., Kimura, R., Inagaki, O., et al. (1994). Comparative-study on alpha(1)-adrenoceptor antagonist binding in human prostate and aorta. *Clinical and Experimental Pharmacology and Physiology, 21*(5), 405–411.

Yamanouchi Pharmaceutical. (1975). Yamanouchi 50 years. Yamanouchi Pharmaceutical.

Yoshida, S., Hideo, T., Hidaka, Y., Okada, K., Akino, Y., Isomatsu, K., Fukuyama, T., et al. (1991) Efficacy and safety in the long-term administration of YM617 for patients with lower urinary tract transit disorder. *Urology of Urology, 37*(4), 421–429.

Chapter 8
Pranlukast (Onon)

An Anti-asthmatic Drug Realized by Intensive Investment in the Arachidonic Acidcascade

Kenta Nakamura

Abstract Pranlukast hydrate (Onon) was the world's first cysteinyl leukotriene (CysLT) receptor antagonist and was developed by Ono Pharmaceutical. The innovation of pranlukast is summarized by its provision of an oral agent with high efficacy against bronchial asthma, and its novel mechanism of action. The case of pranlukast development has the following implications. The first is the importance of the contribution of the underlying scientific knowledge to the development of a drug with a novel mechanism of action, particularly at the conceptualization stage of the project. In the case of pranlukast, the discovery of leukotrienes was the contributing factor. The second is the importance of accumulating drug development experience. Such development of expertise allowed Ono Pharmaceutical to discover such a promising compound ahead of other companies located around the globe that were also in competition to develop an LT antagonist. Pranlukast was the first antagonist of CysLT receptors to be developed and became a highly successful drug in Japan with annual sales of tens of billions of yen. However, the third LT antagonist, montelukast (Singulair), has recorded far greater sales than pranlukast globally. These facts provide two additional insights into a market competition of the pharmaceutical industry. The first is related to patent protection and the subsequent entries into the market. The second is the importance of complementary assets in market competition.

8.1 Introduction

Pranlukast hydrate (Onon) was developed by Ono Pharmaceutical Co., Ltd. as the world's first cysteinyl leukotriene (CysLT) receptor antagonist.[1] On 31 March 1995, Ono Pharmaceutical obtained manufacturing approval for the indications of bronchial

---

[1]Leukotrienes include LTA_4 (leukotriene A4), LTB_4, LTC_4, LTD_4, LTE_4, and LTF_4. Among them, LTC_4, LTD_4, LTE_4 are called cysteinyl leukotrienes (CysLTs) because they have cysteine residues in their structures (Kanaoka 2011).

K. Nakamura (✉)
Kobe University, Kobe, Japan
e-mail: knakamura@econ.kobe-u.ac.jp

© Springer Nature Singapore Pte Ltd. 2019
S. Nagaoka (ed.), *Drug Discovery in Japan*,
https://doi.org/10.1007/978-981-13-8906-1_8

asthma, and in June of the same year, it began selling the new drug as Onon capsules in Japan.[2] On 1 October 1999, the drug was approved as a dry syrup preparation for pediatric patients with bronchial asthma, and Onon Dry Syrup was launched in Japan in January 2000.[3] On 18 January 2000, the indication for allergic rhinitis was added to Onon capsules.[4] Sales of pranlukast, which are calculated as a total of capsules and dry syrup, steadily increased, reaching 399 million USD at its peak around 2006, although most of these sales were in Japan.[5] The indication extension of Onon Dry Syrup for allergic rhinitis was approved in November 2011.

Although pranlukast was also approved for the treatment of allergic rhinitis, this chapter focuses on its development as a bronchial asthma agent. Also, in this chapter, cysteinyl leukotrienes are sometimes referred to as "leukotrienes."

Bronchial asthma is a respiratory disease characterized clinically by a paroxysmal cough, wheezing, and dyspnea; pathophysiologically, it is characterized by chronic airway inflammation, airway hyperresponsiveness, and reversible airflow limitation (Nagase 2009). According to the 2011 Japan Patient Survey conducted by the Ministry of Health, Labour and Welfare, the total number of patients with asthma in Japan was estimated to be 1.04 million.[6] Furthermore, in a large-scale survey of prevalence conducted from 2004 to 2006, the prevalence rate of asthma in adults (ages 20–44) was reported as 9.3% (Fukutomi et al. 2010).[7] The World Health Organization (WHO) recently estimated there are 235 million asthma patients worldwide.[8]

Although it has been pointed out that multiple susceptibility genes and environmental factors are intricately related factors causing bronchial asthma, there remains much to be understood. To elucidate the mechanism of disease development, basic research such as the search for asthma-related genes has been actively carried out. There are many unknown triggers of airway hyperresponsiveness, which is the essence of the pathological state of bronchial asthma; however, it has gradually become apparent that airway inflammation is substantially associated with it. Currently, it is understood that airway inflammation is produced by the complex involvement of physiologically active substances, such as chemical mediators and cytokines released by inflammatory cells and constituent cells of the airway.

As shown in Table 8.1, there are a variety of mediators involved in the pathogenesis of bronchial asthma; among them, pranlukast is an agent that improves asthma

---

[2]Pharmaceutical Interview Form: Onon Capsules 112.5 mg (revised December 2011, 8th edn), p. 1, 62.

[3]Pharmaceutical Interview Form: Onon Dry Syrup 10% (revised January 2012, 8th edn), p. 1, 70.

[4]Pharmaceutical Interview Form: Onon Capsules 112.5 mg (revised December 2011, 8th edn), p. 1.

[5]Ono Pharmaceutical "Supplementary Explanation Materials for Financial Results (Fiscal Year 2007, Consolidated and Non-consolidated)."

[6]"The total number of patients" in the survey refers to those who were continually receiving medical care at the date of the survey, including those who had not received medical treatment at a medical facility at the date of the survey.

[7]The prevalence rate is the ratio of those who responded "with asthma symptoms within the past 12 months" in the questionnaire.

[8]WHO Fact sheet on asthma, http://www.who.int/en/news-room/fact-sheets/detail/asthma, (accessed 31 July 2018).

Table 8.1 Candidate mediators of bronchial asthma

Lipid mediator	CysLTs (LTC$_4$, LTD$_4$, LTE$_4$), LTB$_4$, TXA$_2$, PGD$_2$, PAF
Amine	Histamine, serotonin, etc.
Cytokine	IL-4, IL-5, IL-9, IL-13 (Th2 cytokine), etc.
Chemokine	Eotaxin, RANTES, MCP, lymphotactin, fractalkine, etc.
Peptide	CGRP, AM, tachykinin (e.g., neurokinin A, substance P), etc.

Note Modified from Nagase (2009, p. 3014)

symptoms by regulating the activity of leukotrienes. Leukotrienes, which were initially discovered in 1938 and referred to as the slow reacting substance of anaphylaxis (SRS-A), have been implicated in a wide range of allergic reactions. For bronchial asthma, in particular, the presence of symptoms that are not inhibited by antihistamines had suggested that SRS-A might be more involved than histamine, and that receptor antagonists and biosynthetic inhibitors that regulate leukotriene activity would be a revolutionary treatment for asthma (Hayashi 1983, pp. 56–57; Obata 2001). Against the backdrop of this scientific knowledge, the world's first CysLT receptor antagonist, pranlukast, was developed as a drug with a novel mechanism of action.

Figure 8.1 illustrates pranlukast's mechanism of action; that is, pranlukast binds selectively to and antagonizes the action of leukotriene receptors, which are involved

Fig. 8.1 Pranlukast mechanism of action. Modified from Pharmaceutical Interview Form: Onon Capsules 112.5 mg (revised December 2011, 8th edn), p. 18

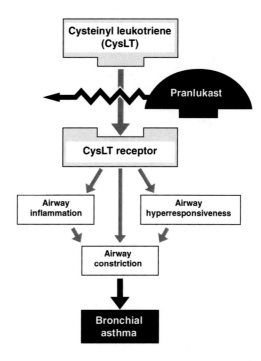

in the underlying pathogenesis of bronchial asthma. Binding of pranlukast to leukotriene receptors inhibits airway constriction, vascular hyperpermeability of airways, edema of airway mucosa, and airway hyperresponsiveness, thereby improving clinical symptoms and lung function in patients with bronchial asthma.[9]

The innovation of pranlukast was that it was an oral agent with high efficacy against bronchial asthma with a novel mechanism of action, working as a cysteinyl leukotriene receptor antagonist. Conventionally, antiallergic agents, oral steroids, inhaled steroids, and β2-agonists had been used for the treatment and management of bronchial asthma. However, steroids have side effects, and inhaled drugs are difficult to handle for children and seniors. Therefore, the introduction of pranlukast, a highly effective drug with a novel mechanism of action, had a substantial impact on patients in terms of substantially eliminating the side effects and difficulty in the use of conventional drugs.

8.2   Timeline from Discovery to Launch

The following is a brief history of the research and development (R&D) toward pranlukast and the scientific discoveries that contributed directly to its development.

1979   Samuelsson at the Karolinska Institute in Sweden discovered several new compounds with conjugated triene structures produced by leukocytes and named them leukotrienes (LTs). These substances were also found to be the main body of SRS-A, which was previously considered to be the causative agent of asthma attacks.

1980   The structures of LTs were determined by identifying the chemical synthesis route of LTs in the laboratory of Elias James Corey at Harvard University. Immediately after that, Ono Pharmaceutical obtained LT synthetic methods from the Corey laboratory.

1981   Ono Pharmaceutical started discovery research on LT receptor antagonists

1985   Ono Pharmaceutical chemically modified the lead compound and obtained ONO-1078 (pranlukast).

1986   Ono Pharmaceutical started Phase I clinical trials.

1995   Ono Pharmaceutical launched pranlukast, the world's first LT receptor antagonist, as Onon in Japan.

As the timeline suggests, the discovery of leukotriene was heavily involved in the conceptualization of the pranlukast development project. It is also worth noting that this development project was supported by Ono Pharmaceutical's continuous R&D on arachidonic acid metabolites, such as prostaglandins (PGs). This implies that it is essential to understand the history of Ono Pharmaceutical before the development of

---

[9]Pharmaceutical Interview Form: Onon Capsules 112.5 mg (revised December 2011, 8th edn), p. 18.

pranlukast. In the following, we present a brief outline of the history and properties of PGs, as well as an overview of PG drug development at Ono Pharmaceutical.

8.3 Prostaglandins and the Arachidonic Acid Cascade

Prostaglandins (PGs) are physiological substances formed from fatty acids (mainly arachidonic acid), and are constituents of phospholipids in biological membranes. They play an essential role in the expression and regulation of cell functions as local hormones (autocoids).[10]

The mechanism of biosynthesis of PGs is as follows. Fatty acids are contained in the meat and fat of animals we ingest as food; among them are unsaturated fatty acids called arachidonic acid, which is used as the raw material for PGs. Arachidonic acid is absorbed from the intestinal tract and enters the blood where it is transported to various tissues. At tissue sites, it is taken up by the plasma membrane. When a cell receives various stimuli, it activates phospholipase A_2 (PLA_2), which releases arachidonic acid from the plasma membrane to the cytoplasm. Free arachidonic acid is metabolized by cyclooxygenase (COX), leading to the production of various physiologically active PG-related substances. Given that all these PG-related substances are produced by COX, the metabolic cascade is called the cyclooxygenase pathway of arachidonic acid.

Another arm of the metabolic system of arachidonic acid involves lipoxygenase (LOX); the lipoxygenase pathway produces various physiologically active substances. Leukotrienes, which are key to the mechanism of action of pranlukast, are produced by the latter system. Combining the two pathways, the total number of arachidonic acid metabolites produced amounts to several dozen, and each has a unique and strong physiological activity. The complete metabolic system using arachidonic acid as a raw material is referred to as the arachidonic acid cascade (Fig. 8.2).

8.4 History of Prostaglandin Research

One of the earliest references of the existence of PG was made by Raphael Kurzrok and Charles Lieb (Table 8.2),[11] two gynecologists in the United States. They observed that during artificial insemination, semen injected into the uterus was immediately expelled, suggesting that semen contains a substance that acts on the myometrium.

In 1934, Ulf von Euler of the Karolinska Institute in Sweden discovered that there are substances in seminal and prostatic fluids that lower blood pressure and induce

---

[10]This section is based on Murota (1988, pp. 41–43) and Tada (1988, p. 1).

[11]Sakamoto and Sato (1983), Hayashi (1983), and Tada (1988) have detailed the history of PG research.

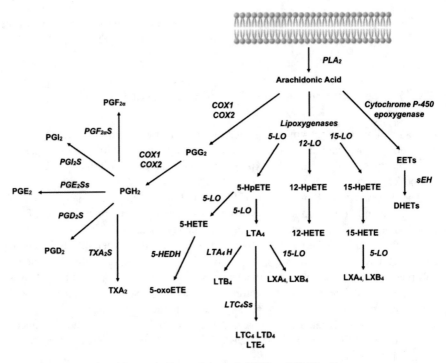

Fig. 8.2 Arachidonic acid cascade *Source* Sala et al. (2018, p. 163, Fig. 1)

Table 8.2 History of prostaglandin studies (major scientific discoveries)

Year	Scientific discovery
1930	Kurzrok and Lieb observed pharmacological effects of PGs
1934	von Euler found an unknown substance and named it prostaglandin
1957	Bergström isolated PGE_1 and $PGF_{1\alpha}$
1962	Bergström identified PGE_2, PGE_3, $PGF_{2\alpha}$, and $PGF_{3\alpha}$
1964	Clarification of the biosynthetic pathway of PGs by Bergström and van Dorp
1969	Synthetic PG methods identified by Corey
1974	Samuelsson identified TXA_2
1976	Vane identified PGI_2

smooth muscle contractions. Furthermore, von Euler identified this stimulator of smooth muscle as a novel physiological agent distinct from known substances, such as acetylcholine and histamine, and named it prostaglandin.[12] von Euler encouraged Sune Bergström to determine the chemical structure of PG, and in 1957, Bergström

---

[12] Kurzrok and Lieb speculated that the seminal fluid contained acetylcholine, which was already known to cause contractions of the myometrium and did not indicate the presence of a new agent. Therefore, the honor of the discovery of prostaglandin does not belong to them.

and his collaborators succeeded in isolating PGE_1 and $PGF_{1\alpha}$. Then in the 1960s, Bergström isolated and identified four additional PGs: PGE_2, PGE_3, $PGF_{2\alpha}$, and $PGF_{3\alpha}$. Consequently, the basic structure of PG was elucidated. Later, in 1974, thromboxane (TX), a metabolite of arachidonic acid and PG was discovered by Bengt Samuelsson, and subsequently prostacyclin (PGI_2) was discovered in 1976.

In 1964, Bergström and David Adriaan van Dorp at the Unilever Research Laboratories in the Netherlands independently elucidated the biosynthesis pathway of PGs. In 1969, E. J. Corey and co-workers reported methods for the synthesis of PG. The technology developed by the Corey laboratory was fundamental to the development of a stable supply of PGs and to the synthesis of synthetic PGs (PG derivatives). Their methodologies have contributed significantly to the development of PG-based drugs since the 1970s.

8.5 Discovery and Development of PG Drugs by Ono Pharmaceutical

Ono Pharmaceutical began researching PGs in 1965 after the strong physiological activity and chemical structure of PG had been characterized. At this time, there was significant expectation that a new PG drug would emerge. Nevertheless, the entire pathway of the arachidonic acid cascade had not yet been elucidated, and because of its uncertainty, some questioned the likelihood of discovering a PG-based drug. Specifically, there were three major concerns at the time (Hayashi 1983, p. 78; Tsuboshima et al. 1992): (1) when PG was externally administered as a drug, it was difficult for it to reach the target organs, and the duration of action could be extremely short; (2) because of its diverse physiological activities, it was difficult to separate the main action from side effects; and (3) because PG is a rather complex molecule, even if chemical synthesis was carried out, a multistage process would be required compared to conventional drug synthesis. Accordingly, there was a possibility that PG-based drugs may have been unprofitable, in addition to the possibility that stability preservation may have been difficult.

Given these challenges, Ono Pharmaceutical was one of a few pharmaceutical companies that began full-scale PG research. In 1965, the president of Ono Pharmaceutical, Yuzo Ono, recognized the rich and diverse potential of PGs as pharmaceuticals during a lecture that Sune Bergström delivered in Japan. Consequently, Ono decided that Ono Pharmaceutical would commit itself to PG research with the aim of developing novel drugs.[13] Furthermore, it was also vital that Prof. Osamu Hayaishi of Kyoto University, who was a prominent biochemist, encouraged Ono Pharmaceutical to enter into the PG field. He also provided opportunities for the company to build a network with world-class scholars, including E. J. Corey of Harvard University, which is discussed later (Noyori 2011, p. 259).

---

[13]Nikkei Daily Morning Edition, 7 February 1984, p. 14.

Table 8.3 History of PG drug development at Ono Pharmaceutical

Year	Development stage at Ono pharmaceutical
1964	(Clarification of the biosynthetic pathways of PGs by Bergström and van Dorp)
1965	Ono Pharmaceutical started PG research
1969	(Synthetic PG methods identified by Corey)
1973	Approval of a $PGF_{2\alpha}$ preparation, Prostarmon-F, as an agent for the induction of labor; the first PG drug in the world
1976	Approval of a PGE_2 preparation, Prostarmon-E, as an oral agent for the induction of labor
1979	Approval of a PGE_1 preparation, Prostandin, as a treatment for peripheral circulatory disorders
1987	Approval of a PGE_1 derivative, Ronok, as an anti-ulcer drug
1988	Approval of a PGE_1 derivative, Opalmon, as a thromboangiitis treatment Approval of a TXA_2 synthase inhibitor, Cataclot, as a treatment for cerebral vasospasm and the associated symptoms of cerebral ischemia after subarachnoid hemorrhage
1992	Approval of a TXA_2 synthase inhibitor, Vega, as a treatment for bronchial asthma

Ono Pharmaceutical had initially procured samples of PGs through enzymatic synthesis using animal organs. However, to convert PGs into pharmaceutical products, it was necessary to provide stable PGs in vast quantities, and in that respect manufacturing by the above method had limitations. Thus, Ono Pharmaceutical contacted Corey with a view to obtaining his group's expertise in PG chemical synthesis. However, Corey's method was for small-scale laboratory level production and was not optimized for large-scale synthesis. Thus, Ono Pharmaceutical developed the technology necessary for mass production on the premise of industrialization (Tsuboshima et al. 1992). The connection with Corey continued and ultimately was critical to the development of pranlukast.

By the 1970s, Ono Pharmaceutical was heavily focused on PG research after having realized its potential and launched various PG drugs (Table 8.3). First, in 1973, $PGF_{2\alpha}$, a natural PG, was approved as Prostarmon-F as a labor-inducing agent and the world's first PG drug. Prostarmon-E, an oral agent for the induction of labor, was approved in 1976, and Prostandin, a treatment for peripheral circulatory disorders, was approved in 1979. After that, in 1987, Ronok, a derivative of PGE_1 with a protective effect on the gastric mucosa, was approved, and Opalmon, a thromboangiitis treatment, was approved in 1988. A a synthetic enzyme inhibitor of thromboxane (TXA_2) was approved in 1988, followed by Cataclot, a treatment for subarachnoid hemorrhage, and, in 1992, Vega, a therapeutic agent for bronchial asthma, was approved.

8.6   Discovery of Leukotrienes

In the early years of research into arachidonic acid metabolism, the metabolites of the COX system, such as PG and TX, attracted a lot of attention because of their potency. In the late 1970s, another pathway, the LOX pathway, was discovered. Then, in 1979, Samuelsson and colleagues found that the LOX-based leukotrienes were identical to those previously identified as SRS-A and responsible for allergic reactions. In 1980, the structure of LTs was determined by Corey.

The origin of the discovery of leukotrienes dates back to the 1930s[14] when Joseph Harkavy reported in 1930 that sputum from patients with bronchial asthma contained an unknown substance that led to a contraction of intestinal smooth muscle cells. In 1938, Wilhelm Feldberg and Charles Kellaway detected a similarly unknown substance in the effluent from perfused guinea pig lungs treated with cobra venom and found that it produced a slow and continuous contraction of smooth muscle, unlike acetylcholine and histamine. Consequently, these materials were named "slow-reacting substance" (SRS).

In the 1950s, it was reported by W. E. Blocklehurst that two kinds of smooth muscle-contractile substances were released into the sensitized lungs of guinea pigs. One was histamine, and the other was a substance similar to SRS that was produced in an anaphylactic condition and therefore called SRS-A (slow reacting substance of anaphylaxis). As the background to the discovery program suggests, from the early stages of research, it was speculated that SRS-A, an unknown substance, was widely implicated as a mediator of allergic reactions, including bronchial asthma. However, SRS-A is an unstable substance, and in the natural world it is synthesized only in trace amounts, so its structure remained unknown for a long time. In the 1970s, it was shown that the stimulation of calcium ionophores induced the production of a significant amount of SRS-A in leukocytes. This improved the availability of SRS-A for experimental systems and contributed to the determination of its structure.

The identity of SRS-A was clarified in 1979 by a collaboration between Samuelsson of the Karolinska Institute and E. J. Corey of Harvard. Samuelsson was a researcher involved in the structure determination of PGs by Bergström in the late 1950s, and, before that, he had been a member of Corey's laboratory as a postdoctoral fellow. Samuelsson and colleagues discovered a new arachidonic acid LOX metabolite from rabbit polymorphonuclear leukocytes and named it leukotriene B4 (LTB_4); they also named one of the production intermediates of LTB_4 leukotriene A4 (LTA_4). Furthermore, under the presumption that LTA_4, which is an intermediate compound in the synthesis of LTB_4 from arachidonic acid, is also an intermediate in the synthesis of SRS-A from arachidonic acid, they began to elucidate the structure of SRS-A. In the structure determination of SRS-A, Samuelsson found that natural SRS-A was identical to a synthetic sample that was provided by Corey; this substance was named leukotriene C4 (LTC_4). Subsequently, Samuelsson and other groups found that SRS contained two other substances; namely, LTD_4 and LTE_4.

---

[14]Hayashi (1983, Chap. 7) and Tada (1988, Chap. 1) describe the history of the discovery of leukotrienes. This section refers to these articles.

Given the results of the above studies, the whole picture of SRS was elucidated. Furthermore, Corey identified methods to chemically synthesize LTA_4, LTB_4, LTC_4, and LTD_4 in 1980.

8.7   Ono Pharmaceutical Started Development of LT Antagonists

The findings of Samuelsson and his colleagues strongly suggested that antagonists and biosynthetic inhibitors that regulate LT activity could become a revolutionary treatment for asthmatics.[15] Therefore, the development of LT antagonists had been of broad interest since the discovery of LT. Ono Pharmaceutical began discovery research in 1981.

At its maximum, the discovery team consisted of eight people in charge of compound synthesis and one or two in charge of pharmacological assessment; at a minimum, the team was made of three people: two in synthesis and one in pharmacological assessment. The core members were Masaaki Toda and Hisao Nakai for compound synthesis, and a pharmacologist, Takaaki Obata, who had been a researcher in the field of drug synthesis. They engaged in discovery research for 3 years until candidate compounds were decided.

The team took two general approaches to discovery research: (1) development of antagonists from LT derivatives, and (2) development of antagonist from an analog of FPL-55712, which was known to be a selective antagonist of SRS-A before the structural determination of LT. However, the former was complicated by the synthesis of an active compound, and the latter could not be expected to have activity through oral administration.[16]

Therefore, these methods were not adopted, and random screening led to the discovery of active compounds.[17] Of course, the success of random screening was not merely reliant on chance but was related to Ono Pharmaceutical's accumulated research on PGs and other arachidonic acid metabolites. Regarding this point, Nakai, who was in charge of synthetic research, reflected in an interview as follows: "The products that arachidonic acid is metabolized to are PG and LT. We have been studying them for a long time, and we have also made many inhibitors. So, we had unknowingly created compounds that were similar to pranlukast. This was because we were specialized in the study of lipids."

---

[15] AA-861, discovered by Takeda Pharmaceutical, was one of the biosynthetic inhibitors that emerged early in the development competition. However, it was not commercialized because of poor pharmacokinetics in humans. In the United States, a 5-lipoxygenase inhibitor zileuton (Zyflo) was commercialized in 1997. Zileuton is hepatotoxic and requires periodic liver function tests (Abiru 2001).

[16] Toda and Arai (1987) and Arai (1988) were detailed about the development approach.

[17] RS-001 was the compound whose antagonistic activity was found by random screening, and ONO-1078 (pranlukast) was synthesized and developed from the same compound. Note that "RS" of RS-001 came from *random screening*.

8.8 Technology Transfer from Corey's Laboratory and Clinical Development Challenges

The main cooperative partner in discovery research was Corey's laboratory at Harvard University. Ono Pharmaceutical had maintained a strong connection with the laboratory since it had entered PG research, which is why it was able to synthesize LTs before the publication of the new method and ahead of other companies. As a result, it became possible to mass-produce LTs with high purity within the company and to construct a screening system in a short period (Mouri 2008). This advantage led to the early discovery of lead compounds and the launch of the world's first LT antagonist in 1995. Thus, the collaboration with Corey's laboratory played a crucial role in the early discovery of pranlukast.

Clinical studies began in 1986 in adult patients with bronchial asthma in Japan.[18] The Phase II and III tests showed the usefulness of pranlukast compared with placebo or an active comparator, namely azelastine (Miyamoto et al. 1993a, b). Pranlukast was approved in March 1995 as the world's first LT receptor antagonist; it was marketed as Onon and sales started in June.

In an interview, the following two points were identified as impediments to the implementation of clinical trials. The first was a problem concerning the physician prejudice towards an LT receptor antagonist for asthma, namely that the drug would not be effective. Such prejudice made it difficult to obtain cooperation in clinical trials. A physician engaged in the clinical studies recalled the following:

> In the late 1980s, clinical trials of anti-leukotriene drugs, such as CysLT receptor 1 antagonists, were started. At the time, I measured leukotriene concentrations in asthmatic patients in a clinical trial, but could not obtain significant results; hence, I was beginning to think that leukotriene was not related to asthma. Therefore, when I was asked to conduct a clinical trial of pranlukast, I replied: "This may not become an anti-asthmatic drug." However, the results of the clinical tests were so striking that my thought was disproved as soon as pranlukast had been administered to asthmatics (Tamura 2004).

The second problem was that the drug substance had strong adhesion and aggregation properties, making it difficult to develop a formulation of a drug that could be supplied for clinical trials. The poor physical properties of pranlukast are also mentioned in the specification of the patent filed by Ono Pharmaceutical (Japanese Patent Application No. 8-149017). The title of the invention is "Granulate containing pranlukast, its production, and improvement in adhering aggregation of pranlukast," which suggests that Ono Pharmaceutical had conducted continuous research on the formulation of pranlukast. However, at the beginning of the clinical study, the company had not possessed an effective formulation technology and overcame this challenge by exploiting fine powder manufacturing technology that was invented by a small start-up company in Japan.

---

[18]Pharmaceutical Interview Form: Onon Capsules 112.5 mg (revised December 2011, 8th edn), p. 1.

Furthermore, Ono Pharmaceutical had succeeded in the development of the thromboxane synthase inhibitor Vega for the treatment of patients with asthma before developing pranlukast. Such experience in the same disease area significantly contributed to the development of pranlukast (Obata 2001).

8.9 Concentrating on PG Research

The most important feature of Ono Pharmaceutical's R&D for pranlukast was that resources had been continuously and intensively invested in the PG field since the 1960s. According to our interview, "there was an air in the company that all works related to arachidonic acid should be undertaken." This implied that the company's drug discovery strategy, which was focused on creating new drugs in the PG field, was well understood. Thus, PG research had been carried out companywide. In fact, in the early 1980s, PG research accounted for 80% of R&D expenditure.[19]

As described above, Ono Pharmaceutical's decision to concentrate on the PG field contributed to its research network, its accumulation of knowledge and techniques in the relevant field, and achieved economy of scope in R&D (e.g., the success of random screening). It is also noteworthy that this strategy resulted in the suppression of influencing activities. Drug development is a lengthy and costly process; therefore, efficient resource allocation is essential. At the same time, there is high uncertainty in R&D, and the asymmetry of information transferred between researchers and executives who make the resource allocation decisions is also significant. Thus, researchers are likely to have incentives to engage in influencing activities to seek more resources for their project. Such activities can prevent a firm's investment in a promising, new R&D field.

In the case of pranlukast, such a problem did not exist. This may be a result of the top-down policy that was strongly focused on arachidonic acid-related metabolites, as mentioned above, and to the fact that Ono Pharmaceutical was a relatively small pharmaceutical company with no originally developed, innovative drug until the early 1970s.

Ono Pharmaceutical's concentration on PGs was also evident in its collaboration with Prof. Corey of Harvard University, who played an important role, from a scientific point of view, in the discovery of pranlukast. After learning of Corey's PG synthesis method, Ono Pharmaceutical continuously sent its researchers to train in the Corey laboratory. Initially, its primary purpose was to absorb cutting-edge knowledge and skills at the laboratory. As Ono's technological capabilities improved, the purpose shifted from information acquisition to information exchange. In addition, Prof. Corey visited Ono Pharmaceutical every 2 years. Thanks to close relations with Corey, Ono Pharmaceutical was able to utilize an LT synthesis method before its publication, and well ahead of other companies. While this technology transfer

---

[19]Nikkei Daily Morning Edition, 29 September 1984, p. 16.

might have been implemented as a result of instructions from Corey, another interesting point is that the synthesis method was delivered by a researcher from another Japanese company who was working in the Corey laboratory.

The Corey laboratory had accepted a relatively large number of Japanese researchers, including the inventors of pranlukast's substance patents: Masaaki Toda, who was in charge of synthesis, and Yoshinobu Arai, a general manager. Furthermore, extraction of the Japanese co-authors from the list of all Corey laboratory publications, that is, more than 1000 articles, identified 57 Japanese co-authors.[20] Among them, 8 authors were employees of Ono Pharmaceutical, more than other pharmaceutical companies or universities in Japan. Hisashi Yamamoto, currently Professor Emeritus at the University of Chicago and Professor at Chubu University, trained under Corey and visited Ono Pharmaceutical about once a month to support synthesis R&D. All of these instances represent a strong link between Ono Pharmaceutical and the Corey laboratory.

8.10   High Efficacy for Bronchial Asthma

One of the major characteristics of pranlukast is that it has high efficacy against bronchial asthma. The efficacy rates of conventional drugs for bronchial asthma were 31.2% (138/443 patients with efficacy or higher) for the anti-allergic agent azelastine hydrochloride (Azeptin), which was approved in April 1986[21]; 42.1% (242/575 patients with moderate or higher improvement) for the thromboxane receptor antagonist Vega, which was approved in March 1992[22]; and 45.7% (266/582 patients with moderate or higher efficacy) for Domenan.[23] In contrast, the efficacy rate of the leukotriene receptor antagonist pranlukast, which was approved in March 1995, reached 65.0% (217/334 patients with moderate improvement or better).[24]

According to a questionnaire survey on asthma symptoms in daily life for 2516 asthmatic patients in the Tohoku region of Japan, administration of pranlukast (Onon) alleviated asthma symptoms (Tamura et al. 2000). Furthermore, it is also reported that the number of deaths cause by asthma attacks had declined since 1997, and this was inversely related to sales of LT receptor antagonists and inhaled steroids (Tamura 2005). These findings suggest that pranlukast was effective in controlling asthmatic symptoms and also contributed to improving patients' quality of life.

---

[20]The Compiled Works of E. J. Corey, http://www.ejcorey.org/corey/publications/publications.php, (accessed 24 February 2014).

[21]Pharmaceutical Interview Form: Azeptin Tablets 0.5 mg, Azeptin Tablets 1 mg, Azeptin Granules 0.2% (revised February 2012, 8th edn), p. 10.

[22]Pharmaceutical Interview Form: Vega Tablets 100 mg/200 mg (revised January 2010, 3rd edn), p. 9.

[23]Pharmaceutical Interview Form: Domenan Tablets 100 mg, Domenan Tablets 200 mg (revised April 2011, 3rd edn), p. 8.

[24]Pharmaceutical Interview Form: Onon Capsules 112.5 mg (revised December 2011, 8th edn), p. 11.

In addition to its effectiveness, pranlukast also had a substantial impact on patients in terms of providing side effect resolution and ease of use. Conventionally, antihistamines, oral steroids, inhaled steroids, and β2-agonists were used for the treatment and management of bronchial asthma; however, steroids have potential side effects. In contrast, in a double-blind, placebo-controlled multicenter Phase II study of pranlukast, asthma symptoms were significantly improved. Moreover, the drug's efficacy in reducing the dose of steroids and bronchodilators and in improving daily life and nocturnal sleeping conditions associated with asthma attacks were confirmed (Obata 2001).

However, there was an issue with the inhalant being difficult for children and the elderly to handle, but this was solved by offering an oral medicine in the form of a capsule and eventually a dry syrup preparation for children that was approved in 1999. In 2008, the Guidelines for the Treatment and Management of Pediatric Bronchial Asthma were revised to include only leukotriene receptor antagonists and sodium cromoglycate (Intal), the stand-alone efficacy of which were validated, and eliminated anti-allergic agents that make patients less likely to recognize efficacy from the basic regimen, for the long-term management of asthma in infants and children aged 2–5 years (Ebisawa 2011). Leukotriene receptor antagonists have been recommended because of inefficiency in inhalational therapy, difficulties in using metered-dose inhalers (MDI), and the safety of leukotriene receptor antagonists (Yoshihara 2009).

Thus, although it is clear that pranlukast has contributed directly to the treatment of asthma, it should also be noted that the launch of an LT receptor antagonist made significant contributions to the understanding of asthma. Namely, the elucidation of the presence and action of LT receptors in asthmatic patients (Mouri 2008).

A clinical trial of pranlukast for allergic rhinitis patients was conducted in 1992, and improvement in the three major symptoms of allergic rhinitis, nasal obstruction, pituita, and sneezing, was confirmed. In particular, the therapeutic effect on nasal obstruction, in which the existing medicine did not have a satisfactory effect, was remarkable. As a result, the extension of indications to allergic rhinitis was carried out in January 2000, in addition to bronchial asthma. Subsequently, Onon Dry Syrup expanded its indication for allergic rhinitis in November 2011, making it possible to use it in children.

8.11   Onon Earned One-Quarter of Ono Pharmaceutical's Sales

The sales of pranlukast peaked between 2006 and 2007, and, at the time, Onon's share of Ono Pharmaceutical's consolidated sales in FY2006 was approximately 27%. For most people with asthma, inhaled corticosteroids are the first choice because they begin to work very quickly. That is to say, Onon has been prescribed mainly for seniors and children who have difficulty using inhalants, and given this limited use, the sales size suggests that the drug was a significant success. Thereafter, the sales declined in 2007, due to the expiration of the patent for bronchial asthma in 2006 and the launch of generic products in July 2007.[25]

8.12 LT Antagonist Development Competition

Once a promising drug target is identified, fierce development competition will occur, such was the case of the LT antagonist. This section describes the competitive conditions of new drug development that were met by Ono Pharmaceutical. For a while after Ono Pharmaceutical entered the area of PG research, few companies, except for Upjohn (Hayashi 1983, p. 88),[26] were able to fully compete. Factors contributing to such a situation included: (1) PG exhibits diverse physiological actions, but is unstable, making drug development difficult; and (2) the earliest PG drugs, which were used in obstetrics and gynecology, did not apply to a large market size, so major pharmaceutical companies were uninterested. However, in line with the progress of PG research, the competition became increasingly intense as many companies entered into PG drug development in the 1970s. In this area, Upjohn was the first entrant, followed by Ono Pharmaceutical, ICI, and Syntex; thereafter, more than 20 companies entered PG development and were competing for invention and patenting in the mid-1970s. Around 1975, PG-related patent applications ranked second in the pharmaceutical field after beta-lactam antibiotics (Hayashi 1983, p. 88). While many companies have developed PG drugs as one of several research fields, Ono Pharmaceutical has invested a large portion of research resources to the field, which is considered to have led to the accumulation of arachidonic acid-related technology by the company.

As the relationship between LTs and disease states became clear, competition in the development of LT antagonists began to intensify. At that time, two approaches were primarily assumed: (1) the development of antagonists by LT derivatives; and (2) the development of antagonists by analogs of the known antagonist FPL-55712. An example of adopting the former approach is the SKF group, which succeeded in the development of the LTD_4 antagonist SKF-104353 by systematically synthesizing LTD_4 analogs; other examples include Ciba-Geigy and Schering-Plough. Development using the latter approach was carried out by many companies such as Eli Lilly, Weiss, Merck Frost, Yamanouchi Pharmaceutical, Roche, and Ciba-Geigy (Toda and Arai 1987; Arai 1988).

---

[25]Ono Pharmaceutical "Notice of Amendment to Forecast of Full-Term Results for the Year Ended March 31, 2008", 8 November 2007.

[26]Upjohn was a company that had been engaged in research since the beginning in the PG field. This may been because of the tight connection between Bergström, who accomplished isolation and identification of PGs, and the company. For such fundamental work, various boosts, including funds were indispensable, and the very person who took care of this was Dr. Weisblatt of Upjohn, an old friend of Bergström (Sakamoto and Sato 1983, p. 4).

As mentioned above, many companies had tried to develop LT antagonists. Nakai stated the following with regard to the situation at the time: "approximately 60 compounds worldwide, including pranlukast (ONO-1078), montelukast, and zafirlukast, were tested in preclinical and clinical settings, but most of them were discontinued for some reason." Thus, the only allogeneic drugs were montelukast and zafirlukast.

Although Ono Pharmaceutical was the pioneer in the synthesis of novel compounds with the discovery of pranlukast (ONO-1078), Merck, which had the best drug development capabilities in the world, was catching up in the clinical stage. However, Merck's development slowed down for an unknown reason, and, as a result, pranlukast was launched as the world's first LT antagonist.[27]

8.13 Exploring Global Expansion

As mentioned above, montelukast and zafirlukast are pranlukast's homogeneous equivalent drugs. Montelukast is a selective and competitive antagonist for the CysLT1 receptor that was found in a series of studies on bronchial asthma and leukotrienes; it is effective with once-daily dosing and is an oral preparation of a therapeutic agent for asthma. The synthesis was done at Merck Frosst Canada in 1991 and development was conducted by Merck. Since it was first approved in Mexico in 1997, it has been approved and marketed in over 100 countries worldwide. In Japan, montelukast was approved in June 2001; it is sold as Singulair (distributor: MSD) or Kipress (distributor: Kyorin Pharmaceutical).[28]

Similarly, zafirlukast is a leukotriene antagonist marketed by AstraZeneca as Acolate. Synthesized by AstraZeneca in the USA in February 1986, full-scale development of this drug as a treatment for asthma began in July 1986. After first gaining approval in Ireland in January 1996, it has been approved in 91 countries, including the USA, and was approved in Japan in December 2000.[29]

In comparing global sales of Onon to its competitors, sales of Onon were $399 million USD in 2006, Singulair (including Kipress) was $3.705 billion USD, and Acolate was $81 million USD. Afterward, while sales of Singulair increased up to about $6 billion USD in 2011, Onon's sales in 2007 were at the same level as the previous year, and this was followed by a declining trend. In other words, although Onon was the first-in-class LT antagonist drug, it has not been able to mark its presence in the global market.

In comparing the daily dosages, only Singulair is taken once a day; Onon and Acolate are both taken twice a day. Therefore, concerning compliance, Singulair has

---

[27]In the interview, it was speculated that Merck's clinical development slowed down, because of some trouble with the clinical compound under development, forcing the company to change to montelukast.

[28]Pharmaceutical Interview Form: Singulair Tablets 5 mg, Singulair Tablets 10 mg, Singulair Tablets 5 mg, Singulair Fine Granules 4 mg (revised in April 2014, 29th edn), p. 1, 68.

[29]Pharmaceutical Interview Form: Acolate Tablets 20 mg (revised in November 2011, 8th edn), p. 1.

an advantage over Onon and Acolate. However, at the same time, some argue that the differences in the efficacy of the three drugs are still unclear (Tamura 2003; Fukui et al. 2008). Thus, it is difficult to explain the small market share of Onon based only on the usefulness of the drug. Instead, the low overseas expansion power of Ono Pharmaceutical at the time and the narrowness of the sales area in the global market may be more meaningful.

According to the pharmaceutical interview form for each drug, Singulair (or Kipres) is approved and sold in over 100 countries and Accolate in over 90 countries worldwide. On the other hand, the main sales regions of Onon are Japan, Korea, and South America, and this does not include the United States and Europe. At the time of Onon's development, Ono Pharmaceutical relied on the domestic market for the majority of its sales and did not have the ability to expand to a global market. For instance, the ratio of overseas sales to consolidated net sales in 1999 was only 1%.[30] In 1993, Ono Pharmaceutical licensed ONO-1078 (Onon) to SmithKline Beecham (SKB) in the United Kingdom, aiming to expand sales channels using SKB's strong sales network. According to media reports at that time, the licensing agreement granted SKB exclusive rights to develop, manufacture, and sell ONO-1078 outside Japan, Taiwan, and South Korea; consequently, SKB initiated a Phase II study in Europe.[31] However, after the licensing agreement was abandoned in 1999, Onon was not released by SKB. This may have been because SKB was one of the world's leading pharmaceutical companies but not necessarily a company with strength in respiratory diseases such as bronchial asthma.[32] Ono Pharmaceutical continued to seek to license Onon in Europe and the United States, but the license was not realized because of the short time remaining on the patent and the advent of competing products.

However, in the 2000s, Onon expanded its sales channels in the Latin American market. This followed Ono Pharmaceutical's licensing of Latin America to Schering-Plough in 2000. In Mexico, sales of Onon began in 2002. The market size in these regions is not large compared with that in Europe and the United States[33,34]; however, the fact that the review and approval period of new drugs is as short as 6 months was thought to be a great advantage for Ono Pharmaceutical because it was eager to expand its leading products overseas.

---

[30]Corporate financial data were obtained from Nikkei NEEDS-FinancialQUEST.

[31]Nikkei Business Daily, 1 April 1993, p. 15.

[32]IMS R&D Focus Drug News, 17 May 1999.

[33]Nikkei Business Daily, 8 November 2000, p. 10.

[34]Nikkei Business Daily, 28 August 2002, p. 11.

8.14 Conclusion

This chapter analyzed the discovery and development process of pranlukast hydrate (Onon), the world's first CysLT receptor antagonist, discovered by Ono Pharmaceutical. The innovation of pranlukast is summarized by two points: it provides an oral agent with high efficacy against bronchial asthma as a cysteinyl leukotriene receptor antagonist, and the drug has a novel mechanism of action. Conventional treatment and management of bronchial asthma had used anti-allergic agents, oral steroids, inhaled steroids, and β2-agonists. However, steroids have side-effects, and inhaled drugs are difficult to handle for children and seniors. Therefore, Onon had a significant positive impact on patients in terms of eliminating side-effects and increasing the ease of use. Moreover, it brought considerable economic benefits to the company.

Finally, we summarize the implications obtained from the case of pranlukast. The first point is the importance of the underlying scientific knowledge in the development of drugs with novel mechanisms of action, particularly at the conceptualization stage of the project. In the case of pranlukast, it was the discovery of leukotrienes that stimulated discovery research. Samuelsson and co-workers revealed that leukotrienes, LOX metabolites, were the same as what was conventionally regarded as a causative agent of allergic reactions, which had been called SRS-A. This strongly suggested that antagonists and biosynthetic inhibitors that regulated LT activity could be an epoch-defining anti-asthmatic agent. Samuelsson's discoveries prompted many companies, including Ono Pharmaceutical, to enter the competition to develop LT antagonists.

The second point is the importance of accumulating competencies in a broad sense for drug development. This is the answer to the question of how Ono Pharmaceutical was able to discover such a promising compound ahead of the global LT antagonist development competition. The most characteristic feature of the company's R&D was that it continued to concentrate its research resources on PGs, metabolites of arachidonic acid, since the 1960s. This led to the accumulation of knowledge in the research field, the construction of scientific networks, and the economy of scope in R&D (e.g., the success of random screening). These features allowed Ono Pharmaceutical to quickly identify drug discovery targets, construct screening systems early, and discover a compound for the first time in the world. The success of this concentration strategy was also attributable to the fact that Ono Pharmaceutical was a relatively small pharmaceutical company with no strong in-house origin drugs until the early 1970s. It also had top-down strategic decisions that prevented competition or congestion in allocating in-house resources within the company.

As we have described, pranlukast is the first-in-class antagonist of CysLT receptor and became a highly successful drug in Japan with annual sales of hundreds of millions of dollars. However, the third LT antagonist, Singulair, has recorded far more sales than pranlukast globally. These facts provide two additional insights into market competition in the pharmaceutical industry.

The first is related to patent protection and subsequent entry into the market. Usually, pharmaceuticals are protected by substance patents, which ensure strong protection for patentees. However, it is undeniable that the chemical structure that achieves the same mechanism of action may be discovered outside the scope of the existing patent. This implies that patent protection does not necessarily guarantee the exclusion of potential competitors. In fact, there were two competitors in the market for LT antagonists.

The second is the importance of complementary assets in market competition. Onon was not able to show its presence on a global basis because it was not sold in the European and US markets. Ono Pharmaceutical at that time did not have the ability to expand its products overseas. However, if a company does not have complementary assets, an alliance is a promising strategy. Ono Pharmaceutical licensed out to a European pharmaceutical company, yet Onon was not successfully expanded to Europe. Thus, this case also appears to suggest that decision-making about timing and choice of alliance partner are very important.

References

Abiru, T. (2001). LT receptor antagonists. In S. Murota & S. Yamamoto (Eds.), *New Development of Prostaglandin Research* (Vol. 38, pp. 202–206). Tokyo: Tokyo Kagaku Dojin (in Japanese).

Arai, Y. (1988). Leukotriene. In S. Yamamoto & S. Murota (Eds.), *Prostaglandin* (Vol. 7, pp. 143–157). Tokyo: Tokyo Kagaku Dojin. (in Japanese).

Ebisawa, M. (2011). Management of infantile asthma in JPGL 2008. *Japanese Journal of Pediatric Allergy and Clinical Immunology, 25*(4), 700–704. (in Japanese).

Fukui, Y., Hizawa, N., Takahashi, D., Maeda, Y., Kobayashi, M., Nasuhara, Y., et al. (2008). Efficacy of leukotriene receptor antagonists in asthma treatment and analysis of background factors evaluated by a questionnaire survey among physicians. *Annals of The Japanese Respiratory Society, 46*(12), 972–980. (in Japanese).

Fukutomi, Y., Nakamura, H., Kobayashi, F., Taniguchi, M., Konno, S., Nishimura, M., et al. (2010). Nationwide cross-sectional population-based study on the prevalences of asthma and asthma symptoms among Japanese adults. *International Archives of Allergy and Immunology, 152*(3), 280–287.

Hayashi, M. (1983). *New bioactive substances: Prostaglandins and related substances.* Tokyo: Kaimeisha. (in Japanese).

Kanaoka, Y. (2011). Cysteinyl leukotriene receptor. *Seikagaku, 83*(7), 609–614. (in Japanese).

Miyamoto, A., Takishima, T., Makino, S., Shida, T., & Nakashima, M. (1993a). Effect of ONO-1078, a selective antagonist to leukotriene C4, D4, and E4 receptors on treatment of bronchial asthma in adults. *Journal of Clinical and Experimental Medicine, 164*(4), 225–247. (in Japanese).

Miyamoto, A., Takishima, T., Makino, S., Shida, T., Nakashima, M., & Hanaoka, K. (1993b). Effect of ONO-1078, a Leukotriene C4, D4, and E4 receptor antagonist on treatment of bronchial asthma in adults. *Journal of Clinical Therapeutics & Medicines, 9*(1), 71–107. (in Japanese).

Mouri, T. (2008). The development and the history of ONON. *Japanese Journal of National Medical Services, 62*(4), 245–246. (in Japanese).

Murota, S. (1988). Advances in prostaglandin research. *Clinician, 35*(367), 40–51. (in Japanese).

Nagase, T. (2009). 3. Topics of asthma-related mediators. *Nihon Naika Gakkai Zasshi, 98*(12), 3013–3018. (in Japanese).

Noyori, R. (2011). *Facts are the enemy of truth.* Tokyo: Nihon Keizai Shinbunsha. (in Japanese).

Obata, T. (2001). Research and development of leukotriene receptor antagonist ONON. *Farumashia, 37*(1), 16. (in Japanese).

Sakamoto, S., & Sato, K. (1983). *Prostaglandin story.* Tokyo: Kodansha. (in Japanese).

Sala, A., Proschak, E., Steinhilber, D., & Rovati, G. E. (2018). Two-pronged approach to anti-inflammatory therapy through the T modulation of the arachidonic acid cascade. *Biochemical Pharmacology, 158,* 161–173.

Tada, M. (1988). *Prostaglandins.* Kyoto: Kagaku Dojin. (in Japanese).

Tamura, G. (2003). Category and class of leukotriene receptor antagonists. In M. Adachi (Ed.), *A complete guide to leukotriene receptor antagonists* (pp. 44–50). Tokyo: SENTA IGAKU-SHA (in Japanese).

Tamura, G. (2004). Special feature leukotrienes and anti-leukotrienes: Introduction. *Allergy and Immunity, 11*(11), 9. (in Japanese).

Tamura, G. (2005). Recent trends and future issues of asthma mortality in Japan. *Allergology International, 54*(2), 223–227.

Tamura, G., Inoue, H., Chihara, J., & Takishima, T. (2000). Population-based open-label clinical effectiveness assessment of the cysteinyl leukotriene receptor antagonist pranlukast. *Allergology International, 49*(3), 189–194.

Toda, M., & Arai, Y. (1987). Biological activities of leukotriene derivatives and development of leukotriene antagonist. *Journal of Synthetic Organic Chemistry, 45*(2), 136–150. (in Japanese).

Tsuboshima, M., Matsumoto, K., Arai, Y., Wakatsuka, H., & Kawasaki, A. (1992). Prostaglandins: Synthetic and pharmacological studies and development. *Yakugaku Zasshi, 112*(7), 447–469. (in Japanese).

Yoshihara, S. (2009). Treatment strategies using leukotriene receptor antagonists. *Allergy and Immunity, 16*(6), 791–793. (in Japanese).

Chapter 9
Tacrolimus (Prograf)

An Immunosuppressant of Global Standards

Kenta Nakamura

Abstract Tacrolimus (Prograf) is an immunosuppressant that was developed by Fujisawa Pharmaceutical. Tacrolimus has a wide range of indications in the fields of transplantation medicine and autoimmune diseases and has had a significant impact on patients. It also generated large sales for Fujisawa as a blockbuster drug. The notable aspect of the discovery and development process of tacrolimus are as follows. First, no compound with immunosuppressive functions beyond tacrolimus has been created, even though over 20 years has elapsed since the total synthesis and elucidation of the mechanism of action of tacrolimus. This suggests that the strategy of searching for compounds from natural products was a good choice. Tacrolimus is an innovative novel drug that was realized by combining Fujisawa Pharmaceutical fermentation technology with the latest immunological theory. Second, tacrolimus development saw the interdependence of drug development and progress of science. The discovery of tacrolimus was substantially driven by scientific advances, but it also triggered academic investigation of the mechanism of action, which later contributed to the creation of system biology. Third, by conducting joint research with Thomas Starzl of the University of Pittsburgh, who was a leader in transplantation medicine, a proof of concept study of tacrolimus was made possible at an early stage. Fourth, when Japanese pharmaceutical companies intended to enter overseas markets, it was common to develop a drug jointly with foreign companies or license it out to foreign companies; the tacrolimus case was rare because Fujisawa Pharmaceutical conducted the whole process by itself, including clinical trials and global sales.

K. Nakamura (✉)
Kobe University, Kobe, Japan
e-mail: knakamura@econ.kobe-u.ac.jp

© Springer Nature Singapore Pte Ltd. 2019
S. Nagaoka (ed.), *Drug Discovery in Japan*,
https://doi.org/10.1007/978-981-13-8906-1_9

9.1 Introduction

Tacrolimus (Prograf) is an immunosuppressant that was developed by Fujisawa Pharmaceutical Co., Ltd. (now Astellas Pharma Inc.). The brand name "Prograf" comes from: "*Pro*tect of *graf*t rejection."[1,2] Tacrolimus was approved in 104 countries around the world by May 2011 and is sold in 96 countries. It is still widely used as one of the major immunosuppressants for renal transplantation. It is also a blockbuster drug that has had annual worldwide sales in excess of $1 billion since the early 2000s.[3]

In March 1984, tacrolimus was discovered in the culture broth of an actinomycete isolated from soil in the suburbs of Mt. Tsukuba and was named "*T*sukuba ma*crol*ide *immunosuppressant*," meaning a macrolide immunosuppressant discovered in Tsukuba. It is also often called "FK506," which was its code number. Before the discovery of tacrolimus, cyclosporine A was the main drug used as an immunosuppressant. Interestingly, tacrolimus is a novel compound and has a completely different chemical structure from cyclosporine A.[4] The development of tacrolimus has had a significant impact on transplantation medicine, and this success has been recognized as a representative achievement in natural drug discovery in Japan, on a par with the discovery of pravastatin by Sankyo Co., Ltd (now Daiichi Sankyo Co., Ltd.).

In April 1993, Prograf (capsule and injection solution) was approved for manufacture in Japan for the control of rejection after liver transplantation, and it was launched in Japan in June of the same year for the first time in the world. Subsequently, suppression of rejection in renal transplantation, heart transplantation, lung transplantation, pancreas transplantation, small intestine transplantation, rejection in bone marrow transplantation, and suppression of graft-versus-host disease were also approved for indications. It was also approved for the treatment of severe autoimmune diseases such as myasthenia gravis, rheumatoid arthritis, lupus nephritis, refractory active ulcerative colitis, interstitial pneumonia associated with polymyositis, and dermatomyositis. Moreover, Fujisawa developed new dosage forms; namely, "Prograf Granules," which facilitated dose adjustments in pediatric use with the same indications as injectables (approved in 2001); and "Graceptor," a sustained-release preparation (approved in July 2008).[5] Tacrolimus was also formulated as an external preparation called "Protopic Ointment," which was approved for the indication

---

[1]Pharmaceutical Interview Form: Prograf Capsules 0.5 mg/1 mg/5 mg, Prograf Granules 0.2 mg/1 mg (revised April 2014, 34th edn), p. 6.

[2]In this chapter, a brand name is used for a drug after marketing, but general names are used otherwise. Tacrolimus hydrate is referred to as "tacrolimus" and Fujisawa Pharmaceutical Co., Ltd. Is referred to as "Fujisawa" throughout the chapter.

[3]Pharmaceutical Interview Form: Prograf Capsules 0.5 mg/1 mg/5 mg, Prograf Granules 0.2 mg/1 mg (revised April 2014, 34th edn), pp. 160–167.

[4]In the following, cyclosporine A is referred to as "cyclosporine."

[5]Pharmaceutical Interview Form: Graceptor Capsules 0.5 mg/1 mg/5 mg (revised April 2014, 14th edn), cover page.

of adult atopic dermatitis in June 1999.[6] Furthermore, Senju Pharmaceutical Co., Ltd., which received a license from Fujisawa, obtained approval in January 2008 for the manufacture and sale of "Tacrolimus Ophthalmic Suspension" for the indication of spring catarrh in cases where an anti-allergic agent is insufficient.[7] Thus, while tacrolimus has been approved for many indications, this chapter focuses on the aspects of immunosuppressants for transplant rejection and summarizes the characteristics of the history and the contribution of science to the discovery and development process of tacrolimus.

9.2 Transplant Immunology

Before explaining the mechanism of action of tacrolimus, we will briefly describe the immune system and a rejection reaction. The immune system is supposed to distinguish between "self" and "non-self" and attack only "non-self" invaders from the outside world such as bacteria and viruses. However, if for some reason the immune system does not function properly and an excessive or abnormal reaction occurs, the immune response may be directed against body tissue. When this happens, it is called an allergic reaction or an autoimmune disease. The immune system is also concerned with graft rejection following organ transplantation. In general, the immune system cannot account for the transplant of cells, tissues, or organs. Thus, it is not at all surprising that even if a transplanted organ is for sustaining life, the immune system recognizes it as non-self and attacks it.

The fact that the immune system rejects a graft suggests that individual cells have self-labeled molecules on their surface that are recognized as transplantation antigens. This antigen is called the major histocompatibility complex (MHC) or the human leukocyte antigen (HLA).

When an organ is transplanted from a donor to a recipient, macrophages find the MHC of the graft and transmit that information to helper T cells. When helper T cells judge the graft as non-self, they release a messenger called interleukin-2 (IL-2), which induces, enhances, and activates cytotoxic T cells and activates natural killer (NK) cells, which attack the transplanted organ.[8] These are the mechanisms of rejection, implying that if there is a drug to control the process, it may become an immunosuppressive agent. In the case of tacrolimus, it exerts immunosuppressive effects by inhibiting IL-2 production in T cells. The detailed mechanism of action is as follows:

---

[6]Pharmaceutical Interview Form: Protopic Ointment 0.1% (revised April 2014, 16th edn), cover page.

[7]Pharmaceutical Interview Form: Tacrolimus Ophthalmic Suspension 0.1% (revised August 2014, 7th edn), cover page.

[8]The foregoing part is based on Kishimoto and Nakashima (2007).

When T cells recognize antigens with T cell receptor (TCR) on the cell surface, signals are transmitted to the nucleus, producing cytokines such as IL-1β, IL-2, IL-4, IL-6, tumor necrosis factor-α (TNF-α), and interferon-γ (INF-γ). Calcineurin (CaN) is a dephosphorylating enzyme involved in signal transduction and dephosphorylates nuclear factor of activated T cells (NFAT) , which is a transcription factor for various cytokines, and then transfers it into the nucleus. Tacrolimus forms a complex with FK506 binding protein (FKBP), which exists in the cytoplasm, and binds to CaN, thereby inhibiting the activity of CaN. As a result, nuclear translocation of NFAT does not occur, transcription of NFAT-dependent genes such as IL-2 is inhibited, and production of cytokines via T cell activation is suppressed (Tani 2012, p. 253).

Figure 9.1 illustrates the mechanism of action. Tacrolimus inhibits the action of calcineurin, thereby suppressing T cell proliferation and inducing immunosuppression; thus, it is called a calcineurin inhibitor. Regarding the mechanism of action, it is important to consider the following two points. First, tacrolimus exerts its effects by an unusual mechanism of simultaneously binding two proteins. That is, tacrolimus itself has no immunosuppressive activity, and the tacrolimus–FKBP complex is the true immunosuppressant. Second, the mechanism of action described above was mostly unknown at the time of its development and became clearer afterward. Thus, the discovery of tacrolimus prompted scientific progress, and the science provided a new tool to understand the mechanism.

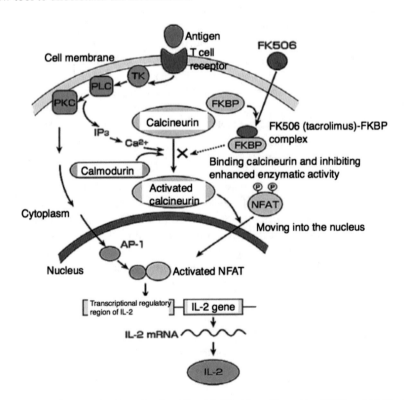

Fig. 9.1 Tacrolimus mechanism of action. *Note* Modified from Yamashita (2013, p. 146, Fig. 6)

The innovativeness of tacrolimus is summarized as being able to actively suppress rejection reactions to transplanted organs (Onozuka 2004, p. 21). In comparison with the existing drug cyclosporine, it was argued that tacrolimus was better in the control of acute rejection, although there was no difference in the survival rate of the transplanted organ, graft function, or the incidence of nephrotoxicity, which is the main side effect. It was also known that hypertension, which is a risk factor of cardiovascular disease, tends to occur with the use of cyclosporine. However, tacrolimus may be useful when cyclosporine, also a calcineurin inhibitor, is ineffective or causes intolerable side effects. While it was evident that cyclosporine presented a novel mechanism of selectively inhibiting T cell activation and was a groundbreaking new drug that fundamentally changed organ transplantations, tacrolimus had an efficacy equal to, or higher than that, of the existing immunosuppressants, contributing to significant improvements in postoperative management and long-term prognosis. In the case of tacrolimus, these points can be described as the core of innovation. Furthermore, tacrolimus was eligible for the first breakthrough premium pricing that was introduced in the 1992 drug pricing reforms, which is another indication of its level of innovation.[9] Fujisawa also received numerous awards for tacrolimus, including the 2004 Prime Minister Invention Award.[10]

9.3 Timeline of Tacrolimus Research and Development

The following is a brief history of the discovery and development of tacrolimus.

1982: At the exploration laboratory under construction in Tsukuba, Japan, a discovery study for active substances derived from natural products was initiated with the aim of developing a new immunosuppressive agent.

---

[9]Prograf received more than 10% price premium (Aoki 2007; Yamashita 2013). The breakthrough premium pricing was certified for new drugs in the National Health Insurance price list that satisfied all of the following requirements: (a) R&D based on entirely new ideas; (b) objective and scientific demonstration of clearly superior efficacy and/or safety compared with existing drugs; and (c) significant impacts on the medical system of the target disease and significant contributions to the improvement and progress of treatment methods are expected. Tacrolimus won the premium because of the following: (a) it was a macrolide immunosuppressant discovered as an actinomyces metabolite; (b) a comparative study with cyclosporine showed a significantly better survival rate for liver transplantation; (c) it was expected that tacrolimus would have fewer toxic side effects and higher immunosuppressive effects than cyclosporine, facilitating liver transplantation, which was previously difficult (Ministry of Health, Labour and Welfare 2005).

[10]According to Aoki (2007) and Yamashita (2013), tacrolimus also received the JSBBA (Japan Society for Bioscience, Biotechnology, and Agrochemistry) Award for Achievement in Technological Research (1995), the Pharmaceutical Society of Japan Award for Drug Research and Development (1997), the SBJ (Society for Biotechnology, Japan) Award (1997), Chunichi Industrial Technology Award (2000), Okochi Memorial Award (2001), and Kinki District Invention Award (2003). In the UK, tacrolimus won the Prix Galien Award (1996).

April 1983: At the same time as the exploration laboratory was launched, the discovery study was gaining momentum.

March 1984: Approximately 1 year after the start of screening, tacrolimus, an immunosuppressant, was discovered in the supernatant of a novel actinomycete species, later named *Streptomyces tsukubaensis* No. 9993. The microorganism was isolated from soils in the suburbs of Mt. Tsukuba. During this period, microorganisms were collected from soils throughout Japan; researchers screened 8000 molds and 12,000 actinomycetes.

December 1984: Isolation and structure determination of tacrolimus. Patent applications covering tacrolimus and its analogs were filed in the United Kingdom three times after December 1984 because there was no internal priority system in Japan at that time. Subsequently, the applications with claims for priority were filed in 28 countries including Japan.[11]

1985: Collaborative research with Takenori Ochiai of Chiba University started. Fujisawa and Ochiai tested the effectiveness of tacrolimus using rat heart transplant and dog kidney transplant models.

August 1986: Preclinical data were presented at the Transplantation Society meeting in Helsinki.

March 1989: Thomas Starzl and colleagues at the University of Pittsburgh first administered tacrolimus to patients with severe rejection after liver transplantation.

May 1990: First human administration in Japan at Shimane University Faculty of Medicine.

June 1990: Liver transplant trial started in Japan.

July 1990: Renal transplant trial started in Japan.

August 1990: A multicenter, large-scale liver transplant trial started in Europe and the United States.

December 1991: An application was filed for approval in Japan for the indication of inhibition of rejection after liver transplantation.

February 1992: Renal transplant trial started in the United States.

September 1992: Renal transplant trial started in Europe.

April 1993: Prograf first approved in Japan as an immunosuppressant. It was launched in June for liver transplantation.

April 1994: Liver transplant approval in the United States (released in June).

---

[11]In 1988, Merck pointed out that one of the tacrolimus analogs (ascomycin) was found to be a known substance. Ascomycin was announced in 1962, and the US patent was granted in 1966 (Takano 1997).

October 1994: Released in the United Kingdom (liver transplant and kidney transplant).

While tacrolimus was discovered in an actinomycetes culture that had been isolated from soil in the suburbs of Mt. Tsukuba, it is essential to understand the history of tacrolimus before it was developed to comprehend the origin of the ability that led to the concept of the research. In the following, we review the history of immunosuppressive agents before the discovery of tacrolimus and the lineage of natural drug discovery at Fujisawa.

9.4 The History of Organ Transplantation is the History of Immunosuppression[12]

The first successful organ transplant was a renal transplantation between identical twins that was conducted by Joseph Murray and co-workers in 1954 in Boston. In the 1950s, total body irradiation (TBI) was initially used for conditioning of organ transplantation-related rejection. For instance, in some renal transplants performed by Jean Hamburger and co-workers in France around 1960, TBI resulted in the suppression of rejection and the long-term engraftment of transplanted organs. However, to achieve immunosuppression with TBI, the irradiation was close to the lethal dose, the control of the dose was difficult, and there were many problems such as frequent infection.

In 1959, Robert Schwarz and colleagues at Tufts University identified an immunosuppressive effect in the antimetabolite 6-MP (mercaptopurine).[13] Although it was already reported that steroids affect immune responses in the early 1950s, the discovery of Schwarz et al. was vital in the sense that it became a turning point from irradiation to drug-induced immunosuppression. Khan, who had studied under Joseph Murray, learned about the findings of Schwarz and demonstrated the efficacy of 6-MP in rejection studies of kidney-transplanted dogs in London in 1960. The Burrows-Welcome Institute also tried to develop derivatives that were less toxic than 6-MP, which led to Gertrude Elion and George Hitchings synthesizing azathioprine in 1960.[14] The effectiveness of azathioprine was also proven by Khan, and the world's first clinical application to renal transplantation was made by Murray in 1963. Moreover, the appearance of azathioprine evoked a transplant boom: lung transplantation and liver transplantation began in 1963, pancreas transplantation in

---

[12]This section refers to Starzl (2001) and Barker and Markmann (2013).

[13]6-MP is an anticancer drug, discovered in 1948 by Hitchings et al. at the Barros Wellcome Research Institute in the United States. It was synthesized in 1952 by Elion et al. See Pharmaceutical Interview Form: Leukerin Powder 10% (revised April 2012, 6th edn), p. 1.

[14]Pharmaceutical Interview Form: Azanin Tablets 50 mg (revised February 2008, 7th edn), p. 1.

1966, and heart transplantation in 1976.[15,16] However, transplant rejection was still an immense obstacle, and the success rate of transplantation at the time remained close to 30%. Therefore, the transplant boom began to gradually fade. In the early 1980s, organ transplantation increased in popularity because of the widespread use of cyclosporine, which was the next generation immunosuppressant.

Cyclosporine is a cyclic polypeptide that was discovered in 1970 by the Swiss pharmaceutical company Sandoz (Novartis) in fungal cultures.[17] In 1972, Jean Borel and colleagues at Sandoz found that the substance had a potent immunosuppressive effect, subsequently leading to its development as an immunosuppressant. In 1978, the first clinical application, cadaveric renal transplantation, was carried out by Khan at Cambridge University, and from 1983 the novel immunosuppressant was launched in various countries as Sandimmun.[18,19]

Conventional immunosuppressive agents, such as 6-MP and azathioprine, are classified as nonspecific immunosuppressive agents. They were limited in efficacy because of their inhibitory effects on cells other than immunocompetent cells, and their side effects were severe. Cyclosporine, on the other hand, acted selectively on T cells, which are the leading players in rejection. Therefore, unlike the existing immunosuppressive agents, it was able to dramatically increase the inhibitory effect without significantly reducing resistance to infection. Cyclosporine is also characterized by fewer side effects, such as myelosuppression, than azathioprine and steroids.[20]

Azathioprine enabled organ transplantation in clinical practices, and the success rate of transplantations rose significantly with the appearance of cyclosporine. For example, Starzl and colleagues reported that the 1-year survival rate after liver transplantation using azathioprine was about 40%, whereas that with cyclosporine improved to 74% (Iwatsuki et al. 1988). Similarly, the introduction of cyclosporine improved the engraftment rate of renal transplants by 15–20% (Markus et al. 1987). However, continued side effects with cyclosporine, such as nephrotoxicity, meant that the demand for a new immunosuppressive agent was increasing.

---

[15]The world's first liver transplant (1963) and the world's first successful liver transplant (1967) were done by Thomas Starzl at the University of Colorado.

[16]In 1968, approximately 100 heart transplants were performed (Kishimoto and Nakashima 2007, p. 94).

[17]The fungus was found in soil collected from the Hardanger Plateau in southern Norway by an employee on vacation.

[18]Pharmaceutical Interview Form: Sandimmun Intravenous Infusion 250 mg (revised March 2015, 10th edn), p. 1.

[19]Cyclosporine as an oral preparation (Sandimmun Oral Solution/Capsule) was liable to have variable efficacy because of the effects of bile acids and meals. Thus, a microemulsion preparation Neoral, which enabled stable absorption, was also developed.

[20]These features are common to tacrolimus, which has a similar mechanism of action.

9.5 Natural Products Drug Discovery in Fujisawa Pharmaceutical

The origin of natural product drug discovery at Fujisawa was based on earlier work at Sanyo Chemical Co., Ltd. in Nagoya.[21] After World War II, through the production of penicillin, Sanyo Chemical had accumulated fermentation technologies and had built a network of external researchers. Fujisawa, which had been responsible for the sales of penicillin produced by Sanyo, absorbed the fermentation technology by acquiring Sanyo in 1956.

The first project to which the technology from Sanyo contributed was the commercialization of cephalosporin C. This research was conducted in collaboration with the National Research and Development Corporation (NRDC) of the United Kingdom, and Fujisawa's fermentation-related technologies were significantly exploited, leading to the discovery of cefazolin and the launch of the first cephem-type antibiotic Cefamedine in 1971. After that, the fermentation research team, which had been collecting soil from all over Japan, continued to search for physiologically active substances, mainly antibiotics, produced by microorganisms.[22]

In the latter half of the 1970s, Fujisawa conducted discovery research for new drugs other than infectious diseases, such as antihypertensive drugs, antiinflammatory drugs, and anticancer drugs, as a foothold to a new field. At one time, the fermentation team of the central research laboratory in Osaka had discovered the immuno-hyperactivity substance FK156; then, FK565, which was chemically decorated FK156, was developed as an anticancer drug, and temporarily an anti-AIDS drug in the United States. It was eventually discontinued. However, the experience of FK565 indicated that it was possible to isolate natural products that act on the immune system depending on the design of the screening system. The concept of a search for immunosuppressants was introduced later. Toshio Goto and Toru Kino, who were the core members of the tacrolimus discovery research team, were also involved in the development of FK565.

9.6 Discovery Research Started at the Newly Established Research Laboratory

At the Tsukuba Discovery Laboratory, which was established in 1983, two research groups, the Biological Science and Fermentation Discovery Departments, began discovery research. The institute was organized around 30 people, mainly young people, and was energetic. While the institute was still new and geographically far from the center of the company, there was a high degree of research freedom (Kino

---

[21] The following explanation relies on Yamashita (2013, pp. 141–142).

[22] Antibiotics such as pyrrolnitrin, thiopeptin (for livestock), trichomycin, and bicozamycin (for livestock) were discovered. Noda Economic Research Institute (1976, p. 159) stated that Fujisawa's most prominent point was to find bacteria that produce antibiotics from bacteria in the soil.

2011). In fact, there was such a high degree of freedom that researchers were able to undertake immunosuppressant discovery research even though this disease area had limited demand at that time in Japan, and the research was not active globally.

Toshio Goto and his fermentation discovery team had set development targets on (1) organ transplantation and autoimmune diseases, (2) inflammation and allergy, and (3) cancer metastasis, for which they possessed assay systems at the time. In addition to having the experimental infrastructure, fundamental advances in immunology and inferences about the mechanism of the immunosuppressive action of cyclosporine were identified as discovery research directions. Regarding this point, Toshio Goto noted as follows:

> [The] rejection response is triggered by thymus-derived lymphocytes called T cells, which were shown to require a T cell proliferator protein, called interleukin-2 (IL-2), for the T cells to increase. Moreover, it was speculated that the immunosuppressive effect of cyclosporine was due to the blocking of IL-2 production (Goto 1993, p. 768).

By searching for a substance that specifically inhibits the production of IL-2, it was considered that an immunosuppressive agent with high safety and excellent efficacy could be found (Kino et al. 1989). However, given the mechanism of action of cyclosporine and the history of its discovery, there was a suspicion that the target compound would not be found in a chemical library, the discovery laboratory started to search for compounds from natural products.[23]

The mixed lymphocyte reaction (MLR) was used as a screening method.[24] MLR is a reaction in which lymphocytes differentiate and proliferate by recognizing a partner as a foreign substance when mixing two lymphocytes with different major histocompatibility complex (MHC). Therefore, it was considered that if substances that specifically inhibit MLR could be screened from mold or actinomycete metabolites, they could be active immunosuppressive agents. After the screening of about 8000 stocks of mold and about 12,000 stocks of Actinomyces by this method, tacrolimus was discovered in March 1984, and the chemical structure was determined in December of the same year.

Transplantation experiments using animals were carried out as joint research with Takenori Ochiai of Chiba University, beginning in April 1985.[25] This collaboration

---

[23] At the beginning of the establishment of the Tsukuba Discovery Laboratory, the facilities and equipment were not in a state where it could operate at full capacity, so researchers were using spare time to collect soils at Mt. Tsukuba, and so forth. For some time after the discovery of tacrolimus, no tacrolimus-producing bacteria were isolated from the soil of the Kansai area in Japan. Given these points, the relocation of the laboratory to Tsukuba may have contributed to the chance early detection of tacrolimus (Yamashita 2013, p. 142).

[24] Prior to screening with the MLR, a mouse spleen lymphocyte-based assay was used. Approximately 3000 samples were screened by this method, but because the target compound could not be found, the evaluation system was reviewed and MLR was adopted (Kino 2011, p. 994; Tsukasaki 2013, pp. 50–51).

[25] Ochiai was engaged in transplant surgery and had experience in research using cyclosporine (Ochiai 1988). Yamashita (2013, p. 146) described that the contact between Ochiai and Fujisawa was as follows. It took place in a cancer immunotherapy workshop held in Osaka in 1984. Hatsuo Aoki, who later became president of Fujisawa, talked to Ochiai about FK506 and that the substance

allowed early confirmation of the effectiveness of tacrolimus in a rat model of heart transplantation and a dog model of kidney transplantation with vascular anastomosis. The results of the transplantation experiments were reported at the International Congress of the Transplantation Society in Helsinki in August 1986 and attracted much attention. This report also triggered collaborative research with Thomas Starzl of Pittsburgh University, who would later conduct the world's first clinical trials of tacrolimus.

The first clinical application of tacrolimus was carried out in 1989 by Starzl at the University of Pittsburgh. Starzl had been acknowledged as one of the world's leading authorities on transplantation medicine; he carried out the world's first liver transplant in 1963 and the world's first successful transplant in 1967. The patient of the first clinical trial was a woman facing severe rejection that was uncontrollable with cyclosporine after the fourth liver transplant, in other words, this clinical trial was salvage therapy, and the administration of tacrolimus succeeded in suppressing the rejection reaction in this patient.[26] At the time, the efficacy, safety, and optimal dose of tacrolimus were unknown while the existing drug cyclosporine had been confirmed to be effective. Moreover, the lives of transplanted patients are immediately at risk if organ rejection cannot be controlled. Under these conditions, monotherapy of tacrolimus was too risky, so the trial focused on the patients in whom cyclosporine had no effect. Cyclosporine and tacrolimus were initially combined, and then the dose of the former drug gradually reduced.

The initial clinical trial and subsequent tests at the University of Pittsburgh confirmed that tacrolimus was clinically as effective as, or better than, cyclosporine. Following these early results, Fujisawa conducted large-scale clinical trials, involving more than 500 patients in the United States and Europe, respectively, from 1990. At the time, when Japanese pharmaceutical companies intended to enter overseas markets, it was common to develop a drug jointly with a foreign company or to license it out to foreign companies; it was rare to conduct the full global research and development (R&D) and sales processes independently. In large-scale clinical trials, it was also confirmed that tacrolimus had an effect equal to or higher than that of cyclosporine in terms of patient and graft survival rates. Furthermore, it was also demonstrated that acute rejection, chronic rejection, and severe rejection, which cannot be controlled by steroids or OKT3 monoclonal antibodies, occurred significantly less frequently (Okuhara et al. 1996; Tanaka et al. 1997; Yamashita 2013).

In Japan, clinical trials were conducted in living-body liver transplantations mainly at Kyoto University in 1990, and in December 1991, an application for approval as a drug for rare diseases was filed (approval was granted in April 1993).[27] Overseas

---

recently found in Tsukuba had 100 times more immunosuppressive effect than cyclosporine. Given that he suspected that it could be used for transplantation, he took about ten residents to the Discovery Research Laboratory in Tsukuba and conducted transplant experiments every Saturday.

[26] Salvage therapy is another treatment option for when the initial treatment fails.

[27] The short review period was attributed to increased attention in Japan during the development stage.

applications were filed in Germany in June 1993, in the United States in July, in the United Kingdom in December, and subsequently applied for in other countries.

9.7   Scientific Contributions to Discovery Research

Scientific contributions to the development of tacrolimus included: (1) the fundamental advancement of immunology, and (2) an increased understanding of cyclosporine's mechanism of action because of the progress of immunology. Since the late 1960s, modern immunology has made remarkable progress. In the early 1980s, it was established that T cells play critical roles in both auto- and transplantation immunity and that in immune diseases, activated T cells can be treated like pathogens in infectious diseases. Furthermore, the advances in the field of immunology set the direction of the search for substances that specifically suppress the production of IL-2 (Kino et al. 1995). Fujisawa had chosen to search for new drugs in natural products because the company had accumulated significant experience in natural-product drug discovery. It is also likely the agricultural chemistry backgrounds of both Toshio Goto and Toru Kino, who played central roles in the discovery research, influenced the discovery strategy for tacrolimus.

Toshio Goto studied at the National Cancer Institute (NCI) in the United States while in his early 30s. His research focus during that time included the significance of NK cells in vivo, the role of NK cells in immunosuppression, and the effects of NK cells on T cell proliferation. The advanced knowledge he obtained in this period significantly contributed to the development of tacrolimus. Furthermore, while Goto was at NCI, clinical trials of cyclosporine were conducted at the nearby National Naval Medical Center. Goto's international studies allowed him to build a network with cutting-edge researchers; this was evident with the participation of the National Naval Medical Center staff in the tacrolimus research workshop.

The use of MLR for compound screening can also be regarded as a scientific contribution. MLR is an excellent tool because of its simplicity, quantitative nature, and ability to rapidly process large numbers of specimens. It is currently the gold standard for immunosuppressant screening; however, in 1982, an effective screening method for immunosuppressants was unknown.

It is important to note here cyclosporine and tacrolimus differ significantly in their discovery history. Notably, cyclosporine was initially considered as an antifungal agent, but its development for this application was abandoned. Fortunately, immunosuppressive effects were found in routine tests for the properties of substances (Borel et al. 1995). In other words, targeted screening of substances with immunosuppressive action was not conducted in the case of cyclosporine, although this was done for tacrolimus.

MLR made it possible to screen over 20,000 specimens in a short amount of time, resulting in the discovery of tacrolimus. The idea behind the screening strategy is credited to Toru Kino who had extensive experience in the use of MLR, which was gained as a visiting researcher at Tokai University. Indeed, before participating in the discovery research of tacrolimus, he was placed on secondment to the laboratory of Sonoko Habu at Tokai University to acquire knowledge and skills that could be used to elucidate the mechanism of action of FK565 (Kino 2011).

9.8 The World Authority in Liver Transplantation Led Clinical Trials

At the time of tacrolimus development, kidney transplantation was rarely carried out in Japan; it was an extremely unsuitable environment for the development of an immunosuppressive agent for organ transplantation. Therefore, clinical development in Europe and the United States where transplantation was active was indispensable, and tacrolimus was developed in a partnership with Thomas Starzl of the University of Pittsburgh.[28] Starzl was famous for the first liver transplantation and the first successful liver transplantation, and was a global authority in organ transplantation. He had been involved in a number of liver transplantations using cyclosporine in the early 1980s, including the successful use of it in combination with steroids. Thus, he was familiar with the efficacy and side effects of cyclosporine.

As previously mentioned, the proof of concept study was carried out in patients receiving salvage treatment by Starzl's team. The University of Pittsburgh, which can be regarded as a mainstay of transplantation medicine, was the best place to demonstrate the concept of tacrolimus. The presence of Starzl also served as a signal to external researchers and the US Food and Drug Administration (FDA) to demonstrate the quality of clinical trials, enabling Fujisawa to independently implement large-scale clinical trials. An article on the front page of the *New York Times* (18th October 1989) detailed the performance of a transplant using tacrolimus at the University of Pittsburgh, presumably because Starzl was in the lead position in transplant medical care. These public endorsements likely played some role in the early launch of tacrolimus.

9.9 Tacrolimus Is More Effective Than Cyclosporine

Organ transplantation is one of the predominant treatments for the dysfunction of vital organs caused by chronic diseases; however, its success depends on the control of rejection. Since the 1960s, azathioprine and steroidal anti-inflammatory drugs had been used as immunosuppressive agents, but their effects were not satisfactory.

---

[28]Fujisawa had also provided a sample of tacrolimus to Khan of Cambridge University, who had contributed to the clinical application of azathioprine and cyclosporine. However, Khan insisted that toxicity occurred when he administered tacrolimus to dogs; thus, he referred to tacrolimus as "Fujitoxin" (toxin made by Fujisawa). Therefore, joint research with Khan at the clinical stage never took place.

Cyclosporine was developed in the 1980s, and its use dramatically increased the success rate of the organ transplantation; nevertheless, the immunosuppressive effect of tacrolimus was even more potent than that of cyclosporine. In 2002, a comparison study of tacrolimus and the oral microemulsion formulation of cyclosporine (Neoral) was published in *The Lancet*, one of the world's most highly regarded medical journals. According to the article, tacrolimus was superior to cyclosporine in both liver transplantation and kidney transplantation; thus, it was recommended to use tacrolimus.[29] Because of the influence of such academic research, tacrolimus has since been established as a global standard for immunosuppressants. The *Merck Manual* explains tacrolimus as the most commonly used drug in kidney, liver, pancreas, and intestinal transplants.[30] It has also been reported that tacrolimus is used in about 90% of liver transplants and 60% of kidney transplants in Japan and the United States.[31]

The advent of more potent immunosuppressive agents was valuable in terms of improved graft survival and survival of transplant patients. Depending on the type of immunosuppressant used, the history of transplantation can be divided into three periods: azathioprine in the 1970s, cyclosporine in the 1980s, and tacrolimus since the 1990s. For example, in liver transplantation, the 1-year survival rates for each period were 35%, 70%, and 80%, respectively, and the 5-year survival rates were 20%, 60%, and 70%, respectively (Todo 1999). Although the impact of cyclosporine was excellent, tacrolimus further improved the success rate of transplantation.

Tacrolimus was originally approved as capsule and injectable preparations (Prograf Capsules and Prograf Injection) for the prevention of transplant-related rejection, and subsequently, by adding formulations and application areas, its dominance of the market has expanded. Specifically, it gained approval for granules, which were initially developed for pediatric use,[32] and sustained-release drugs, which could be administered once a day while maintaining the same safety and efficacy as Prograf. It is considered that tacrolimus enhanced long-term transplant performance by improving the quality of life of transplant patients who take several drugs, and has improved the compliance of patients with regard to drug administration.[33]

---

[29] See O'Grady et al. (2002) for liver transplantation and Margreiter (2002) for kidney transplantation.

[30] Merck Manual Japanese Online version "General Principles of Transplantation" http://merckmanual.jp/mmpej/sec13/ch166/ch166b.html (accessed August 2015).

[31] Fujisawa Pharmaceutical Industry (2004) "Notice of Prime Minister Invention Award 2004: Immunosuppressant Tacrolimus." http://www.astellas.com/jp/corporate/news/fujisawa/pdf/040426.pdf (accessed August 2015).

[32] Liver transplantation in Japan is mostly completed as biological segmental liver transplantation from living donors and is often applied to children. However, in the case of children, it may be difficult to take capsules, and dosage adjustment is difficult because the dosage is small. Refer to Pharmaceutical Interview Form: Prograf Capsules 0.5 mg/1 mg/5 mg, Prograf Granules 0.2 mg/1 mg (revised April 2014, 34th edn), p. 1.

[33] Pharmaceutical Interview Form: Graceptor Capsules 0.5 mg/1 mg/5 mg (revised April 2014, 14th edn), p. 1.

In addition to suppressing transplant-related rejection, the use of tacrolimus has been extended to autoimmune diseases (systemic myasthenia gravis, rheumatoid arthritis, lupus nephritis, and refractory active ulcerative colitis), vernal keratoconjunctivitis, and atopic dermatitis (Protopic Ointment). In particular, Protopic was classified into a new drug category called topical immunomodulators (TIM), and atopic dermatitis treatment was launched in Japan in 1999. It was the first new drug to be released for this disease in the 40 years since steroids were commercialized. Indeed, the significant contribution of tacrolimus to patients is based on the strength of its immunosuppression and a broad spectrum of indications.

9.10 Contributing to the Creation of Chemical Biology

The contributions of tacrolimus to patients have already been described, but the role of tacrolimus in the development of science cannot be overlooked. To begin, the mechanism of action of tacrolimus became clear at a very late stage of drug development. Although Fujisawa studied the mechanism of action at the cellular and protein levels, studies at the genetic level were conducted mainly in academia, where tacrolimus samples were provided to collaborators.[34] Fujisawa did not actively investigate the mechanism of action because it was considered that in-house resources should be invested in the R&D of new drugs rather than elucidating the mechanism of drugs that had already been launched. However, the characterization of the tacrolimus mechanism of action brought about a significant contribution in the form of expanding the indication of the drug to atopic dermatitis.

In terms of mechanistic studies, the contributions of Stuart Schreiber and colleagues at Harvard University were significant. Schreiber discovered FK506 binding protein (FKBP) in 1988 and subsequently found that tacrolimus itself has no effect as a pharmaceutical agent, rather tacrolimus binds to FKBP to form a complex with immunosuppressive activity.

This finding had a crucial impact on subsequent studies. It led to the elucidation of the signals in T cells upstream of IL-2 production, which is required for T cell proliferation, and the early stages of T cell signaling cascades were connected. It also led to the rediscovery of the immunosuppressive activity of rapamycin, which is

---

[34]Specifically, Harvard University, Stanford University, Merck, Tonen (a Japanese chemical company), and others (Kobayashi 1995).

structurally similar to tacrolimus, and the analysis of the later stages of signal transduction.[35] Importantly, a more general impact was the creation of a new discipline
of chemical biology through the mechanistic studies of tacrolimus.[36]

9.11   Annual Sales Exceed 2 Billion Dollars

This section presents an overview of the sales trends and discusses the competition
faced by Fujisawa in relation to tacrolimus. Tacrolimus was launched in 1993 as
Prograf in Japan. Figure 9.2 shows the global trends of tacrolimus sales, including
Prograf, Graceptor, and Protopic, since 1999.[37] In April 2008, the US substance
patent of tacrolimus expired; sales of tacrolimus had increased until then, yet the
introduction of generics to the United States in 2009 was associated with a decline in
sales. However, Prograf is still a blockbuster drug with annual sales of approximately
2 billion USD.

Figure 9.3 compares the sales of tacrolimus (Prograf, Graceptor, and Protopic)
and cyclosporine (Sandimmun and Neoral).[38] Initially, the sales of cyclosporine were

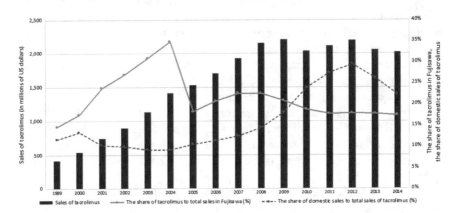

Fig. 9.2 Trends in tacrolimus sales

---

[35]Rapamycin was an analog of tacrolimus produced by actinomycetes contained in the soil of
Easter Island in 1965 and was initially developed as an antifungal agent. Although it strongly
binds to FKBP, it exerts immunosuppressive action using a completely different mechanism from
tacrolimus (Kobayashi 1995).

[36]Chemical biology was defined as a new discipline to elucidate biological phenomena from a
chemical point of view and is narrowly defined as approaches to elucidate the functions of biopolymers by probing with small molecules that interact specifically with biopolymers such as DNA,
RNA, and proteins (Hagiwara 2008, p. 4). Chemical genetics is also a similar concept.

[37]Astellas Pharma "Documents relating to its settlements of accounts." https://www.astellas.com/
jp/ir/library/results.html (accessed August 2015). Sales are calculated based on the value of the yen;
therefore, it may be affected by fluctuations in exchange rates.

[38]Data sources are Pharma Future (Uto Brain Division, Cegedim Strategic Data).

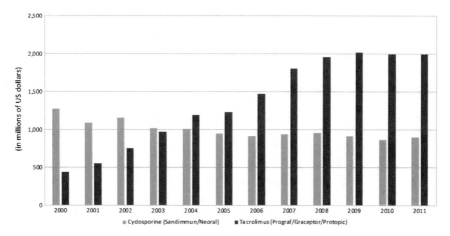

Fig. 9.3 Sales trends for tacrolimus and cyclosporine

more significant than the sales of tacrolimus; however, in 2003 (the year following the publication of *The Lancet* paper recommending tacrolimus), the sales of were almost equal. From 2004 onwards, the sales of tacrolimus increased, while the sales of cyclosporine declined.

In Fig. 9.2, the solid line in the figure shows the tacrolimus-related share in consolidated sales of Fujisawa Pharmaceutical and Astellas. Astellas was formed in April 2005 by the merger of Yamanouchi Pharmaceutical and Fujisawa Pharmaceutical. In the Fujisawa era (up to 2004), sales of tacrolimus accounted for up to 35%, and since the merger, tacrolimus has accounted for nearly 20% of Astellas sales. The dashed line in Fig. 9.2 shows the share of domestic sales relative to the global sales of tacrolimus. Focusing on the period before the expiration of the US patent in 2008, the share of sales in Japan only accounted for about 10–15%, indicating that the majority of sales were in Europe and the United States. It is suggested that such a high rate of foreign sales was realized because the overseas market is much larger than the Japan market, and that Fujisawa independently entered foreign markets rather than in collaboration with foreign pharmaceutical companies.

9.12   R&D Competition and the First-Mover Advantage

During the search for tacrolimus, the global development of immunosuppressive agents was limited. Among the companies aiming to discover immunosuppressive agents, Fujisawa was the only company to intensively search for compounds derived from natural products. While more than 20 years have elapsed since the total synthesis of tacrolimus and full understanding of the mechanism of action, the creation of compounds with more potent immunosuppressive functions has not been realized. This suggests that the Fujisawa strategy was an astute choice.

After discovering tacrolimus, Fujisawa readily responded to sample requests from other companies.[39] Hence, during the period when it engaged in compound optimization and development research based on tacrolimus, competitors such as Merck, Glaxo, and Sandoz, entered the field. The tacrolimus development team had the following to say about the discovery competition: "Fujisawa began derivative synthesis around 1987; however, in 1988, Merck was exploring similar avenues with 150 staff." Thus, it can be inferred that the competitors attempted to catch up with Fujisawa by introducing large-scale resources. Given these circumstances, an analog of tacrolimus, ascomycin, was found to be a known substance in 1988, based on Merck's information provided to the US Patent and Trademark Office. Ascomycin was reported in 1962, and a US patent was granted in 1966. Thus, if Merck had discovered an immunosuppressant from ascomycin derivatives, a competing product directly against tacrolimus would have emerged. In reality, however, neither Fujisawa nor other companies have discovered a compound superior to tacrolimus. To date, only pimecrolimus (Elidel), a treatment for atopic dermatitis from Novartis, has been launched as an analog of tacrolimus.[40]

Given that the strategy of seeking substances with immunosuppressive effects from natural products was not common, Fujisawa was able to precede in drug development against Merck and others. On the sources of such first-mover advantage, two things are clear. First, Fujisawa possessed natural product-derived drug discovery capabilities based on agricultural chemistry. At the time, many pharmaceutical companies had scaled down or abolished fermentation divisions because of inefficiencies in creating microbial-derived pharmaceuticals. However, Fujisawa Pharmaceutical continued to focus on the development of microorganism-derived drugs because of its past achievements and the understanding of top executives. The technology for natural product-derived drug discovery itself was not uncommon but continuing to accumulate technological expertise led to the discovery of a microbial-derived immunosuppressant.

Second, Fujisawa had accumulated research on biological response modifiers (BRMs) and was able to acquire and apply new immunological techniques and concepts by placing corporate researchers on secondment to domestic and overseas research facilities, such as universities. It was in the mid-1970s that the company began full-fledged entry into the immunology field because it considered that there was a limit to the search for conventional antibiotics. However, before that time, Fujisawa had been researching BRMs, including joint research with academia in the cancer field.

---

[39] When applying for a patent on an invention relating to microorganisms, it is necessary to deposit microorganisms in a given institution, and a third party may receive microorganisms under certain conditions. The existence of this system seems to have influenced the background in which Fujisawa provided the sample to other companies.

[40] Pimecrolimus (Elidel) is a derivative of ascomycin; it has not yet been launched in Japan.

9.13   Conclusion

This chapter analyzed the R&D process for tacrolimus (Prograf), an immunosuppressant developed by Fujisawa Pharmaceutical. The chapter is primarily based on interviews with the researchers that were involved in the project.

Tacrolimus was innovative in its ability to potently suppress rejection of transplanted organs. It demonstrated immunosuppressive activity that was equivalent to or higher than that of the calcineurin inhibitor cyclosporine and contributed to the improvement of both post-operative and long-term management of transplant patients. The drug has a wide range of indications in the fields of transplantation medicine and autoimmune diseases and has had a significant impact on patients from both fields. It also generated large sales for Fujisawa as a blockbuster drug.

The notable aspects of the discovery and development process of tacrolimus are as follows. First, no compound with immunosuppressive functions beyond tacrolimus has been created, even though over 20 years has elapsed since the total synthesis and elucidation of the mechanism of action of tacrolimus. This suggests that the strategy employed by Fujisawa to search for compounds from natural products was a good choice. Of course, we do not merely endorse drug discovery from natural products. It is important to understand that tacrolimus was an innovative drug that was realized by combining Fujisawa Pharmaceutical's accumulated fermentation technology with the latest immunological theory, experimental infrastructure, and so forth.

Second, there was the interdependence of drug development and scientific progress. On the one hand, drug development is substantially driven by scientific advances. On the other hand, scientific advancements may develop in the process of investigating the mechanism of action of new drugs. Tacrolimus was used to elucidate the signal transduction mechanism in T cells and also contributed to the birth of the new discipline of chemical biology. Moreover, research into the tacrolimus mechanism of action contributed to the expansion of its indications for use.

Third, by conducting joint research with Thomas Starzl of the University of Pittsburgh, who was a leader in the field of transplantation medicine, a tacrolimus proof of concept study was made possible at an early stage. Furthermore, Fujisawa actively cooperated with academia and obtained information that was hard to gain from inside the company, such as experiments that involved transplant physicians and elucidation of mechanisms not directly linked to drug discovery.

Fourth, when Japanese pharmaceutical companies intended to enter overseas markets, it was common to develop a drug jointly with foreign companies or license it out to foreign companies. Fujisawa was a pioneer in that it independently conducted the whole process from clinical trials to global sales.

References

Aoki, H. (2007). "History and Future of Medical Care," the material for the November meeting of Medical Journalists Association of Japan (in Japanese).

Barker, C. F., & Markmann, J. F. (2013). Historical overview of transplantation. *Cold Spring Harbor Perspectives in Medicine, 3*(4), a014977.

Borel, J. F., Kis, Z. L., & Beveridge, T. (1995). The history of the discovery and development of cyclosporine (Sandimmune). In V. J. Merluzzi & J. Adams (Eds.), *The search for anti-inflammatory drugs: Case histories from concept to clinic* (pp. 27–64). Boston: Birkhäuser.

Goto, Toshio. (1993). Medicine of microbial origin: From penicillin to FK-506. *Chemistry and Chemical Industry, 46*(5), 765–768. (in Japanese).

Hagiwara, Masatoshi. (2008). Overview of chemical biology. *Folia Pharmacologica Japonica, 132*(1), 4–6. (in Japanese).

Iwatsuki, S., Starzl, T. E., Todo, S., Gordon, R. D., Esquivel, C. O., Tzakis, A. G., et al. (1988). Experience in 1,000 liver transplants under cyclosporine-steroid therapy: A survival report. *Transplantation Proceedings, 20*, No. 1, Suppl. 1, 498–504.

Kishimoto, T., & Nakashima, A. (2007). *Modern immunology*. Tokyo: Kodansha. (in Japanese).

Kino, T. (2011). An immunosuppressant, Prograf (FK506). *PHARMA TECH JAPAN, 27*(6), 993–997. (in Japanese).

Kino, T., Goto, T., Hosoda, J., & Okuhara M. (1995). Discovery and development of an immunosuppressant, FK506 (Tacrolimus). *Nippon Nogeikagaku Kaishi, 69*(Special issue), 471–473 (in Japanese).

Kino, T., Hatanaka, H., Goto, T., & Masakuni, O. (1989). FK506, a novel immunosuppressant. *Nippon Nogeikagaku Kaishi, 63*(2), 224–227. (in Japanese).

Kobayashi, M. (1995). Macrolide antibiotic tacrolimus (FK506), rapamycin (Rapa) and T-cell signaling. *The Japanese journal of antibiotics, 48*(Special issue), 48–55 (in Japanese).

Margreiter, R. (2002). Efficacy and safety of tacrolimus compared with ciclosporin microemulsion in renal transplantation: A randomised multicentre study. *The Lancet, 359,* 741–746.

Markus, B. H., Hakala, T. R., Tzakis, A., Mitchell, S., Marino, I. R., Gordon, R. D., et al. (1987). Kidney transplantation in Pittsburgh: Experience and innovations. *Clinical Transplantation*, 141–54.

Ministry of Health, Labour and Welfare. (2005). "Relevant Documents of Drug Pricing Rules," the material for Special Committee on Drug Prices of Central Social Insurance Medical Council, April 20, 2005 (in Japanese).

Noda Economic Research Institute (Noda Keizai Kenkyujo). (1976). *The challenge to 1/3000.* Tokyo: Noda Keizai Kenkyujo. (in Japanese).

Ochiai, T. (1988). An immunosuppressant FK506. *Journal of Adult Disease, 18*(4), 539–543. (in Japanese).

O'Grady, J. D., Burroughs, A., Hardy, P., Elbourne, D., & Truesdale, A. (2002). Tacrolimus versus microemulsified ciclosporin in liver transplantation: The TMC randomised controlled trial. *The Lancet, 360,* 1119–1125.

Okuhara, M., Goto, T., Kino, T., & Hosoda, J. (1996). Discovery and development of an immunosuppressant, tacrolimus (FK506). *Nippon Nogeikagaku Kaishi, 70*(1), 1–8. (in Japanese).

Onozuka, S. (2004). "The Vales of Medicines," Research Paper Series No. 20, the Office of Pharmaceutical Industry Research, the Japan Pharmaceutical Manufacturers Association (in Japanese).

Starzl, T. E. (2001). The birth of clinical organ transplantation. *Journal of the American College of Surgeons, 192*(4), 431–446.

Takano, N. (1997). Particulars on practice in patent application of fermentation products: A case study of tacrolimus (FK506). *Bioscience and Industry, 55*(6), 22–24. (in Japanese).

Tanaka, H., Nakahara, K., Hatanaka, Hiroshi, Inamura, N., & Kuroda, A. (1997). Discovery and development of a novel immunosuppressant, tacrolimus hydrate. *Yakugaku Zasshi, 117*(8), 542–554. (in Japanese).

Tani, S. (2012). From discovery to clinical applications of tacrolimus (FK506). *Inflammation and Immunity, 20*(3), 252–259. (in Japanese).

Todo, S. (1999). Immunology through liver transplantation in Japan and the U.S. *JSI Newsletter,* 7(2), 11. (in Japanese).

Tsukasaki, A. (2013). *Japanese scientists who challenged new drugs.* Tokyo: Kodansha. (in Japanese).

Yamashita, M. (2013). Tacrolimus (FK506) developmental story. *Seibutsu-kogaku Kaishi, 91*(3), 141–154.

Chapter 10
Pioglitazone (Actos, Glustin)

A Drug That Transformed Diabetes Therapy

Naoki Takada and Koichi Genda

Abstract Pioglitazone was an early member of a new category of diabetes treatment drugs called insulin sensitizers, pioneered by Takeda Pharmaceutical. It is used for the treatment of Type 2 diabetes, and acts by lowering blood glucose levels. The thiazolidinedione backbone discovered by Takeda Pharmaceutical was widely adopted in industry. Research on pioglitazone began as a result of serendipitous discovery of its lead compound during research on lipid-lowering agents. New animal models made a major contribution to understanding the pathogenesis of diabetes in basic research, but also led the researchers to the discovery of compounds with thiazolidinedione backbone that improve insulin sensitivity. Although ciglitazone, the first compound studied in clinical trials, failed, pioglitazone was discovered and selected as a candidate compound based on its potent activity, and low propensity to induce side effects. Pioglitazone is a breakthrough drug generated by the culmination of basic research, including the development of animal models, careful screening techniques, and excellent synthesis technologies.

10.1 Introduction

Pioglitazone hydrochloride (Actos) is used for the treatment of Type 2 diabetes and was developed by Takeda Pharmaceutical Co., Ltd.; it was one of the first insulin sensitizers developed for the treatment of diabetes. Pioglitazone (Development number: AD-4833) was selected as a candidate for clinical development in 1986; clinical studies in the USA began in 1989, and approval was obtained in the USA in 1997. The clinical trial in Japan began in 1991, and the drug was put on the market in 1999. Pioglitazone is sold in approximately 90 countries worldwide, and in 2012, it had the largest market share of insulin sensitizers.

N. Takada (✉)
Institute of Advanced Sciences, Yokohama National University, Kanagawa, Japan
e-mail: takada-naoki-jx@ynu.ac.jp

K. Genda
Career Development Department, Shionogi Career Development Center Co., Ltd., Osaka, Japan
e-mail: koichi.genda@shionogi.co.jp

© Springer Nature Singapore Pte Ltd. 2019
S. Nagaoka (ed.), *Drug Discovery in Japan*,
https://doi.org/10.1007/978-981-13-8906-1_10

   The innovation of Actos lies in its ability to lower blood glucose levels using a mechanism of action that was different from other diabetes medications available at the time. Furthermore, at the time of development, the conventional pharmacotherapy of diabetes mellitus included oral administration of hypoglycemic drugs, and insulin injection for lowering the glucose level. Existing drugs, such as sulfonylureas (SUs), stimulated basal insulin secretion; however, excessive stimulation was associated with excessive insulin secretion, resulting in hypoglycemia. In contrast, pioglitazone improved the body's insulin sensitivity (reduced insulin resistance), and because as it does not increase insulin secretion, there was little concern of drug-induced hypoglycemia. Additional benefits included the ability to lower neutral blood fat, which is an additional favorable effect for patients with obesity and hyperlipidemia.

10.2   Pioglitazone Mechanism of Action

According to the Diabetes Diagnostic Criteria of the Japan Diabetes Society, diabetes mellitus is defined as "a group of diseases with chronic hyperglycemia caused by insufficient insulin action and with dysbolism" (Seino et al. 2010). Diabetes is classified into four categories according to its origin: (1) Type 1 diabetes, (2) Type 2 diabetes, (3) gestational diabetes, and (4) diabetes brought about by other causes, such as disease-induced, drug-induced, or chemical-induced diabetes. Type 2 diabetes mellitus, the target disease of pioglitazone, is a lifestyle-related disease to which both genetic factors and environmental factors, such diet and exercise, contribute. According to the Ministry of Health, Labour and Welfare, over 95% of Japanese patients with diabetes mellitus have Type 2. Type 2 diabetes is caused by either decreased insulin secretion or decreased insulin action, also known as insulin resistance, which prevents the liver, muscle, and other cells from adequately taking up glucose. Insulin resistance-improving drugs, such as pioglitazone, treat diabetes by improving the sensitivity of these cells to insulin.

   Pioglitazone reduces insulin resistance by binding and activating the transcription factor peroxisome proliferator-activated receptor gamma (PPAR-γ), a type of nuclear receptor expressed in PPAR cells. It is closely involved in intracellular metabolism and cell differentiation. Activation of PPAR-γ results in the following actions, which reduce insulin resistance (Lehmann et al. 1995):

a. Glycerol kinase is expressed in adipocytes.
   As the number of small adipocytes increases following PPAR-γ activation, the increased secretion of adiponectin by these cells favors the binding of adiponectin to its receptor. Receptor-associated adiponectin activates AMP-kinase (AMPK), which stimulates glucose uptake into muscle cells.
b. Hypertrophic adipocytes are reduced by apoptosis.
   Hypertrophic adipocytes and the surrounding macrophages secrete substances that cause insulin resistance, such as free fatty acids and tumor necrosis factor alpha (TNF-α); the reduction of hypertrophic cells improves insulin resistance.

c. Enhances insulin signaling by acting on insulin receptors.
    Insulin binding to the insulin receptor stimulates tyrosine phosphorylation of
    the β-subunit of the receptor and insulin receptor substrate 1 (IRS-1), thereby
    promoting phosphatidylinositol-3 kinase (P13K) activity, glycolysis, and glyco-
    gen synthesis. Insulin decreases blood glucose levels through this pathway, but
    pioglitazone enhances this signaling mechanism by improving insulin resistance.

10.3   Timeline of Pioglitazone Development

1960s: Takeda Pharmaceutical began basic science diabetes research.
1968: Takeda basic science research team began using KKAy mice.
1973: Takeda chemistry laboratory started research on the synthesis of lipid-lowering
agents.
1975: Discovery of AL-321 prompted the company to switch to diabetes drug
research.
1979: Ciglitazone (ADD-3878) was selected as a candidate compound.
1983: Ciglitazone was discontinued after the clinical trial failed.
1986: Pioglitazone (AD-4833) was selected as a candidate compound.
1989: Takeda's joint research partner, Upjohn, began pioglitazone clinical trials in
the USA.
1991: Upjohn suspended clinical development in the USA, while Takeda started its
clinical trials in Japan.
1995: Takeda American Research and Development Center resumed its development
in the USA.
December 1996: An application for manufacturing approval was filed in Japan.
January 1999: A New Drug Application was filed with the US Food and Drug Administration (FDA) .
July 1999: Takeda acquired approval to commence marketing in the USA.
September 1999: Drug approval was acquired in Japan
December 1999: Pioglitazone launched in Japan.

10.4   Discovery of Lead Compounds Through Animal
Models

Pioglitazone discovery research began in the Takeda Pharmaceutical biological laboratory, and lead compounds were discovered during research into lipid-lowering
drugs in the chemistry laboratory (Takeda Pharmaceutical 2008) Takeda Pharmaceutical. Takeda had begun basic research on obesity and related diseases in the early
1960s in an attempt to discover new drug targets. In anticipation of the increased

prevalence of lifestyle-related diseases in Japan and the USA, researchers at Takeda explored the pathogenesis of obesity-associated hyperglycemia. The aim of this early stage research was to improve understanding of the pathogenesis of diabetes. In 1963, Hisashi Iwatsuka and Takao Matsuo introduced the KK mouse model to Takeda Pharmaceutical. KK mice are spontaneously diabetic, as reported in 1962 by Kyoji Kondo (Nagoya University). Iwatsuka used the KK mouse to study the pathogenesis of diabetes. However, KK mice did not develop hyperglycemia, which is one of the main symptoms of diabetes, despite their genetic predisposition (Ikeda 1998). The investigation into the cause of this strange phenotype revealed that body weight (obesity) was involved in the development of diabetes. This finding was shared with researchers at Nagoya University and was used for the development of the next animal model of diabetes. Accordingly, Masahiko Nishimura of Nagoya University introduced obesity-onset genes, Ay, into the KK mouse, producing the KKAy mouse model of Type 2 diabetes mellitus. The KKAy mouse was immediately introduced to Takeda Pharmaceutical, where it has been used for compound screening and other studies since 1968 (Ikeda and Sugiyama 2001).

In laboratory studies, Takeda used KKAy mice, as well as other animals and mouse lines to study disease. Among them, Takeda's Wistar fatty rats were made widely available to academia and other companies, with the aim of diffusing it as a well-known standard research tool. Data were collected through contracted studies on those animal models not owned by Takeda Pharmaceutical, including WHHL rabbits produced by Yoshio Watanabe (Kobe University).

Meanwhile, synthesis research into lipid-lowering agents began in 1973. Yutaka Kawamatsu, a key investigator of lipid-lowering drugs, advanced synthetic research using clofibrate as a lead compound and found that a compound, AL-54, lowered cholesterol and triglycerides in rats. In addition, Matsuo found that AL-54 had a hypoglycemic effect, which was fortuitous because blood glucose was usually measured only as reserve or reference data. Regarding the discovery of this compound's hypoglycemic effect, Kanji Meguro wrote the following:

> Dr. Matsuo, who became aware of [the hypolipidemic effect of AL-54], visited Dr. Kawamatsu in order to administer AL-54 to KKAy mouse, and they found that the compound not only lowered [triglycerides] but also significantly lowered blood glucose levels. This result demonstrated the presence of a drug that would lower blood glucose levels in these mice for which a potent hypoglycemic agent, the sulfonylurea agent, was [sic] completely ineffective. (Meguro 1991, p. 900)

Although hypoglycemic agents were well known, AL-54 was a compound originally found in lipid-lowering drug studies. This was during a time when hypolipidemic drugs, including fenofibrate, dominated the markets. Therefore, Takeda researchers decided to develop AL-54 as a hypolipidemic agent rather than as a treatment for diabetes.

Synthetic studies were carried out with the aim to develop a compound with a stronger lipid-lowering effect, and the ester-derivative AL-294 was discovered in 1973. AL-294 and its related derivatives were the first compounds to be reported as insulin sensitizers. However, AL-294 continued to be developed as a hypolipidemic agent because the hypoglycemic effect was found to be too weak for it to be developed as a therapeutic agent for diabetes (Sohda et al. 2002).

10.5   Balancing Potency and Side Effects

Discovery research using AL-294 as a lead compound was conducted mainly by Takashi Sohda and Katsutoshi Mizuno. They synthesized a derivative of AL-294, considering the ease of crystallization, and found AL-321, a 2,4-thiazolidinedione derivative. Once it was discovered that AL-321 was several times more potent than AL-294 in lowering blood glucose levels, they decided to restart the synthetic research as an AL-321-based diabetes treatment project.

Further exploration of AL-321 led to the discovery of ciglitazone (ADD-3878) in 1979. Results of the pharmacological and mechanism-of-action studies by Takeshi Fujita demonstrated that ciglitazone increased cell sensitivity to insulin by acting on the insulin receptor, improving the efficiency of insulin. This meant that while there was a risk of hypoglycemia with significant doses of existing diabetes drugs, ciglitazone was very unlikely to cause hypoglycemia (Sohda et al. 2002).

The patent for ciglitazone was filed with the US Patent Office on 27 July 1979 and registered on 1 September 1981. In 1981, a clinical trial of ciglitazone was initiated in Japan. However, results from Phase II clinical trials carried out in 1983 showed that the drug efficacy in humans was insufficient. At this point, the development of ciglitazone was stopped, and the search for more potent compounds resumed (Kawamatsu and Fujita 1981). In the discovery laboratory, it was suggested that the attenuation of side effects would make it possible to administer more of the drug, resolving the issue of drug efficacy. Therefore, the activities of metabolites of ciglitazone were reexamined. At this time, joint research with Upjohn was initiated to select candidate compounds for clinical use.

In 1986, pioglitazone (AD-4833) was selected as a clinical candidate from more than 1000 synthesized compounds. It was reported that AD-5061 and AD-5075, which were synthesized after pioglitazone, had stronger hypoglycemic effects than pioglitazone. However, pioglitazone was chosen despite this because its balance between potency and side effects was the best among the various synthesized compounds (Sohda et al. 2002).

When pioglitazone was newly synthesized in 1982, it was originally planned to be developed as an acetic acid adduct. However, acetic acid adducts of pioglitazone caused problems in uniformity because of the release of acetic acid in the manufacturing process (Takeda Pharmaceutical 1999) Takeda Pharmaceutical. Therefore, when pioglitazone was selected as a clinical candidate compound, it was decided that it should be developed as a hydrochloride salt. Although further development then found that pioglitazone hydrochloride was a racemic mixture, development was continued because the racemic mixture was found to easily interconvert in the rat and there was no difference in drug efficacy between the optical isomers.

After the selection of pioglitazone, pharmacological and safety studies for the initial preclinical evaluation, synthetic method and basic physical property examinations, and industrialization research were carried out. During these studies, it was shown that pioglitazone caused increases in heart weight, which indicated the possibility of serious side effects. After investigation, it was proven that the increase in heart weight was benign, and that pioglitazone was found to have a similar effect to exercise-induced cardiac hypertrophy.

10.6 Clinical Development

In 1989, clinical trials for pioglitazone began in the USA with Upjohn as a collaborative partner. However, Upjohn did not approve of the results demonstrating the increase in heart weight, despite the high degree of activity, and withdrew from the project in 1991, suspending the US clinical trial. Nevertheless, Takeda imitated a clinical trial in Japan. Development in the USA was restarted in 1995. Following the demonstration of efficacy in clinical trials, an application for manufacturing approval was filed in December 1996 in Japan, January 1999 in the USA, and March 1999 in Europe. In the USA, pioglitazone became a priority review drug, and entered the market in August 1999, 7 months after the manufacturing approval application. It was launched in Japan in December 1999 and in Europe in October 2000.

10.7 Diabetes Treatment Before Pioglitazone

The history of diabetes is long, and it is said that the disease state was first described about 2000 years ago (Makino 2006). Until the seventeenth century, only polyuria was recognized as a cardinal symptom of diabetes mellitus, meaning that diabetes mellitus and diabetes insipidus remained indistinguishable. However, a technique to scientifically identify diabetes was established in the latter half of eighteenth century. In 1776, Dobson found that the sugar in the blood of diabetics was higher, and in 1815 Chevreul discovered that the sugar in the urine was glucose.

In 1889, Oskar Minkowski and Joseph von Mering found that diabetes was associated with the pancreas, and that dogs that had their pancreas removed became diabetic. However, this finding itself did not suggest a treatment of diabetes. Minkowski and contemporary researchers also studied diabetes by giving pancreatic extracts to dogs that had their pancreas removed, but they did not find a substance that could block diabetes (Sakai 1998).

Frederick Banting and Charles Best were the first group to succeed in developing a treatment for the prevention of diabetes when they administered a pancreas extract named "Iletin" to a human in January 1922. The results were reported at the American Medical Association in May 1922, and the extract was named "Insulin". Thanks to this achievement, Banting and John Macleod, who supported his work, received the

Nobel Prize in Physiology and Medicine in 1923. However, although insulin therapy became possible, insulin preparation and test methods were underdeveloped, which imposed large burdens on patients (Ninomiya 1996).

In 1956, Frederick Sanger established a method to determine the amino acid sequence of proteins and elucidated the amino acid sequence of insulin. Chemists then tried to make artificial insulin on the basis of this chemical structure, which led to Zahn's synthesis of sheep insulin in 1965, and Katsoyannis' synthesis of human insulin in 1967.[1]

10.8 Insulin Mechanism of Action

The demonstration of the relationship between diabetes mellitus and insulin suggested that diabetes originated from insulin shortage. However, Harold Himsworth proposed the concept of insulin resistance, defining diabetes caused by insulin resistance as "a condition in which the presence of insulin in the blood fails to produce an adequate effect in the target tissues of the liver, muscle, and adipocytes," and distinguished diabetes caused by insulin resistance from the other diabetes (Himsworth 1936). At that time, however, it was not possible to carry out a clinical trial based on the measurement of insulin, because the technology to measure insulin in the blood was not yet established. Thus, although the existence of several types of diabetes had been described, it had not yet been established what caused the difference.

The problem with measuring blood insulin was solved by Solomon Berson and Rosalyn Yalow in 1960 when they established a radioimmunoassay of insulin. As a result, the existence of "diabetes mellitus," in which the blood insulin concentration is low, and "diabetes mellitus of obesity," which presents as disease in spite of the high blood concentrations of insulin, was identified. However, it remained unclear whether the two types were attributable to differences in severity or etiology. It was not until the 1970s that scientists were able to show that diabetes could be classified according to its etiology. In 1979, the National Diabetes Data Group (NDDG) in the USA developed a systematic categorization of diabetes. The NDDG classified diabetes into three categories: (1) Type 1 diabetes (insulin-dependent diabetes), (2) Type 2 diabetes (non-insulin-dependent diabetes), and (3) miscellaneous diabetes. In the following year, the World Health Organization (WHO) also announced diagnostic criteria that followed the NDDG classification system, and this classification became the worldwide standard (National Diabetes Data Group 1979; WHO 1980).

As a part of the research effort to identify the origin of diabetes mellitus, attention also gathered surrounding the mechanism of the action of insulin. Although the hypothesis of insulin acting through the cell membrane was proposed in 1939 by Einar Lundsgaard, it was difficult to isolate insulin's action at the cell membrane

---

[1]Novo Nordisk Pharma Limited. Advances in Insulin Preparations 2: Looking for Human Insulin. http://www.novonordisk.co.jp/documents/article_page/document/CO_tcinsulin_011.asp (accessed 7 August 2014).

until 1972, when the protein that binds to insulin, the insulin receptor, was extracted. Then in 1982, Masato Kasuga discovered phosphorylation of the β subunits of the insulin receptor. This discovery together with the subsequent findings was critical to elucidating the mechanism of action of pioglitazone (Maruyama 1992).

At this time, the relationship between obesity and Type 2 diabetes was reported to be caused by a variety of sources. It was advocated by Gerald Reaven in 1988 that three main components of lifestyle-related diseases, namely hypertension, anomalies in glucose metabolism, and disorders in lipid metabolism, combined with insulin resistance to cause disease. Gökhan Hotamisligil confirmed the relationship between inflammation, obesity, and insulin resistance in 1993 by discovering the role of inflammation through TNFα, and Kliewer demonstrated the action of thiazolidinediones on PPAR-γ in 1995 (Kawai 2007).

10.9 Scientific Basis for Research and Development

Research and development of pioglitazone at the preclinical stage, including basic research, was conducted at the Takeda Pharmaceutical Central Laboratory. Clinical development and approval were obtained by the company's Drug Research Laboratory in Japan and the Takeda America Research and Development Center in the USA. The Takeda American Research and Development Center was established in 1997, and its origin goes back to the joint research agreement with Abbott in 1972. This agreement developed into a joint venture agreement (Takeda Abbott Product Partnership, TAP) with Abbott in 1977 and TAP Pharmaceuticals (TAP) in 1985; the establishment of the US arm of Takeda Pharmaceutical began with TAP. In 1993, the Takeda American Research and Development Center was established as a division of Takeda America. This division developed pioglitazone (Obayashi 2008).

Throughout the basic research process, animal models introduced from universities and firms greatly contributed to the accumulation of diabetes mellitus knowledge and to the screening of candidate drugs. For example, the KKAy mouse and Wistar fatty rats were the primary means to evaluate the efficacy of pioglitazone (Meguro 2014). It is very likely that had the hypoglycemic effects not been discovered, diabetes drug studies would not have been initiated. This suggests that the contribution of the animal models was significant.

Several players contributed to the elucidation of the pioglitazone mechanism of action and related research. The first player was Upjohn, a joint research and development partner, although the contribution of Upjohn itself was indirect. More specifically, joint research with Upjohn promoted insulin resistance research in the USA by attracting attention to ciglitazone and pioglitazone in the USA. The second player was academia. Takeda Pharmaceutical identified the mechanism of action and side effects through joint research programs with domestic academic institutions. Cooperative research on the measurement of the insulin signal was conducted with Masashi Kobayashi (Shiga Medical College), where insulin signaling was measured before and after administration of pioglitazone. The researchers also collaborated with Takashi Kadowaki (University of Tokyo) on the miniaturization of adipocytes, Yuji Matsuzawa (Osaka University) on the relationship between adiponectin and diabetes, and Kazuwa Nakao (Kyoto University) on adipose turnover, including mitochondrial uncoupling protein.

10.10 Market Success

Pioglitazone was launched in 1999, and, as of 2015, was sold in 90 countries around the world. In the USA, Takeda Pharmaceuticals North America, a wholly owned subsidiary of Takeda, and Eli Lilly co-marketed Actos. Takeda Pharmaceutical and Eli Lilly had co-promotion agreements for Actos between 1999 and 2006 to overcome Takeda Pharmaceutical's lack of sales experience in the US market. Sales were approximately 5 billion USD in the highest year, mostly in foreign markets. Figure 10.1 shows the sales of Actos from 2000 to 2012. Sales declined significantly in 2012, but this was a consequence of the expiration of the patent, not because the diabetes therapeutic market was shrinking.

Takeda Pharmaceutical was the world's leading company in basic diabetes research, likely because of early initiation of diabetes research. Even when the company stopped developing ciglitazone in 1983, Yasuo Sugiyama, then the director of the second drug discovery laboratory of Takeda Pharmaceutical, said: "Because there were no competitors, we could have started it once again." However, the situation changed following the ciglitazone publication as many pharmaceutical companies began the development of thiazolidinedione compounds, and fierce competition

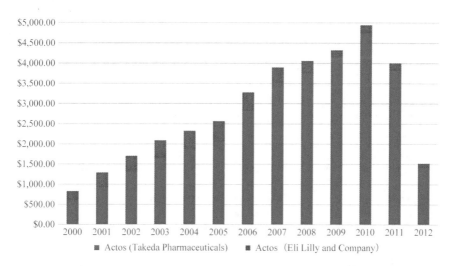

Fig. 10.1 Annual sales of pioglitazone (unit: 1 M USD) *Source* Pharma Future (Cegedim Strategic Data)

ensued. Although a large number of analogs have been reported since the publication of ciglitazone, two of the analogs were on the market by 2000, troglitazone (Noscar) of Sankyo and rosiglitazone (Avandia) of SmithKline Beecham.

Among them, Sankyo and SmithKline Beecham recognized the usefulness of thiazolidinedione derivatives, and started the development of insulin resistance-improving drugs. Sankyo changed the direction from the development of lipid peroxide lowering therapeutics, and the research program of the insulin resistance-improving drugs began in 1983. Clinical trials were started in 1987, and the first insulin resistance-improving drug was put on the market in March 1997. SmithKline Beecham started the clinical trial of rosiglitazone in 1993, and the second insulin resistance-improving drug in the world was put on the market in the USA in June 1999. In the meantime, Takeda Pharmaceutical, which ran the top in basic research, started the clinical trial of pioglitazone in 1991, and the third insulin resistance-improving drug was put on the US market in August 1999.

Sankyo and SmithKline Beecham were not the only companies involved in the research and development of insulin sensitizers. Table 10.1 summarizes the insulin resistance-improving drugs among the new drug candidates for the treatment of diabetes as of 2003.

However, none of these second-generation insulin sensitizers were ever launched. KRP-297 of Kyorin (Development Code at Merck: MK-767) was discontinued in November 2003 because of the presence of neoplastic lesions in a mouse carcinogenicity study (Kyorin Pharmaceutical 2003). Mitsubishi Pharma's MCC-555 (Development Code at Johnson & Johnson: RWJ-241947) was discontinued in 2009. In 2005, Fujisawa decided to discontinue the development of FK614 because of a failure to demonstrate superiority relative to existing drugs following completion of Phase II trials in Japan and the USA (Astellas Pharma 2005) . ONO-5816 of Ono

Table 10.1 Development of insulin sensitizers by Japanese Pharmaceutical Companies as of 2003

Firm	Development code	Origin	Clinical stage		
			Japan	US	Europe
KYORIN Pharmaceutical Banyu Pharmaceutical	KRP-297 MK-767	KYORIN Pharmaceutical	Ph II	Ph III (Merck)	Prestage (Merck)
Mitsubishi Pharma	MCC-555 RWJ-241947	Mitsubishi Pharma	Ph II	Ph II (J&J)	Ph II (J&J)
Fujisawa Pharmaceutical	FK614	Fujisawa Pharmaceutical	Ph II	Ph II	–
Ono Pharmaceutical	ONO-5816 SP-134101	Shaman Pharmaceutical	Ph II	–	Ph II (Merck KGaA)

Source March 2003 issue of Nikkei Bio-business, pp. 131–133

Pharmaceutical (Development Code at Merck KGaA: SP-134101) was also with-drawn in 2003 because of poor performance in experimental animals (Ono Pharmaceutical 2003). In 2006, Jeri El-Hage of the US FDA stated that "more than 50 insulin-sensitizing drugs have been accepted for IND applications since 1999, but none has reached the market" (El-Hage 2006).

10.11 World's Largest Market Share of Insulin Sensitizers

Biguanides, such as metformin, were launched prior to pioglitazone and the other thiazolidinediones were released to improve insulin sensitization. However, because thiazolidinediones and biguanides have different mechanisms of action, only the thiazolidinedione antidiabetic drugs troglitazone and rosiglitazone are considered as competing with pioglitazone in this chapter. Figure 10.2 shows the changes of the market shares of these insulin sensitizers up to 2012. Of note, annual sales have decreased over time because of the entry of generic drugs, and the switch to combination therapies.

As of 2011, Actos made up the world's largest share of insulin sensitizers. In fact, many non-Actos insulin sensitizers were withdrawn from the market over concerns of side effects. Noscal, the world's first insulin sensitizer, reported 130 cases of liver damage, including 6 deaths, as of autumn 1997, the year of launch. This led the Japanese Ministry of Health, Labour and Welfare to issue an urgent safety information (yellow letter) at the end of 1997. GlaxoSmithKline, which sold Noscal in the UK, withdrew the drug in early 1997, and Sankyo and Park Davis, as well as Warner Lambert, completely withdrew sales from Japan and the USA by March 2003. Avandia, which once had the highest share of insulin sensitizers in the world, was also forced to withdraw from the market in 2010, when cardiovascular side effects became evident (Tsutani 2007).

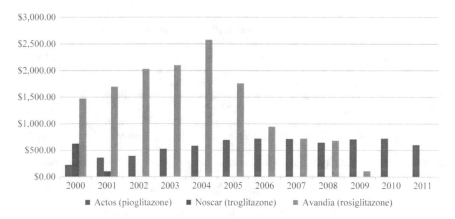

Fig. 10.2 Annual sales of insulin sensitizers (unit: 1 M USD). *Source* Medtrack database

10.12 Conclusion

Pioglitazone is a drug that took approximately 40 years, from the beginning of basic research, to launch. A careful investigation of this development process suggests that pioglitazone can be viewed as a breakthrough product that was generated by the combination of basic research, animal model development, careful screening techniques, and excellent synthesis technologies.

Takeda Pharmaceutical began diabetes mellitus research in the early 1960s, and it grew with the availability of various animal models that were obtained through cooperation with academia. Not only did these animal models make a major contribution to the understanding of diabetes, but they also led the researchers to the discovery of lead compounds with a thiazolidinedione backbone, which could be utilized to improve insulin sensitivity. Ciglitazone, the first compound studied in clinical trials, failed, but pioglitazone has become a major therapeutic for people with Type 2 diabetes worldwide. In addition, the thiazolidinedione backbone discovered by Takeda Pharmaceutical was also adopted by numerous other companies as the basis for insulin sensitizing drug development. Such spillover effects to other companies shows the scientific benefit of Takeda's discovery of a novel therapeutic mechanism of action.

Some may say that pioglitazone was not wholly successful in terms of competition for product development, because Takeda Pharmaceutical was the third company to launch a thiazolidinedione insulin sensitizer. However, the two drugs launched before pioglitazone have since been withdrawn from the market because of side effects. This provides evidence that Takeda Pharmaceutical's strategy to minimize the risk and uncertainty of side effects outweighed the "first mover" advantage, which may apply to other situations where the mechanism of action of a drug is not fully understood.

References

Astellas Pharma Inc. (2005). *Summary of fiscal 2005 interim summary.* http://www.astellas.com/jp/ir/library/pdf/h_pre2006_jp.pdf. Accessed in July 3, 2014.

El-Hage, J. (2006). *Peroxisome Proliferator-Activated Receptor (PPAR) agonists preclinical and clinical cardiac safety considerations.* http://www.fda.gov/downloads/AboutFDA/CentersOffices/CDER/ucm119071.pdf. Accessed in May 16, 2014.

Himsworth, H. P. (1936). Diabetes mellitus: Its differentiation into insulin-sensitive and insulin-insensitive types. *The Lancet, 227*(5864), 127–130.

Ikeda, H. (1998). Introduction to the study of diabetes mellitus by Takeda. *The Japanese Society of Diabetes and Obese Zoology News Letter 2*(2). http://jsedo.jp/news/nl2-2.html. Accessed in May 16, 2014.

Ikeda, H., & Sugiyama, Y. (2001). Insulin-resistance improving effects of pioglitazone. *Journal of Science, Japan drug, 117*(5), 335–342.

Kawai, K. (2007). *Insulin-sensitizing drugs and "frosted" carcasses.* http://www.somos.co.jp/solution/d/076.html (accessed in May 16, 2014).

Kawamatsu, Y., & Fujita, T. (1981). Useful as remedies for diabetes and hyperlipemia, US4287200A.

Kyorin Pharmaceutical Co., Ltd. (2003). Discontinuation of development of KRP-297, a diabetes treatment at Merck, U.S.A. (Phase 3 clinical trial).

Lehmann, J. M., Moore, L. B., Smith-Oliver, T. A., Wilkison, W. O., Willson, T. M., & Kliewer, S. A. (1995). An antidiabetic thiazolidinedione is a high affinity ligand for peroxisome proliferator-activated receptor γ (PPARγ). *Journal of Biological Chemistry, 270*(22), 12953–12956.

Makino, H. (2006). Diabetes mellitus and Kappadkia. *News Letter of the Japanese Society for Diabetes and Obesity, Zoological Society 10*(2). http://jsedo.jp/news/nl10-2.html. Accessed in February 28, 2014.

Maruyama, K. (1992). *The New Insulin Story.* Tokyo Dojin.

Meguro, K. (1991). Seeking new treatments for diabetes. *Pharmacia, 27*(9), 899–902.

Meguro, K. (2014). Luck, stolidity, and perseverance in drug discovery. In *Kinki Division of Chemistry Lecture Meeting*, June 19, 2014. http://pe-eco.jp/articles/show/378. Accessed in August 7, 2014.

National Diabetes Data Group. (1979). Classification and diagnosis of diabetes mellitus and other categories of glucose intolerance. *Diabetes, 28*(12), 1039–1057.

Nikkei Bio-Business. (2003, March). *Competition for insulin resistance improvement drug* (pp. 131–133).

Ninomiya, R. (1996). Insulin-finding banting for $1. *Medicina, 33*(3), 612–613.

Obayashi, M. (2008). From our experience in new drug development in the United States-centering on the development of Actos. In The Japanese Society of Pharmacopoeia (Ed.), *The Pharmaceutical Expert Workshop Series 9: How to Learn from Practical Cases-How to Develop New Drugs Overseas-To Find Topics of Actos Success* (pp. 2–34).

Ono Pharmaceutical Co., Ltd. (2003). *ONO-5816, Diabetes Treatment.* http://www.ono.co.jp/jpnw/news/pdf/2003/n03_0219.pdf. Accessed in July 3, 2014.

Sakai, S. (1998). Fighting diabetes from a history perspective. *Diabetes, 41*(2), 89–93.

Seino, S., Nanjo, K., Tajima, N., Kadowaki, T., Kashiwagi, A., Araki, E., et al. (2010). The Committee Report on *Classification and Diagnostic Criteria of Diabetes Mellitus 53*(6), 450–467.

Sohda, T., Kawamatsu, Y., Fujita, T., Meguro, K., & Ikeda, H. (2002). Creation of pioglitazone, an insulin-sensitizing drug. *Yakugaku Zasshi, 122*(11), 909–918.

Takeda Pharmaceutical Co. Ltd. (1999). *Data on Actos (Pioglitazone hydrochloride).* http://www.info.pmda.go.jp/shinyaku/g990913/01ctdp_1-354.pdf. Accessed in February 10, 2014.

Takeda Pharmaceutical Co. Ltd. (2008). Birth and Evidence of Japanese World Pharmaceuticals (23) Pioglitazone (Actos®). *Medicine 62*(11), 637–638.

Tsutani, K. (2007). Drug withdrawal. *Pharmacia, 43*(1), 1097–1102.

World Health Organization. (1980). *WHO Expert Committee on Diabetes Mellitus* [meeting held in Geneva from 25 September to 1 October 1979]: second report. http://whqlibdoc.who.int/trs/WHO_TRS_646.pdf. Accessed in May 16, 2014.

Chapter 11
Donepezil (Aricept)

The World's First Drug for Alzheimer's Disease

Yasushi Hara and Hideo Kawabe

Abstract Donepezil is a symptomatic drug for Alzheimer's disease that was discovered and developed by Eisai Co., Ltd. It has a temporary suppressive effect on the pathology of Alzheimer's disease. Donepezil increases the concentration of acetylcholine, a key neurotransmitter in the brain, by inhibiting the degradative enzyme acetylcholinesterase. While other pharmaceutical groups focused on other mechanisms of action, such as cerebral metabolism improvement or cerebral blood flow improvement, as targets of anti-dementia drug development, the research group at Eisai focused on the acetylcholine hypothesis, for which scientific verification had not been sufficiently carried out. Throughout the research and development process, the difficulty of low bioavailability of lead compounds was encountered and the project was officially suspended within the firm. However, researchers continued to work on the drug. Donepezil was virtually the first drug in its class for Alzheimer's disease in the world, and has long dominated the marketplace as the best-in-class drug, primarily because no better second-generation drugs have been developed. The diffusion of donepezil into the market was also helped by the popularization of MRI scanning, which is effective for early diagnosis and detection of Alzheimer's disease. This has resulted in the accurate diagnosis of patients with Alzheimer's-type dementia, who otherwise may have been diagnosed as having other dementias, thereby expanding the market for donepezil.

11.1 Introduction

Donepezil hydrochloride (Aricept) is a drug for Alzheimer's disease (AD) that was discovered and developed by Eisai Co., Ltd. (Eisai). It was the world's second AD

Y. Hara (✉)
CEAFJP/EHESS, Paris, France
e-mail: yasushi.hara@r.hit-u.ac.jp

Faculty of Economics, Hitotsubashi University, Tokyo, Japan

H. Kawabe
Patent Attorney in Hiraki & Associates/IP counsel in Manufacturing Technology Association of Biologics, Tokyo, Japan

© Springer Nature Singapore Pte Ltd. 2019
S. Nagaoka (ed.), *Drug Discovery in Japan*,
https://doi.org/10.1007/978-981-13-8906-1_11

drug, after tacrine, for Alzheimer's-type dementia, and was the first in Japan. Given that tacrine was withdrawn from the market because of its side effects, donepezil is the world's first AD drug within the AD drugs currently on the market (Ono 2003).

Alzheimer's disease was first identified in 1907 when Dr. Alois Alzheimer, a neuropathologist, presented his findings of pathological changes in the brains of women who died with progressive dementias to the German Psychiatric Association (Sugimoto et al. 1998). In AD, abnormal shrinkage occurs in the brain, and the patients have symptoms such as memory disorders, depression, and persecution delusions (Ishii 2005). It has been identified that abnormal protein of the filament state is deposited in the nerve cells. Alzheimer's disease can be divided into three major clinical stages: (1) cognitive deficits are limited to memory development; (2) the symptoms of the memory disorders are clear, mental status becomes unstable, and symptoms such as delusions, restlessness, and depression occur; (3) patients suffer further declines in health and physical capability and may become bedridden. Death eventually results from aspiration pneumonia or urinary tract infection associated sepsis (MHLW 2014). Cognitive dysfunction, such as memory disorder, disorientation, and impaired judgment are considered to be the hallmark symptoms of AD (Takahashi 2001). Psychiatric symptoms and behavioral abnormalities summarized as "action and psychiatric symptom group of the dementia" are defined as accessory symptoms. To improve the quality of life (QOL) of patients and the others involved, the development of AD drugs was strongly desired.

The number of patients with AD is increasing annually; in the United States, the number of AD patients was approximately 4.68 million in 2002, rising to 5.24 million in 2014. According to a survey of Connecticut State, the incidence of AD is 720 per 10,000 patients in people aged 65 years or older.[1] The number of dementia disease patients diagnosed as having AD has also been increasing in Japan, with an incidence of 380 cases per 10,000 patients in people aged 65 years or older (Sekita et al. 2010).

11.2 Mechanism of Action of Donepezil

An exact understanding of the cause and pathogenesis of AD has not yet been elucidated; therefore, there is no preventative therapy. However, donepezil is a symptomatic drug that only has a temporary suppressive effect on the pathology of AD, and it increases acetylcholine concentrations by inhibiting the degradative enzyme acetylcholinesterase (Ono 2003).

Acetylcholine is a neural transmitter that acts between neuronal synapses. The signal is transmitted from the presynaptic cell where the acetylcholine, synthesized by choline acetyltransferase (ChAT), interacts with the acetylcholine receptor of the

---

[1] Medtrack database, Epidemiology Report.

postsynaptic cell. Under normal conditions, acetylcholine is degraded by acetyl-cholinesterase (AChE) to choline, which prevents excessive signaling by the receptors. Donepezil blocks the action of AChE, thereby increasing acetylcholine in the brain and facilitating signal transduction (Mizoguchi 2003).

Although the pathogenesis of Alzheimer's disease remains largely unknown, the β-amyloid hypothesis, which postulates that the disease arises from the deposition of β-amyloid protein, is the most popular hypothesis as of 2015. Because amyloid plaques accumulate before the onset of dementia, it remains difficult to develop drugs based on the inhibition of amyloid accumulation.

11.3 Timeline of Donepezil R&D

1973–1981: Eisai initiated discovery research targeting AD after multiple failed research and development (R&D) efforts for dementia drugs that were associated with cerebrovascular disorders.

Around 1983: Study of tacrine derivatives began.

1983–1985: Discovery research based on the new lead compound that was accidentally found to have acetylcholine elevation activity.

March 1986: Eisai determined the new lead compound had low bioactivity.

April 1986: Dr. Iimura joined Eisai and was put in charge of compound synthesis. Dr. Kawakami placed in charge of computer-aided drug design (CADD) at Eisai.

December 1986: Donepezil hydrochloride (Aricept) was discovered and successfully synthesized.

March 1987: Donepezil was selected as a drug candidate for clinical testing.

Initiation of clinical development

Japan

1989: Started Phase I clinical trial.
1991: Started Phase IIa clinical trial.
1996: Started Phase III clinical trial.

USA

1991: Started Phase I clinical trial.
1992: Started Phase IIa clinical trial.
1994: Started Phase III clinical trial.

Submitted application for approval of a new drug:

USA

March 1996: New Drug Application.
November 1996: New Drug Approval.

Japan

　1998: New Drug Application.
　1999: New Drug Approval.

11.4　Science Basis for R&D

When basic research for donepezil began, the conventional view was that dementia was caused by cerebrovascular disease in the majority of patients, and only a minority of dementia cases were caused by AD. Therefore, drugs for dementia were mainly those that improved the cerebral blood flow such as Avan (Idevenon) and Hopate (Calcium hopantenate hydrate), which improved ischemic cerebral dysfunction; only a few pharmaceutical companies focused on AD drugs specifically.[2]

　In the early 1980s, Bartus et al. (1982) proposed the cholinergic hypothesis, in which the failures of cerebral cholinergic nerves themselves constitute an essential pathology in AD. Based on this hypothesis, there are three possible ways to increase brain acetylcholine: (1) facilitating the release of acetylcholine from presynaptic neurons (synaptic vesicles); (2) inhibiting the action of acetylcholinesterase, an enzyme that metabolizes acetylcholine, thereby increasing the concentration of acetylcholine released between synapses; and (3) introducing an agonist for muscarinic receptors located at postsynapses (Fig. 11.1).

　For the first method (1), there were no compounds known to exert potent effects in clinical studies (Mohs et al. 1979; Renvoize and Jerram 1979). However, it had

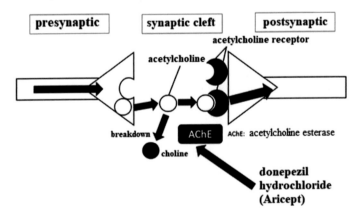

Fig. 11.1 Mechanism of action of donepezil

---

[2]Later, the indication for cerebral blood flow-improving drugs for dementia was questioned, and the approval was respectively canceled in 1998 for Calcium hopantenate hydrate and Idevenon. Subsequently, changes in diagnostic methods, including the introduction of MRI, identified a number of dementias as being of the AD type (Box 11.1).

been reported that the development of a candidate by DuPont (DuP-996) to promote the release of acetylcholine was discontinued (Nickolson et al. 1990; Sugimoto et al. 1999). For the second approach, cholinesterase inhibitors such as physostigmine and tacrine were also studied. For the third approach, arecholine, oxtremorine, RS-86, and others were developed as muscarinic acetylcholine receptor agonists, although these drugs caused peripheral side effects, such as decreases in blood pressure. Clinical development of these candidates failed (Sugimoto et al. 1999).

In the 1970s, Eisai's Hachiro Sugimoto conducted a 4-year joint project with Prof. Shimadzu of the Department of Neurology, Saitama Medical University, on a drug to treat a sequela of cerebrovascular disorders; the project's objective was to develop a drug that would improve cerebral blood flow. The synthesis was carried out by Eisai, and the effect on cerebral blood flow in monkeys was investigated at Saitama Medical University. As a result of this joint development effort, a chemical compound that increased cerebral blood flow in monkeys was discovered. However, concerns of liver functional impairment arose in Phase I clinical trials, and product development was abandoned. This process lasted 8 years and cost 800 million yen from the start of development to the end (Sugimoto 2004c).

11.5 Discovery Research of Tacrine Derivatives

In April 1983, a proposal for a drug-development project targeting senile dementia by activation of the acetylcholine system was initiated by the Clinical Development Department and the Research Planning Department at Eisai. The initial assessment of a three-member team, led by Dr. Sugimoto, concluded that there was a plausible rationale for testing the hypothesis. This conclusion was based on their prediction made from a review of the literature that predicted future developments in diagnostic methods would increase the number of patients diagnosed with AD.[3] It was decided that the development program would have two main goals: (1) a drug with long-lasting effects, and (2) a drug that was highly localized in the brain rather than in the blood. These goals would be the focus of a discovery program investigating drugs capable of increasing the concentration of acetylcholine between synapses by inhibiting the function of AChE (Sugimoto et al. 1999).

Derivatives of tacrine with AChE inhibitory effects were synthesized, using the clinical paper on tacrine by Summers et al. (1981) as a guide, and pharmacological assessments were carried out. The reasons for making derivatives of tacrine included: (1) its relative ease of synthesis compared with physostigmine, (2) its relatively high level of concentration in blood compared with physostigmine, and (3) clinical studies that were archived by Prof. Summers (Umeda 2002). However, despite the preparation of more than 30 tacrine derivatives over about 6 months, it was not possible to obtain an adequate lead compound (Sugimoto 2004c).

---

[3]From interview with Dr. Sugimoto.

After failing to develop a tacrine derivative, a researcher from another research group at Eisai found, after monitoring in vivo screens, that a compound synthesized as a drug candidate for hyperlipidemia had an acetylcholine-elevating effect. Administration of the compound in mice resulted in symptoms such as salivation, miosis, and convulsions, all of which are symptoms caused by increased acetylcholine. Then, when the level of the AChE activity was measured, strong activity was shown. This compound was selected as the candidate compound, and about 200 derivatives of the compound were synthesized. Through these synthetic works, it was also identified that there was a species difference in the pharmacological activity of enzymes derived from the rat brain and those derived from the electric eel.

11.6   Lead Compound Derivation and Project Discontinuation

The research team led by Sugimoto synthesized about 300 derivatives over 3 years by using the compound that was developed as a therapeutic agent for hyperlipidemia as the base molecule. The change of piperazine of the lead compound to piperidine confirmed a more than 70-fold increase in activity. Also, the AChE source was changed to enzymes derived from rat brain homogenate, which was more commonly used in in vivo studies than eels. Subsequently, the team confirmed that the activity was enhanced by changing the ether linkage of an ether derivative to an amide linkage, making the amide derivative a lead compound. Examination of the structure–activity relationship of this lead compound revealed that: (1) introduction of a bulky functional group at the *para*-position of the benzamide improved the activity; (2) introduction of a methyl, ethyl or femyl group at the nitrogen atom of the amide group improved the activity; and (3) introduction of a substituent to the benzyl group or conversion to another functional group did not improve activity. Through these chemical syntheses, the research team discovered a substance with very potent AChE inhibitory activity by introducing a benzylsulfonyl group at the *para*-position of benzamide and a methyl group at the nitrogen atom of the amide group (Sugimoto et al. 1999).

However, characterization of the substance revealed that bioavailability in dogs was low, below 2%. Furthermore, the example that irreversible AChE inhibitor "organophosphate" had been used as pesticide, might have given the peers in Eisai a negative image of reversible AChE inhibitor development. At a subsequent meeting, comments such as the following predominated:

> It is a pesticide but not a drug for humans, or
>
> It is not possible to increase only local acetylcholine concentration in the brain, because the acetylcholine receptor is widely distributed in the whole body, and the inhibitor is carried to the whole body by the blood. The side effects cannot be avoided if acetylcholine increases in the whole body. (Sugimoto 2004b)

As a result, the compound was removed from the company's development program.

11.7 *Yami* Research, Project Restart, and Derivation of Donepezil

Subsequent research continued as *yami* research (unauthorized research) and focused on improving drug bioavailability. The work was initiated by Dr. Sugimoto and was carried out by two or three researchers in the Eisai Laboratory. Dr. Sugimoto described the conditions when conducting the *yami* research as follows:

> Does the management know whether unauthorized research is going on? The department lead would not know. However, the unit manager often knows that it is. Otherwise, the resources would not come to the research group for *yami* research.

After the progress made by Dr. Sugimoto's team, the director of the research institute approved the resumption of the donepezil R&D process. Dr. Iimura was then hired by Eisai and worked within the synthetic chemistry group. In addition, the CADD research support group, which included Dr. Kawakami, were included to provide insight into chemical structures based on accumulated stability and brain absorption data (Sugimoto 2005a). In this way, the chemical synthesis of over 200 compounds was carried out. The main synthesis target was the benzamide derivative, because it was originally considered to be the basic structure having AChE inhibition activity, but it was found that the indanone derivative also showed activity.

Finally, donepezil hydrochloride (E2020), a derivative of 5, 6-dimethoxyindanone, was synthesized on 12 December 1986 during the development based on CADD information (Nikkei Business 2001) (Fig. 11.2). Dr. Iimura was in charge of chemical synthesis (Umeda 2002). Dr Sugimoto said the following with respect to Iimura:

> This compound was synthesized by a new employee from the university, Dr. Youichi Iimura. Despite his young age, his synthetic ability was very high and he had the youthful capability of synthesizing all the compounds he wanted. The structure–activity relationships used were certainly based on knowledge accumulated over the last 3 years, but the subsequent scoping and limitation for donepezil was made by his sophistication, certainly.

Subsequent examinations identified the structure–function relationship of the indanone derivate (Sugimoto et al. 1999):

(1) When the five-membered ring of the indanone skeleton was expanded to a six-membered or seven-membered ring, the activity of the compound was reduced. Carbonyl groups were essential for the expression of activity, and that activity was enhanced by the introduction of a substituent at the *para*-position of the carbonyl.
(2) Methylene chain lengths of one or three were desirable.
(3) The basic nature of the nitrogen atom was essential for activity, and 1-benzylpiperidine exhibited the strongest activity.
(4) The substituent effect of the benzene ring showed a slight increase in activity in the case of *meta*-substitution with both electron-donating and electron-withdrawing groups.

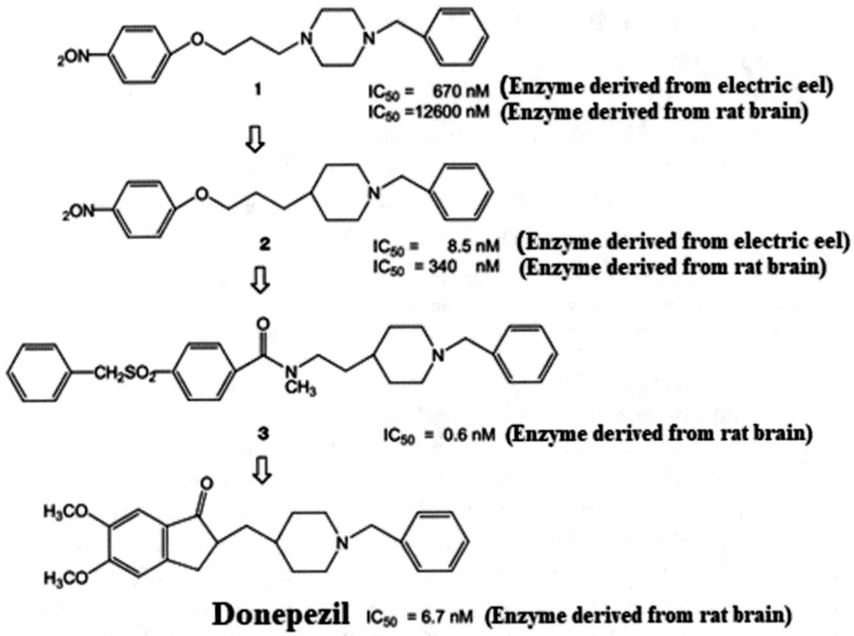

Fig. 11.2 Development from early lead compounds to donepezil hydrochloride. *Source* Sugimoto (2014)

When donepezil hydrochloride showed efficacy in pharmacological tests in vivo, pharmacokinetics experiments were initiated, and the bioavailability was confirmed. Investigation of the pharmacological effects revealed that the drug had a higher selectivity for AChE than physostigmine and tacrine (Yamanishi et al. 1991). It was also shown that donepezil hydrochloride had the most potent inhibition of the AChE enzyme, and had excellent localization into the brain (Sugimoto et al. 1999).

Because synthesized donepezil contained an asymmetric carbon, optical resolution was performed by HPLC fractionation using a chiral column and an asymmetric reduction method using BINAP (patent H4-21670,[4] H4-187674[5]). It was discovered that the asymmetric center was in the α-position to the carbonyl group, and easily racemizes. As a result, donepezil hydrochloride was developed as a racemic mixture.

---

[4]Japanese Patent Number: H4-21670, Inventor: Iimura, Y., Kajima, T., Araki, N., Sugimoto, H. Title: Optically Active Indanone Derivatives. Assignee: Eisai Co., Ltd.

[5]Japanese Patent Number: H4-187674, Inventor: Iimura, Y., Kajima, T., Araki, N., Sugimoto, H. Title: (−)-1-Benzyl-4-[(5,6-dimethoxy-1-indanon)-2-yl] methylpiperidine. Assignee: Eisai Co., Ltd., Takasago Kogyo Co., Ltd.

11.8   Clinical Trials

In Japan, the Phase I clinical trial was started in January 1989, and was completed in July 1989. In the United States, it was started in 1991. The subsequent Phase II clinical study was initiated in Japan in May 1990. Open studies, double-blind studies, dose-escalation studies, and dose-finding studies were conducted at doses of 0.1, 1, 2, 3, 4, and 5 mg, respectively, and completed in January 1995 (Sugimoto et al. 2002). The Phase II clinical trial in the United States was started in 1992.

The following two approaches were used to evaluate drug efficacy in these clinical trials. The Alzheimer's Disease Assessment Scale Cognitive Subscale (ADAS-Cog) method was used as an evaluation method to determine improvements in the memory disorder of patients, and the Clinician's Interview-Based Impression of Change Plus Caregiver Assessment (CIBIC-Plus) method was used as an index evaluation method of the patients quality of life (Sugimoto 1995). Donepezil showed statistically significant improvements in both tests. Clinical studies in the United States were conducted by Pfizer, the United States licensee of donepezil. It was estimated that the drug had the ability to delay the progression of AD by roughly 1 year, although such characterization was acquired only through implementing the multifaceted clinical research programs described above.[6]

Phase III clinical trials in Japan were initiated in 1996, 2 years later than the Phase III clinical trials in the United States. This was because: (1) there were significant differences in the numbers of AD patients in the United States and Japan, and (2) the doses used in United States clinical trials were larger (5 and 10 mg) than the doses used in Japan (3 and 5 mg). It is believed that this difference in dose was an effort to evaluate efficacy of the drug more clearly in the United States. However, this made it difficult to bridge the results of the two studies. Afterward, in the United States, a new drug application was made to the US Food and Drug Administration (FDA) in March 1996, and the approval was obtained in November 1996. In the United Kingdom, the sales started in April 1997 (Sugimoto 2001).

> When we were going to do a clinical trial in Japan, US clinical investigators advised us that the level of the dose in Japan was too low, so we changed the drug concentrations being tested in Japan. That's the key of clinical development success. In the case of clinical trials in the USA, development will proceed even if a few side effects occur. In America and Japan, the way of thinking for clinical trials was totally different.[7]

A Japanese Phase III study testing a 5 mg dose and structured as a double-blind, placebo-controlled study, was initiated in September 1996 and completed in January 1998. In Japan, an application for a new drug was filed in 1998, and the approval was received in 1999. As of January 2013, it was sold in 97 countries.[8]

The development of the drug dosage of fine granules was also carried out in an attempt to improve the ease of drug administration, because the targeted patients

---

[6]From interview with Hachiro Sugimoto.

[7]From interview with Dr. Hachiro Sugimoto

[8]Aricept Interview Form.

were elderly. This was approved in March 2001. Later, an oral disintegrating tablet was developed for patients with ingestion difficulty, and it was approved in February 2004. Oral jelly and dry syrup were approved in July 2009 and February 2013, respectively.[9]

11.9 Science Basis of Donepezil: The Acetylcholine Hypothesis

In 1976, the acetylcholine hypothesis was formed by three independent research groups in the UK, including Prof. Bowen (Bowen et al. 1976; Davies and Maloney 1976; Perry et al. 1977). The activity of ChAT, the enzyme responsible for the synthesis of acetylcholine (ACh), was significantly reduced in the cerebral cortex or hippocampus of AD patients. In AD, the temporal lobe of the cerebral cortex and the adjacent limbic system are most substantially affected (Bowen et al. 1976). Davis, Perry and others hypothesized that increasing brain acetylcholine would be one mechanism to improve memory function (Perry et al. 1977).

Then, in 1982, Whitehouse et al. (1982) reported that large neuronal cell groups were markedly sloughed in the Meynert's nucleus, which is present in the substantia inominata of the basal forebrain, in AD brains. The contributions of this paper include the discovery of neuronal loss under the cerebral cortex in AD, as well as the finding of a specific cholinergic neuronal disorder.

11.10 Development of Tacrine Based on the Choline Hypothesis

From these studies, the drug development of anti-Alzheimer's disease drugs based on the choline hypothesis was initiated. Although clinical trials of the drug candidates were carried out with the aim of activating cholinergic nerves using lecithin and choline, the effectiveness could not be confirmed. Subsequently, cholinesterase inhibitors were studied, such as physostigmine and tacrine. However, physostigmine was associated with a short half-life in the blood and serious side effects at off-target sites.

In the meantime, a pilot study was carried out in 12 patients by Prof. Summers in 1981. It was discovered that during the administration of tacrine (originally developed as an antibacterial agent) to Alzheimer's patients, the drug had an inhibitory effect on acetylcholinesterase. In 1986, a study was conducted involving intravenous administration of tacrine, and oral administration of tacrine might be at least temporarily useful in the long-term palliative treatment of patients with AD (Koopmans et al. 1986). Based on these results, Prof. Summers conducted a double-blind study

---

[9]Aricept Interview Form.

and showed the potential of cholinergic drug as a symptomatic drug for Alzheimer's disease. Professor Summers filed a method-of-use patent in the United States in 1986.[10] Thereafter, tacrine was approved as an AD drug by the FDA in the United States, and, in 1993, Warner-Lambert launched it on the market. However, it became known that the drug led to a high probability of hepatic dysfunction. Given that the drug's half-life in the blood was 2–4 h, the drug required administration four times a day, and because of the extremely high rate of hepatic dysfunction and gastrointestinal symptoms, regular monitoring of hepatic function was mandated (Takahashi 2001). Eventually, tacrine was withdrawn as an AD drug (Tumiatti et al. 2010).

In 1980, the National Institutes of Health invested 13 million USD to support research on AD, and the Alzheimer's Association was established in the same year to start research support.

11.11   Donepezil R&D Team Members

The following members played central roles in the development of donepezil (Umeda 2002):

- Hachiro Sugimoto: Senior synthetic scientist, donepezil R&D team leader and synthetic group leader.
- Youichi Iimura: Research Fellow in the field of synthetic chemistry. Responsible for the synthesis of donepezil.
- Yoshiyuki Kawakami: Supported donepezil synthesis in CADD.
- Noboru Araki: Biologist, in charge of biochemical experiments.
- Hiroo Ogura: Biologist, contributed to the development of amnesia models.
- Norio Karibe: Synthetic researcher, synthesized pyrazine derivatives that remained as final candidates.
- Atsuhiko Kubota: Biologist.
- Takashi Osasa: Biologist.
- Michiko Otake: Biological Researcher.
- Jun Sasaki: Synthetic Researcher.
- Yutaka Tsuchiya: Synthesized "the world's strongest compound" before donepezil.
- Higurashi Kunizou: Synthetic Researcher.
- Yoshiharu Yamanishi: Senior Biologist, donepezil sub-investigator, and leader of biological groups.

Dr. Sugimoto played a central role as the principal investigator of the R&D group, and led the research in the early stages. Sugimoto had discovered and developed the blood pressure-lowering drug Detantol before the development of the donepezil

---

[10]Administration of monoamine acridines in cholinergic neuronal deficit states, http://patft. uspto.gov/netacgi/nph-Parser?Sect1=PTO1&Sect2=HITOFF&d=PALL&p=1&u=%2Fnetahtml% 2FPTO%2Fsrchnum.htm&r=1&f=G&l=50&s1=4,816,456.PN.&OS=PN/4,816,456&RS=PN/ 4,816,456 (accessed 24 October 2014).

(Sugimoto 2005b). Iimura and Kawakami played key roles as specialists in synthetic chemistry and CADD, respectively, in the process of identifying donepezil as the lead compound.

When the research of donepezil began, Eisai had just established the Tsukuba Research Institute in 1982, and the Senior Managing Director, Mr. Naito (later President and CEO), strongly supported in-house drug-discovery development. Chemical synthesis and pharmacology researchers were designated based on research area, given that Eisai's research institute was divided into six disease-specific laboratories: infectious diseases, cranial nerves, digestive organs, cardiovascular, asthma/allergy, and blood. A discovery research team was established within each laboratory. However, among the newly established laboratories, interlaboratory researchers were able to communicate closely, which allowed for collaboration between chemical synthesis and pharmacology specialists at the discovery stage. At the same time, it also created a sense of competition for new drug development between laboratories (Ono 2004). Director Naito encouraged such competition by stating: "It's no use returning home before nine o'clock in the evening... researchers should go to work on Saturdays," and "synthesize at least five samples per week."

Eisai also actively hired new graduate researchers when the Tsukuba Research Institute was established (Ono 2003). In parallel with strong leadership by Director Naito, strong lead authority was given to the project leader, Dr. Sugimoto, for donepezil research. Eisai had a bottom-up R&D framework at the time.

Eisai was a relatively late entrant among major pharmaceutical companies in Japan. However, the company put an emphasis on developing its own products rather than licensing from overseas companies, which was the most popular practice in the pharmaceutical industry at the time. As a result, Eisai's product lineup was primarily based on the products developed in-house. Eisai was good at developing products that faced high hurdles to approval, such as coenzyme Q10.

11.12   Interactions Between Science and Donepezil

The discovery team of donepezil had identified compounds with strong AChE inhibitory activity but poor bioavailability. To overcome the shortcomings of these compounds, Kawakami enrolled at the University of Tsukuba as a research student, while engaging in the work for Eisai. This was to expedite the acquisition of CADD technology to help design a drug that not only had strong enzymatic activity, but also had good stability and high bioavailability. To improve CADD methods, Kawakami learned the molecular orbital method at the University of Tsukuba and reflected the opinions of young researchers at Eisai into the application of the method. He then made various suggestions on the direction of the discovery research through use of the unique CADD method.

> Fortunately, a course in computational science was offered at the University of Tsukuba. I enrolled as a research student for about 2 years. I learned the molecular orbital method and utilized it for electronic structural analysis of (Aricet) amide. I went to the laboratory once

or twice every week to discuss the structure–activity relationship in depth with the professor of the computational science laboratory.[11]

The key relationship between brain acetylcholine and Alzheimer's dementia was proposed 5–6 years before the start of research by Eisai. However, the choline hypothesis had been more or less abandoned when Sugimoto's team started the R&D process in 1983 (Sugimoto 2004a). However, recognizing that "If an overseas pharmaceutical company enters AD drug development, more than 50 R&D members will be hired the next day following that decision. Eisai cannot win against such a company if it competes on the same ground." Because of this, Sugimoto decided to pursue this "niche" research trend before other companies entered the competition (Sena 2004).

Donepezil was utilized for structural analysis of AChE in the brain and molecular analysis of the reaction mechanism, which greatly contributed to the study of the brain. Furthermore, molecular structural studies of donepezil were conducted in collaboration with Anton Hopfinger, a professor at the University of Illinois, since the beginning of the clinical studies (Cardozo et al. 1992a, 1992b).

11.13 Sales of Aricept: Over 3 Billion USD Sold Worldwide Annually

Today, AD has a large number of patients, but that number was believed to be much smaller before the launch of donepezil development. Because of this, few pharmaceutical firms were interested in the development of AChE inhibitors from a market standpoint. However, advances in diagnostic imaging techniques such as CT and MRI scanning have contributed to the detection of AD, and the number of patients diagnosed with AD has dramatically increased. These circumstances have led to an increase in the importance and first-mover advantage of donepezil as a pioneer drug.

Aricept's annual sales in Japan in 2011 were 1.297 billion USD (144 billion yen), and total worldwide sales in 2010 before the patent expiration were 3.209 billion USD (356.1 billion yen). Afterward, sales declined sharply because of substance patent expiration. The worldwide sales of donepezil from 2005 to 2012 are shown in Fig. 11.3. The peak in annual sales of donepezil was about 3.5 billion USD in 2010.

The effects of donepezil on patient cognitive ability are shown in Fig. 11.4. Donepezil has the ability to slow the progression of dementia by roughly 6 months to 1 year. Considering that the progression of AD increases the cost of care, donepezil contributes to reduced expenditure (and labor) by patients, patient families, and public institutions, and to improvements in QOL of patients and their families. Cost-effectiveness analyses using Quality Adjusted Life Years (QALY) have shown that donepezil was more effective than other AD medications (Ikeda et al. 2000).

---

[11] From interview with Kawakami.

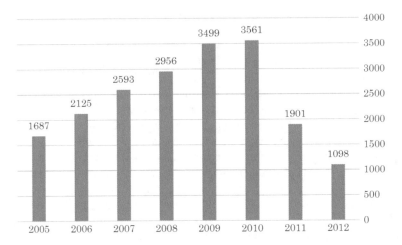

Fig. 11.3 Global sales of aricept (unit: 1 M USD) *Source* Pharma Future Sezidem Strategic Data, Uto Brain Division

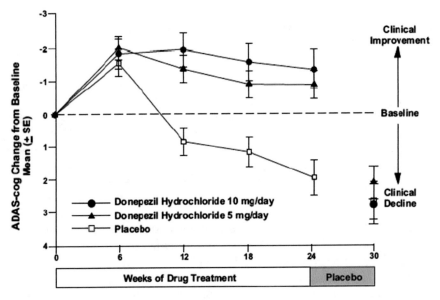

Fig. 11.4 Effects of donepezil on patient cognitive ability. *Source* drugs.com (Donepezil, drugs.com, https://www.drugs.com/pro/donepezil.html, accessed 26 August 2018)

11.14 Entries and Failures of Other Companies in R&D of Anti-alzheimer Drugs

In addition to acetylcholine agonists such as donepezil, R&D of serotonin agonists and cerebral circulation metabolism drugs were explored. After the synthesis of donepezil, it became the benchmark target, and a number of similar structural compounds were developed, including TAK-14 from Takeda and CP-118954 from Pfizer. As shown in Fig. 11.5, all of these compounds have an N-benzylpiperidine partial structure, but only donepezil has been successfully developed and marketed in Japan until recently. In many cases, development programs were discontinued because of a lack of significant difference from placebo through Phase II and III clinical trials (see Table 11.1). Many of the product patents for these compounds cite the basic substance patent (US4895841) for donepezil (Takahashi 2001). Among such companies, Pfizer chose to partner with Eisai after abandoning their own development.

> Pfizer had candidates for Phase II clinical trials of Alzheimer's drugs, but Aricept was already in Phase III trials. Perhaps Pfizer synthesized Aricept and compared it with its own candidate. As a result, Pfizer stopped developing in-house candidates and selected sales partnership with Eisai. An alliance with Pfizer was a great opportunity for Aricept to become a blockbuster drug.[12]

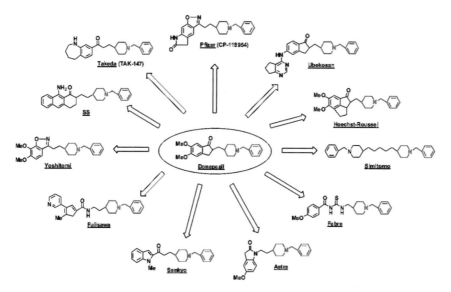

Fig. 11.5 Donepezil hydrochloride-related compounds. N-Benzylpiperidine moiety is highlighted. *Source* Sugimoto (2004c)

---

[12]From interview with Dr. Sugimoto.

Table 11.1 Research and development of donepezil hydrochloride-related compounds (benzylpedilimine derivatives) by other companies

Company name	Study number, generic name	Development status	US Patent number	Publication date	Application date	Direct reference to donepezil basic patents
Takeda Chemical Industries	TAK-147, zanapezil	Discontinuation after Phase III clinical study (2009)	US5273974	28 Dec 1993	22 Nov 1991	Yes
Pfizer	CP-118954, icopezil	Discontinuation after Phase II clinical study (1995)	US6498255	24 Dec 2002	19 Sept 2001	Yes
Ube Industries	UR1827	Discontinuation of development in preclinical studies	US5610303	11 Mar 1997	1 Oct 1993	Yes
Hoechest-Roussel	Hextol	Discontinuation of AD development after Phase III clinical trials (2000)	US6028083	22 Feb 2000	25 Jul 1997	Yes
Dainippon Sumitomo Pharma	AC3933	Discontinuation after Phase II clinical study (2009)				No
Pierre Fabre SA	F14413	Discontinuation of development in preclinical studies	–	–	–	–
Astra	AZD0328	Discontinuation after Phase II clinical study (2010)	US6110914	29 Aug 2000	10 Jul 1998	No

(continued)

Table 11.1 (continued)

Company name	Study number, generic name	Development status	US Patent number	Publication date	Application date	Direct reference to donepezil basic patents
Sankyo Pharmaceutical	BGC200406	Discontinuation of development in preclinical studies	–	–	–	–
Fujisawa Pharmaceutical	FK960, FK962	Discontinuation of study in AD after Phase II clinical trial (2003)	US5250528	5 Oct 1993	8 Oct 1992	Yes
Yoshitomi Pharmaceutical Industries	Unknown	Unknown	US5141930	25 Aug 1992	8 Jul 1991	Yes
Arena Pharmaceuticals	T-82	Discontinuation after Phase II clinical study (2004)	US6218402B1	17 Apr 2001	9 Jun 1999	No

Development status from Medtrack Database; Patent information from USPTO Database

11.15 Competition in the Anti-Alzheimer's Drug Market

Figure 11.6 shows the changes in sales over time for the main drugs used to treat AD. The first drug approved for AD in the United States was tacrine (1993), but it was withdrawn from the market. Afterward, donepezil was approved in 1996. Following this, galantamine, a compound discovered in the plant *Galanthus woronowii*, which was traditionally used to treat myasthenia, was approved in 2001. Johnson & Johnson sells galantamine under the trade name Razadin, and Shire and Takeda Pharmaceutical sell it as Remenyl. A controlled-release formulation of memantine hydrochloride, which has an antagonistic effect on N-methyl-D-aspartate (NMDA) glutamate receptors, was approved in 2002 by the European Medicines Agency and in 2003 by the FDA for moderate to severe Alzheimer's-type dementia. It is currently marketed as Namenda in the USA, Evixa in Europe, and, in June 2011, was launched by Daiichi Sankyo under the name Memary. Rivastigmine, which has a chemical structure similar to that of physostigmine, was approved in 2007.[13] Rivastigmine is marketed by

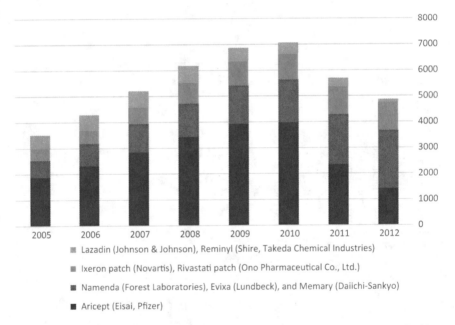

■ Lazadin (Johnson & Johnson), Reminyl (Shire, Takeda Chemical Industries)

■ Ixeron patch (Novartis), Rivastati patch (Ono Pharmaceutical Co., Ltd.)

■ Namenda (Forest Laboratories), Evixa (Lundbeck), and Memary (Daiichi-Sankyo)

■ Aricept (Eisai, Pfizer)

Fig. 11.6 Sale trends of major Alzheimer drugs (unit: 1 M USD). *Source* Pharma Future Sezidem Strategic Data, Uto Brain Division

---

[13]Medicinal Interview Form "Memary."

Novartis Pharma as the Exelon Patch[14] and by Ono Pharmaceutical as the Rivastach Patch[15] (Sugimoto 2006).

Box 11.1 Innovations in Diagnosing Alzheimer's Disease by Magnetic Resonance Imaging

The diffusion of cerebral examination by magnetic resonance imaging (MRI) greatly contributed to the increase in the number of patients recognized as having Alzheimer's disease in recent years. Studies on the use of functional magnetic resonance imaging (fMRI) to diagnose Alzheimer's disease had been reported since the mid-1980s, and in 1990, a study on cholinergic central innervation in rats was published (Kikutani 1990). Furthermore, the use of MRI for early detection of AD was reported in the Lancet in 1994 (Minoshima et al. 1994).

The significance of measuring cerebral blood flow using MRI is threefold: (1) the disease cases can be extracted relatively early from the distribution of the decreases of such flow, (2) the classification of the disease type is possible, and (3) the depth of tissue damage can be measured. By distinguishing Alzheimer's disease from psychiatric and other dementias, the risk of exacerbations of the patients with frontotemporal dementia could be prevented by donepezil (Aricept) (Shimosekawa 2011).

Shimadzu Corporation, which led the R&D process of MRI, started discovery research on NMR-CT in 1980, and launched the project the following year. However, attention at that time was focused on diseases such as brain tumors, and the diagnosis of Alzheimer's disease was not considered a major target of MRI. At that time, the main themes of the technology were the improvement of the image and the accumulation of expertise in brain content measurement.

Afterward, technological improvements made it possible to acquire the cerebral function image using MRI at high resolution (Takeo et al. 1996). Today, these techniques, known as non-invasive functional imaging of the brain, utilize two techniques: (1) the measurement of neural currents caused by neural activity; and (2) the measurement of regional cerebral blood flow changes (Shimizu 2010).

These MRI technological developments have led to the use of MRI in AD diagnostics. In the diagnosis of juvenile Alzheimer's disease, it is possible to characterize the change from the initial stage of the diseases by combining neuropsychological tests such as the Revised Hasegawa Dementia Scale (HDS-R), head CT, and MRI. This is because MRI can be used to image the hippocampus in the frontal and sagittal planes, thereby detecting hippocampal atrophy (Kimura 2001). The diagnosis of senile Alzheimer's disease also uses

---

[14]Exelon® Patch, Highlights of Prescribing Information, https://www.accessdata.fda.gov/drugsatfda_docs/label/2007/022083lbl.pdf. Accessed: Jul 4 2019

[15]Medicinal Interview Form "Rivastach Patch", https://www.ononavi1717.jp/drug_info_files/drug_info/rivastach/if/10030057/RIV_IF_6.pdf. Accessed: Jul 4 2019

MRI at diagnosis to distinguish Alzheimer's disease from other dementias, which may be curable (Kawano 2004). In addition, Dai Nippon Printing and Eisai jointly developed the "Early Alzheimer's Dementia Diagnosis Support System (VSRAD)," a diagnostic support solution combining MRI and analytical software, for the early treatment of Alzheimer's disease using AD drugs such as donepezil (Goto et al. 2006).

The above paragraphs mainly focusing on MRI; however, the development of donepezil occurred when image analysis of the brain began to diffuse into medical institutions. Importantly, image analysis of the brain of patients with dementia by X-ray, CT, and MRI showed abnormal atrophy of the brain, that is, Alzheimer's disease, but did not show the cause of cerebral metabolism disorder (Wahund 1996).

With such advances in brain image analysis technology, the number of patients diagnosed with Alzheimer's disease increased sharply, and as mentioned earlier, the Ministry of Health, Labour and Welfare (MHLW) canceled the entire approval of cerebral metabolism improving drugs, which had blood flow-improving effects after cerebral infarction, that were used for dementia while donepezil was being developed between 1989 and 1998.

In other words, during the development period of donepezil, the innovation in diagnostic imaging in clinical practice led to an increase in the number of patients receiving donepezil, and pre-existing drugs that had been classified as effective for dementia had been withdrawn from the market.

Donepezil was launched in Japan in 1999 and by 2001 MRI was being successfully used to diagnose Alzheimer's disease approximately 80% of the time. The MRI diagnosis of Alzheimer's was based on the morphology of the cortex from the MRI image and the evaluation of atrophy in the para-hippocampal gyrus using the Z score image (Imabayashi 2010). Furthermore, by using MRI image analysis, differential diagnoses of dementia and grading of dementia have been made possible. MRI image analysis can determine the morphological character of local parts of the brain and locate blood flow decrease in the brain (Uno 2002). The combination of drugs with a clear mechanism of action for AD therapy and advanced diagnostic imaging techniques such as MRI has promoted research on the diagnosis and treatment of Alzheimer's disease (Uno 2003).

11.16   Conclusion

While market competitors focused on other mechanisms of action, such as cerebral metabolism improvement or cerebral blood flow improvement, as the target of anti-dementia drug development, the research group led by Dr. Sugimoto focused on the acetylcholine hypothesis. In the R&D process, the difficulty of low bioavailability of lead compounds was encountered, and, once the project was suspended as an official R&D project within the firm, the research was continued as *yami* research, before finally being resumed.

As a characteristic feature of the donepezil development team, Dr. Sugimoto, the project leader, was passionately engaged in a consistent way from the start of the R&D project to the start of clinical development. At the same time, Sugimoto appointed young researchers to draw on their strengths to drive the R&D team and to successfully lead the project. For example, Dr. Iimura, who ultimately succeeded in synthesizing donepezil hydrochloride, was a new doctoral program graduate who had experience in synthesizing antibiotics and introduced synthetic processes that took advantage of his experience. Dr. Kawakami also contributed to expanding the area of chemical synthesis by learning CADD technology at the University of Tsukuba.

A major characteristic of the development environment at the newly established Tsukuba Research Laboratories of Eisai was that chemical synthesis and pharmacology scientists were integrated in each disease area and engaged in the development of new drugs. As a result, a close relationship between the two groups was realized at an early stage. At the same time, the slogan "Pressure Makes Diamonds" prompted a strong competition for development between research groups.

Donepezil was virtually the first-in-class drug for AD, and it has long been able to dominate the marketplace as the best-in-class drug. Approval of drugs for improvement of cerebral blood flow were discontinued, and many pharmaceutical companies have withdrawn from development of drugs for dementia. At the same time, subsequent development of AChE-inhibitory drugs has been unable to demonstrate significant improvements over donepezil or placebo in clinical trials.

Furthermore, diffusion of donepezil into the market has been helped by the popularization of MRI scanning, which has facilitated early diagnosis and detection of Alzheimer's disease. MRI scanning can easily detect brain shrinkage, especially in the hippocampus, so that early diagnosis and detection of AD has become possible. This has allowed for the correct diagnosis of Alzheimer's-type dementia (these patients had been often diagnosed as having other dementias) and has consequently expanded the market for donepezil.

References

Bartus, R. T., Dean, R. L., 3rd, Beer, B., & Lippa, A. S. (1982). The cholinergic hypothesis of geriatric memory dysfunction. *Science, 217*(4558), 408–414.
Bowen, D. M., Smith, C. B., White, P., & Davison, A. N. (1976). Neurotransmitter-related enzymes and indices of hypoxia in senile dementia and other abiotrophies. *Brain, 99*(3), 459–496.
Business, N. (2001). 21st century Edison: Technology and Innovation Prof. Hachiro Sugimoto, who develop Alzheimer drug which leading the world; Modern alchemist who bet on new medicine creation (in Japanese). *Nikkei Business*, Issue March 12, 2001.

Cardozo, M. G., Kawai, T., Iimura, Y., Sugimoto, H., Yamanishi, Y., & Hopfinger, A. J. (1992a). Conformational analyses and molecular-shape comparisons of a series of indanone-benlpiperdine inhibitors of acetylcholisterase. *Journal of Medical Chemistry, 35,* 590–601.

Cardozo, M. G., Iimura, Y., Sugimoto, H., Yamanishi, Y., & Hopfinger, A. J. (1992b). QSAR analyses of the substituted indanone and benzylpiperidine rings of a series of indanone-benzylpiperidine inhibitors of acetylcholinesterase. *Journal of Medicinal Chemistry, 35*(3), 584–589.

Davies, P., & Maloney, A. J. (1976). Selective loss of central cholinergic neurons in Alzheimer's disease. *Lancet, 25*(2), 1403.

Goto, M., Aoki, S., Abe, S., Masumoto, T., Watanabe, Y., Satake, Y., et al. (2006). The usefulness of horizontal cross-sectional images in early Alzheimer-type dementia support systems (VSRAD) (in Japan Radiological Society's 34th Autumn Medical Congress). *Japan Radiological Society of Radiological Technology Journal, 62*(9), 1339–1344.

Ikeda, T., Yamada, Y., & Ikegami, T. (2000). Economic evaluation of donepezil, an anti-dementia drug. *Medicine and Society, 10*(3), 27–38.

Imabayashi, E. (2010). Advances in diagnostic imaging for dementia and regional medical collaboration news. *Saitama Medical School International Medical Cooperation Center, 3,* 16–17.

Ishii, M. (2005). *Research of researchers and engineers who are responsible for the development of innovative products.* NISTEP Discussion Paper.

Kawano, K. (2004). Usefulness of clock drawing test (The Clock Drawing Test, CDT) in dementing clinics. *Journal of the Biomedical Fuzzy Systems Society, 6*(1), 69–79.

Kimura, Y. (2001). Attempts at an outpatient clinic specializing in juvenile Alzheimer's disease. *Juntendo Medical, 47*(1), 15–22.

Kinutani, K. (1990). Basic research on nuclear medical imaging of cholinergic central innervation. *Journal of Kanazawa University Juzen Medical Association, 99*(2), 342–359. (in Japanese).

Ministry of Health, Labour and Welfare [MHLW]. (2014). *From knowledge to begin mental health-dementia.* http://www.mhlw.go.jp/kokoro/speciality/detail_recog.html. Accessed February 23 2014.

Minoshima, S., Foster, N. L., & Kuhl, D. E. (1994). Posterior cingulate cortex in Alzheimer's disease. *Lancet, 344*(8926), 895.

Mizoguchi, A. (2003). *Global-level technology from Japan* (p. 135). Shogakukan (in Japanese).

Mohs, R. C., Davis, K. L., Tinklenberg, J. R., Hollister, L. E., Yesavage, J. A., & Kopell, B. S. (1979). Choline chloride drug of memory deficits in the elderly. *The American Journal of Psychiatry, 136*(10), 1275–1277.

Nickolson, V. J., Tam, S. W., Myers, M. J., & Cook, L. (1990). DuP 996 (3, 3-bis(4-pyrindinylmethyl)-1-phenylindolin-2-one) enhances the stimulus-induced release of acetylcholine from rat brain in vitro and in vivo. *Drug Development Research, 19*(3), 285–300.

Ono, Y. (2003, Spring). Business case: Eisai Alzheimer's disease development process and organizational management. *Hitotsubashi Business Review,* 133–145.

Ono, Y. (2004) A discussion of the relationship between the division of roles of leadership and leadership and the activation of team activities, from the case of Aricept, a drug for Alzheimer's disease from Eisai Co., Ltd. *Management Behavioral Science, 17*(3), 185–196.

Perry, E. K., Perry, R. H., Blessed, G., & Tomlinson, B. E. (1977). Necropsy evidence of central cholinergic deficits in senile dementia. *Lancet, 22*(1), 189.

Renvoize, E. B., & Jerram, T. (1979). Choline in Alzheimer's disease. *New England Journal of Medicine, 301,* 330.

Sekita, A., Ninomiya, T., Tanizaki, Y., Doi, Y., Hata, J., Yonemoto, K., et al. (2010). Trends in prevalence of Alzheimer's disease and vascular dementia in a Japanese community: The Hisayama study. *Acta Psychiatrica Scandinavica, 122,* 319–325.

Sena, H. (2004). Invitation to Science 5th Hachiro Sugimoto. *Weekly Toyo-Keizai, July 31 2014,* pp. 93–95 (in Japanese).

Shimizu, K. (2010). Non-invasive cerebral function imaging. *Journal of the Society of Imaging Information and Media, 64*(6), 794–798.

Shimosekawa, Eihisa. (2011). Measuring cerebral blood flow: Significance and limitations in the diagnosis of pathological conditions (abstract before presentation in the basic course of the 63rd Nuclear Medicine Subcommittee). *Journal of the Nuclear Medicine Subcommittee, 63,* 10–15.

Sugimoto, H., Ogura, H., Arai, Y., IImura, Y., & Yamanishi, Y. (2002). Research and development of donepezil hydrochloride: A new type of acetylcholinesterase inhibitor. *The Japanese Journal of Pharmacology, 89(1),* (7–20).

Sugimoto, H. (1995). *One hundred soul of inventors: Developing ARICEPT, a drug for Alzheimer's disease* (pp. 47–55) (in Japanese).

Sugimoto, H. (2001). Discovery of drug discovery drug Donepezil for Alzheimer's disease, representing the 20th century (37, 1, p. 17).

Sugimoto, H. (2004a). Donepezil: Advances in development and drug. *Japanese Journal of Pharmacology, 124,* 163–170.

Sugimoto, Hachiro. (2004b). R&D strategies for donepezil hydrochloride: Its lights and shadows (pharmaceuticals produced by pharmacists"-collaboration between pharmaceuticals, pharmaceuticals, and science). *Summary of Annual Meeting of the Japanese Society of Medical Pharmacology, 14,* 146–159.

Sugimoto, H. (2004c). Condition of successful development of new drug. *R&D Leader, 1*(1), 14–19.

Sugimoto, H. (2005a). Chasing the dreams of developing drugs for Alzheimer's disease. *Bioscience and Industry, 63,* 153–157. (in Japanese).

Sugimoto, Hachiro. (2005b). Interview innovators Dr. Hachiro Sugimoto. *Rinkuru Chuo Houki Shuppan, 4,* 23–25. (in Japanese).

Sugimoto, H. (2006). Chasing the dream of developing Alzheimer's disease drug. *Kagaku to Kyoiku, 54*(3), 130–133 (in Japanese).

Sugimoto, H. (2014). Chasing the dreams of developing drugs for Alzheimer disease. http://www.rsihata.com/pdf/070108cre2.pdf. Accessed 29 October 2014.

Sugimoto, H., Yamanishi, Y., Ogura, H., Iimura, Y., & Yamazu, Y. (1998). Research and development of donepezil hydrochloride for Alzheimer's disease. *Journal of Pharmaceutical Sciences, 119*(2), 101–113.

Sugimoto, H., Yamanishi, Y., Ogura, H., Iimura, Y., & Yamazu, Y. (1999). Research and development of donepezil hydrochloride, a drug for Alzheimer's disease. *Journal of Pharmaceutical Sciences, 119*(2), 101–113.

Summers, W. K., et al. (1981). Use of THA in drug of Alzheimer-like dementia: Pilot study in twelve patients. *Biological Psychiatry, 16*(2), 145–153.

Summers, W. K., Majovski, L. V., Marsh, G. M., Tachiki, K., & Kling, A. (1986). Oral tetrahydroaminoacridine in long-term drug of senile dementia. *Alzheimer Type, 315*(20), 1241–1245.

Takahashi, K. (2001). Drug for cognitive dysfunction in Alzheimer's disease-current status of the development of anti-dementia drugs. *Juntendo Medical, 47*(2), 184–190.

Takeo, K., Shimizu, K., & Shimizu, K. (1996). EPI with a MRI device for clinical use (application to cerebral function imaging). In *Japanese Society of Radiological Technology. General Research Abstract of the 52nd Annual Scientific Congress* (p. 1177).

Tumiatti, V., Minarini, A., Bolognesi, M. L., Milelli, A., Rosini, M., & Melchiorre, C. (2010). Tacrine derivatives and Alzheimer's disease. *Current Medicinal Chemistry, 17*(17), 1825–1838.

Umeda, E. (2002). New drug development project for miracle. In *Kodansha Plus Alpha* (in Japanese).

Uno, Masanari. (2002). The significance and potential of the forgotten outpatient. *Psychiatry Therapeutics, 17*(3), 269–274.

Uno, M. (2003). Mental health and the forgetfulness clinic. *Elderly Mental Health, 22*(4), 409–411.

Wahund, L. O. (1996). Magnetic resonance imaging and computed tomography in Alzheimer's disease. *Acta Neurologica Scandinavica Supplementum, 168,* 50–53.

Whitehouse, P. J., Price, D. L., Struble, R. G., Clark, A. W., Coyle, J. T., & Delon, M. R. (1982). Alzheimer's disease and senile dementia: Loss of neurons in the basal forebrain. *Science, 215*(4537), 1237–1239.

Yamanishi, Y., Ogura, H., & Kosaka, T. (1991). In T. Nagatsu (Ed.), *Basic, clinical and therapeutic aspects of Alzheimer's and Perkinson's diseases* (Vol. 2, pp. 409–413). New York: Plenum Press.

Chapter 12
Candesartan (Blopress, Atacand)

Antihypertensive Drug with a New Mechanism of Action

Naoki Takada, Koichi Genda and Akira Nagumo

Abstract Candesartan (Blopress) is an antihypertensive drug known to act as an angiotensin-II receptor blocker (ARB). It was discovered and developed by Takeda Pharmaceutical. Compared to earlier antihypertensive drugs, candesartan had a new mechanism of action, fewer side effects, and demonstrated efficacy even at low doses. The research project started as exploratory research on diuretics. Takeda discovered CV-2198, the first nonpeptidic ARB, by making full use of spontaneously hypertensive rats (a pathological animal model) provided by academia, and through capturing serendipitous event brought about by cutting-edge chemical synthesis methodology. However, the project was suspended because of weak efficacy, while DuPont developed losartan, building on the Takeda's invention. Although preceded by DuPont, Takeda quickly responded, and successfully launched candesartan shortly after. The case of candesartan is a good illustration of the role of unique capabilities and the overall organizational strength for innovative drug discovery, as well as the difficulty of go/no-go decisions and indicates the importance of flexible allocation of resources for grasping opportunities.

12.1 Introduction

Candesartan cilexetil (Blopress) is an antihypertensive medicine discovered by Takeda Pharmaceutical Co., Ltd. It is an angiotensin-II (AII) receptor antagonist, also known as an AII receptor blocker (ARB). The unique feature of ARBs is their

N. Takada (✉)
Institute of Advanced Sciences, Yokohama National University, Kanagawa, Japan
e-mail: takada-naoki-jx@ynu.ac.jp

K. Genda
Career Development Department, Shionogi Career Development Center Co., Ltd, Osaka, Japan
e-mail: koichi.genda@shionogi.co.jp

A. Nagumo
Medical Affairs, Medical Science Liaison, MSD K.K., Tokyo, Japan
e-mail: akira.nagumo@merck.com

© Springer Nature Singapore Pte Ltd. 2019
S. Nagaoka (ed.), *Drug Discovery in Japan*,
https://doi.org/10.1007/978-981-13-8906-1_12

ability to lower blood pressure by a different mechanism than pre-existing antihypertensive drugs. Above all, ARBs lead to fewer side effects than seen in pre-existing drugs, such as nervous system depression and dry cough, by demonstrating positive effects even at low doses. ARBs with these characteristics have greatly contributed to increased options for treatment of hypertension and to increased quality of life of hypertensive patients.

CV-2198, the lead compound of Blopress, was originally created by Takeda and has a basic structure similar to all nonpeptidic ARBs currently on the market. CV-2198 was discovered during the research and development (R&D) process for the discovery of new diuretic drugs, and Takeda's full-scale ARB R&D began after the discovery of CV-2198. However, because early ARBs in clinical trials were ineffective in humans, ARB R&D was suspended before they reached the market. In the meantime, DuPont USA promoted the R&D based on chemical compounds synthesized by Takeda, and published a patent for losartan, the first nonpeptidic ARB in the world. Takeda resumed ARB research after the publication of losartan, and candesartan cilexetil was advanced to clinical trials in 1991. As a result, Blopress was approved in the UK in 1997, and in Japan in 1999 as the second ARB in Japan.

The R&D of candesartan involved the culmination of scientific knowledge on the renin-angiotensin system (RAS). The history of RAS research began in 1898, when renin was discovered. The complete picture of the present RAS was established in the 1960s, when R&D into RAS drugs began. However, at the time, peptide ARBs and angiotensin-converting enzyme (ACE) inhibitors were insufficient to develop as a medicine, and the diuretics that were used to treat hypertension were concerning because of their side effects. These unresolved problems led Takeda's researchers to begin R&D for candesartan.

12.2   Mechanism of Action

Candesartan and other ARBs lower blood pressure by pharmacologically affecting the RAS through blocking of AII and AII receptor binding. The RAS, which is a hormonal system involved in the regulation of blood pressure and extracellular volume, elevates blood pressure by the pathway shown in Fig. 12.1. The first substrate in the RAS is angiotensinogen, which is secreted by the liver and hypertrophic adipocytes. Angiotensinogen released into the blood is partially degraded to angiotensin I (AI) by renin produced by juxtaglomerular cells of the kidney. AI is converted to AII by ACE in the endothelial cells of the pulmonary capillaries. Binding of AII, a bioactive vasopressor, to its receptor increases blood pressure.

AII has two receptors, AT1 and AT2. Most AIIs bind to the AT1 receptor and initiate the following cascade (a) to (e), thereby increasing blood pressure: (a) constricting blood vessels by influx of calcium ions into smooth muscle cells, (b) stimulating biosynthesis and secretion of aldosterone in the glomerular layer of the adrenal cortex, (c) promoting sense of thirst and release of antidiuretic hormone (ADH) through the hypothalamus, (d) stimulating Na^+ reabsorption in the proximal tubule, and (e)

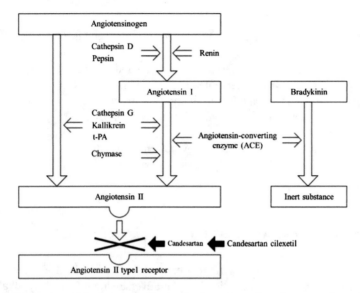

Fig. 12.1 Mechanism of blood pressure elevation by the renin-angiotensin system. *Source* Blopress Drug Interview Form (2012), pp. 21, translated by author

suppressing renin secretion. Specifically, vasoconstriction increases peripheral vascular resistance (the pressure exerted by blood as it flows through blood vessels throughout the body) and blood pressure increases. In addition, when water and sodium ions are reabsorbed, the volume of blood circulating in the body increases, resulting in an increase in blood pressure (Shimamoto 2011).

12.3   Timeline of the Candesartan R&D Process

1970s: Initiation of diuretic research; screening using spontaneously hypertensive rats.

1978: Discovery of CV-2198, which provided the basic structure of ARB drugs. The full-scale R&D of ARBs began.

1981: Establishment of AII inhibitory effect of CV-2973, a related compound of CV-2198.

September 1981: Clinical trial of CV-2973 started.

1982: Research of CV system was suspended as CV-2973 failed to demonstrate efficacy in humans.

October 1982: Filed patent application for CV-2973-related compounds (US4355040).

February 1989: DuPont published losartan referring to Takeda's CV-2961 as the sole lead compound.

1990: Takeda resumed ARB research.

April 1990: TCV-116 (candesartan) was selected as the final candidate for clinical trials.

1991: Clinical trials of candesartan began in Japan and Germany at the same time. Subsequently, the trial was implemented in Japan, USA, and EU.

August 1996: Applied for manufacturing approval in Japan.

September 1996: Applied for manufacturing approval in EU.

November 1996: Launched in EU.

October 1998: Launched in the USA by AstraZeneca as Atacand.

June 1999: Launched in Japan as Blopress.

12.4 Start as Diuretic Research

In the 1970s, thiazide diuretic antihypertensives were prescribed to hypertensive patients. However, thiazide diuretics had the adverse side effect of hypokalemia. Therefore, there were unmet medical needs for antihypertensive drugs that could replace thiazide diuretics. Kohei Nishikawa, who conducted the antihypertensive research project, discussed the process that led to the initiation of diuretic R&D:

> The problem with thiazide diuretics was that they have "a stronger excretion of Na than water, a loss of K," and the so-called water diuretics; that is, a diuretic with less potent K excretion was needed, according to a nephrologist. We therefore thought that finding a new basic chemical structure different from thiazide diuretics was important and initiated a random screen to find diuretics with new basic structures, and tested various compounds synthesized in the synthesis department of the Takeda Drug Research Institute in rat animal models (Nishikawa 2007, p. 72).

Initially, a small team of five people in the pharmacological group studied metabolites of Alinamin, which is a vitamin B1 derivative. At the time, advice from a nephrologist spurred the team to start exploring new diuretics with a basic structure different from thiazide diuretics. With full-scale diuretic research, the drug synthesis department synthesized test compounds, and the pharmacological department carried out compound screening using rats. Approximately 7 years after starting screening with animal models, CV-2198, the lead compound of ARB, was discovered.

12.5 A New Synthesis Method and Serendipitous Findings

CV-2198 was discovered by the basic research team in the drug synthesis group. The teams had been synthesizing compounds using a new synthetic process called vapor-phase pyrolysis that was devised by Katsura Morita. The gas-phase pyrolysis reaction had been based entirely on a theoretical idea, as Morita mentioned: "What would happen if we exposed a material for a very short time to a high temperature of more than 1000 °C and cooled it immediately? This was my original idea." When the reaction apparatus was set up and put into operation for the chemical reaction, it gave rise to a larger number of compounds than the basic research team had expected, because the team did not have specific R&D disease targets. Therefore, the compounds synthesized by gas-phase pyrolysis were sent to Nishikawa. As a result, CV-2198 was discovered among the compounds synthesized by Koichi Matsumura (Morita 2000, pp. 245–246).

CV-2198 showed potent diuretic activity comparable to that of furosemide, which had been considered first-line therapy among existing diuretics. CV-2198 was administered to spontaneously hypertensive rats (SHRs) that were treated with furosemide, and then the RAS was stimulated by placing the rats on a reduced salt diet. CV-2198 reduced blood pressure even at low doses (3 mg/kg) and did not cause diuretic effects. Based on the hypothesis of the vasoconstrictive effect of CV-2198, rabbit aortic vessels were hung in Magnus tube and used to investigate the inhibitory effects of CV-2198 on vasoconstrictive responses induced by various stimuli including epinephrine, prostaglandin F2α, and AII. CV-2198 suppressed only the vasoconstrictive response induced by AII. Based on the results of these tests, the close collaboration between the drug synthesis group and the pharmacology group found that the antihypertensive effect of CV-2198 was caused by AII antagonism. This meant that the first nonpeptidic ARB in the world had been identified. Nishikawa noted that the discovery of the AII inhibitory effect of CV-2198 was "serendipity" (Imura 2010). Nishikawa wrote the following:

> This inhibitory effect of CV-2198 on AII activity, although significant, was weak and could have been missed by a researcher who was not paying attention. The finding arose from the researcher's careful observation and it can be called serendipity. However, we should not say just serendipity for this finding. Daily persistent pharmacological experiments and strong interest and enthusiasm to research brought us this finding. (Nishikawa 2007, pp. 73–74)

Thus, CV-2198 was produced, but its potency was too weak to be developed as an antihypertensive agent. Therefore, further drug synthesis efforts were conducted using CV-2198 as the base molecule. After about 200 derivatives were synthesized, it was decided that a clinical candidate be selected from the following four compounds: CV-2196, CV-2961, CV-2973, and CV-3382. CV-2973, which had diuretic and AII-inhibiting effects, was selected, and the clinical trial was conducted in September 1981 (Inada and Naka 2000).

12.6 Drug Lacked Efficacy in the Clinical Trial

In the clinical trial, CV-2973 did not prevent higher blood pressure and did not show a diuretic effect. Therefore, the development of CV compounds was suspended after the US patent for CV compounds was issued in 1982. In an interview, Yoshimi Imura mentioned the process of decision making for the termination of ARB R&D as follows, even though the potential basic structure for ARB had been discovered by them:

Because the costs of clinical trials are very high, the failure of its clinical trial meant many people had negative feelings towards the continuation of the project. The drug synthesis group was dissolved, and the pharmacology group shifted to the research of calcium antagonists and ACE inhibitors, which attracted more attention in the world at the time.

As Imura mentioned, the drug synthesis group was immediately dissolved with the termination of CV compound research. The pharmacology group was assigned to Adecut, an ACE inhibitor, and Calslot, a calcium channel blocker. The experience that the pharmacology group would develop throughout the R&D of these other drugs would become useful in the development of candesartan later.

12.7 The Discovery of Losartan

Research on CV compounds, which was suspended after its patent application, led to the discovery of losartan, the first ARB launched in the world. When the Takeda clinical trials of CV-2973 failed, DuPont was struggling to develop a high molecular weight peptidic ARB. In the early 1980s, when DuPont decided to focus its resources on priority areas, young researcher John Duncia was assigned to be the ARB research lead. The original strategy adopted by Duncia was to develop a more potent and novel peptidic ARB by amino acid substitution of pre-existing peptidic ARBs. However, the research did not progress as much as Duncia had expected, and a promising candidate was not found until 1982 (Bhardwaj 2006).

Under these circumstances, DuPont discovered Takeda's US patent on CV compounds in 1982 during its exploration of ARB-related patents. David Carini synthesized the sodium salt (S-8308) of CV-2961 described in Takeda's patent and Andrew Chiu and Pancras Wong confirmed its antihypertensive effect in a verification experiment. Afterward, DuPont shifted their resources and efforts from peptidic ARBs to designing new compounds based on CV-2961. To fit the molecule to the AII receptor, the molecule was altered to become bifunctional with the aim of increasing the bulkiness, resulting in a 10-fold increase in activity. The "bulk-up" of this moiety was achieved by attaching a phenyl residue, which resulted in a 100-fold increase in activity, but the compound still failed to show oral bioavailability. Then, by connecting the amide bond to the phenyl residue, the resultant molecule showed oral bioavailability without any reduction in activity. Moreover, the introduction of tetrazole increased the activity 1000-fold compared to the original candidate. DuPont developed this compound as losartan (Kuno and Sato 2011). After the publication of the losartan patent, many ARBs were developed by modifying the chemical structure of losartan. The chemical structures of the ARBs are shown in Fig. 12.2.

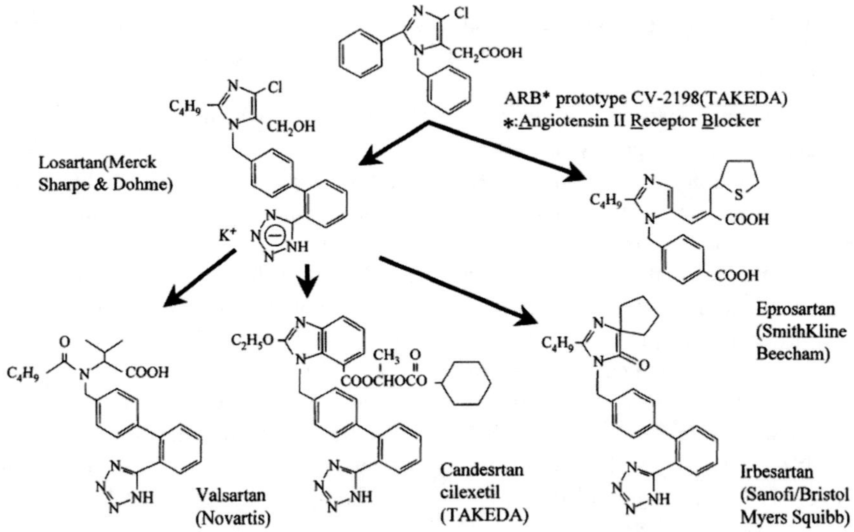

Fig. 12.2 Structural comparison of CV-2961 with losartan, and approved ARBs. *Source* Takeda Pharmaceutical (2007, p. 828)

12.8 Resumption of Takeda's ARB Research and the Discovery of Candesartan

DuPont first published losartan at the Gordon Research Conference held in February 1989. In the conference, the poster presentation by DuPont stated that losartan was created with reference to CV-2961 of Takeda. The researchers of Takeda were informed of the discovery of losartan by a person who participated in the conference and realized the potential of the ARB. Takeda had already completed the launches of Adecut and Calslot and was searching for new antihypertensive drugs to be developed; the R&D group was immediately reorganized to resume ARB research. At that time, the pharmacology group was composed of previous members of the project, while the drug synthesis group was comprised of the members recalled from the other groups, to be reorganized around Takehiko Naka.

The reorganized drug synthesis group designed and synthesized a variety of heterocyclic compounds, focusing attention on the methyl biphenyl tetrazolyl moiety discovered by DuPont. The structural design was mainly carried out by Keiji Kubo, who was previously in charge of the synthesis of the proton pump inhibitor lamprazole. After resuming the research and synthesizing 429 compounds, Kubo discovered candesartan (CV-11974). Given that the pharmacology group had experience in supporting the launches of Calslot and Adecut to their clinical stages, the group was able to immediately start acquiring the data needed for the clinical development of the newly discovered candesartan. Candesartan was selected as the lead candidate compound, and advanced to clinical trials in April 1990 (Kubo 2006; Nishikawa 2007).

Further investigations demonstrated that candesartan, which should be a potent AII antagonist, was poorly absorbed by the body and had no hypotensive effect following oral administration. Therefore, prodrug modification was conducted to enhance efficiency of absorption from the gastrointestinal tract. After examining the introduction of various substituents, the orally bioavailable prodrug of candesartan was prepared using the cilexetil residue. Fortunately, the safety evaluation of the substituent cilexetil residue had already been completed in the research for the prodrug modification of the antibiotic pansporine (Kubo et al. 1993).

The clinical trial of candesartan cilexetil (TCV-116) was started in 1991 in Japan and Germany at the same time. Afterward, clinical development in other countries were conducted in collaboration with AstraZeneca, and clinical trials started in Japan, the USA, and Europe. The manufacturing approval application was filed in Japan in August 1996. Candesartan launched in Japan in June 1999, in the UK in April 1997, and in the USA in October 1998.

12.9   The Scientific Basis of Candesartan

ARBs, including candesartan, are AII receptor antagonists. The following discussion presents the history of drug discovery and development targeting the RAS, from identification of the RAS to the development of ARBs.

The history of RAS research began in 1898, when Robert Tigerstedt and Per Bergman discovered renin. They found that rabbit kidney extracts had an unknown vasopressor effector and named it renin. However, they were unable to artificially reproduce chronic hypertension by renin infusion. Although renin was found to play a role in the development of hypertension, the actual vasopressor agent was not known at the time (Abe and Tsunoda 2002; Marks and Maxwell 1979).

In 1934, Harry Goldblatt, a pathologist at the Cleveland Clinic, succeeded in inducing chronic hypertension by ligation of the renal arteries in dogs (Goldblatt et al. 1934). Prior to this it had been reported that stenosis of the renal artery had also been shown to cause hypertension. The experimental hypertension animal model by Goldblatt was significant in identifying a link between the kidney and hypertension, leading to later research on the RAS (Kagami 2010).

In 1940, Irvine Page at the Cleveland Clinic found that renin itself had no pressor effect, but instead there was a substance formed by renin following enzymatic cleavage of an unknown plasma substrate. The resultant molecule was responsible for the pressor effect (Page 1990). At about the same time, similar discoveries were made by Eduardo Braun-Menéndez at the University of Buenos Aires. After discussions between Page and Braun-Menendez, the substance was named angiotensin in 1958. In 1954, Leonard Skeggs, a biochemist at the Cleveland Clinic, identified the existence of AI and AII and proposed a hypothesis that there were substances that converted AI to AII, which he called ACEs (Skeggs et al. 1954). Furthermore, in 1956, Skeggs and co-workers partially purified angiotensinogen and officially proposed the renin-ACE-angiotensin system (Skeggs et al. 1956).

In 1958, Franz Gross, a pharmacologist at Harvard University, completed the picture of the present RAS (Gross 1958). However, the research by Gross was carried out in horses and cattle, and therefore the purification of human angiotensin was the next target. In the 1960s, purification of various substrates and enzymes of the RA system proceeded, and competition for human angiotensin purification intensified. This competition was won in 1966 by Kikuo Arakawa of Japan, who purified human AI and determined the structure (Arakawa et al. 1967). In 1968, Kunio Hiwada at Osaka University synthesized tetrapeptide analogs of rabbit renin with competitive inhibitory activity and reported the results in *Nature* (Kokubu et al. 1968). This finding suggested the possibility of treating hypertension by blocking the action of renin.

12.10 Drug Development Era

At the time that the RAS was proposed, Skeggs predicted that there would be three ways to inhibit the RAS: renin inhibitors, ACE inhibitors, and ARBs. This sentiment led to research into renin inhibitors and ARBs beginning in the 1960s (Arakawa 2006; Yanagisawa 2006).

In particular, ARB research became a very active field of investigation after Skeggs isolated AII in 1956 and Merlin Bumpus and co-workers established a purification method for AII in 1957 (Bumpus et al. 1957). After the purification method was established and it became possible to use AII as a raw material, many AII analogs were synthesized, and the structure-activity relationships were investigated. In 1970, Philip Khairallah at the Cleveland Clinic reported that Ala-8 AII had inhibitory effects on the binding of angiotensin and its receptor (Khairallah et al. 1970). However, because the peptide had serious problems with oral absorption, it was not possible to use the peptidic ARB, which was made from the AII base molecular structure itself, in humans.

Miguel Ondetti at Squibb discovered the world's first ACE-inhibitor, captopril, in 1977. However, captopril was associated with a very high incidence of adverse effects, including rash, and the drug was only approved for use in high-renin hypertension in the USA (Ogihara and Kumahara 1982). In addition, Yoshihiro Kaneko at the University of Tokyo pointed out in 1985 that dry cough was also one of the side effects of ACE inhibitors. The problems with captopril became one of the driving forces in the development of nonpeptidic ARBs by Takeda, as well as eventually stimulating ACE inhibitor research (Arakawa 2006).

In the early 1980s, as genetic engineering began to develop, attempts were made to isolate the genes for renin and angiotensinogen (Fukamizu 2011). In 1983, Kazuo Murakami at the University of Tsukuba determined the basic structure of human renin, which enabled large-scale preparation of pure renin (Imai et al. 1983). Tadashi

Inagami at Vanderbilt University isolated the AT1 receptor, a subtype of angiotensin receptor, in 1991, and Masatsugu Horiuchi at Harvard University isolated the AT2 receptor in 1993 (Kambayashi et al. 1993). Takehiko Naka, who synthesized candesartan, mentioned that ARBs such as candesartan played a crucial role as research tools to study the angiotensin receptor, including identification of receptor subtypes and clarification of physiological functions (Naka et al. 1995).

12.11 Contributions of Scientific Knowledge

The contributions of scientific knowledge to the basic research of candesartan were particularly important in the following two aspects. The first is the tool of the disease model animal. At Takeda, the efforts to improve breeding environments of rats and mice began from around 1953, and since the central research laboratory was established at Juzo, Osaka, in 1958, the breeding and research facilities had been expanded to introduce various disease model animals. In the research of antihypertensive drugs, SHRs, which could not be freely used even in the national and public hospitals at the time, could be readily used for the diuretic research because the facilities for the disease model animals had been improved and they were grown in-house.

The SHRs, produced by Kouzo Okamoto at Kyoto University in 1963, spontaneously developed high blood pressure, and Takeda was the first to recognize the usefulness of SHRs for antihypertensive agent screening. Given that it was well known that there was a purchase order delay for SHRs, Takeda decided to establish a breeding system to breed SHRs in-house. Takeda also participated in the preparation of stroke-prone spontaneously hypertensive rats (SHRSPs) when Akinobu Nagaoka of the biological laboratory collaborated with the department of pathology at Kyoto University (Nagaoka 1987).

The second contribution was the scientific discoveries by universities in Japan and abroad. Such knowledge had been transferred to Takeda by Takeda researchers studying abroad. First, Kohei Nishikawa, who was the leader of the pharmacology group during diuretics research, studied in 1975–1976 under Philip Needleman at the Department of Pharmacology at the University of Washington School of Medicine, and was able to apply his knowledge of the RAS to the research at Takeda (Morrison et al. 1977). In addition, an experimental method used in the university project to understand the relationship between thromboxane A2 production and the development of hypertension was launched at Takeda immediately after Nishikawa's return to Japan and helped to discover the AII antagonism of CV-2198. Other information on losartan from academic researchers in Japan led to the early resumption of ARB research by Takeda, as described above.

Although the examples given above are examples of direct knowledge transfer, knowledge acquired from published papers and books was also useful for R&D. The active metabolites of candesartan and losartan were known to selectively bind to the AT1 receptor and antagonize the action of AII. The subtyping of angiotensin receptors greatly contributed to the investigation of how candesartan suppresses pharmacological action, specifically vasoconstriction by AII.

12.12 Impact of Candesartan

Candesartan was sold through a licensing agreement with AstraZeneca, a co-developer, in addition to Takeda. Since its launch in Japan in 1999, candesartan has been used in many countries around the world and is currently sold in 50 countries worldwide. Figure 12.3 shows the sales of candesartan from 2000 to 2012. The sales declined significantly in 2012, which is presumed to be a result of the expiration of patents in Europe, as well as the decline in demand for single-agent drugs and the increased sale of combination drugs with diuretics.

The benefits of ARBs to hypertensive patients were significant because they avoided side effect problems found in previous drug classes, such as nervous system depression and dry cough, and were effective at low doses. Nevertheless, the major contribution of ARBs, including candesartan, may be its novel mechanism of action, which generated a new option of treatment for hypertension. Indeed, ARBs continue to be mainstays in combination therapy for the treatment of hypertension.

The current guidelines for antihypertensive treatment are to administer three drugs: an ARB, a Ca blocker, and a diuretic, in the most appropriate combination for each patient. Alternatively, ACE inhibitors or sympatholytics (β-blockers,

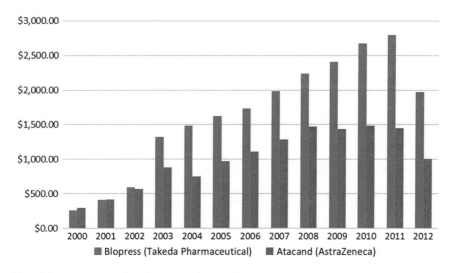

Fig. 12.3 Annual sales of candesartan (unit: 1 M USD). *Source* Pharma Future (Cegedim Strategic Data)

α-blockers) may be used. Because each of these medications has different characteristics, it is possible to provide ideal hypertension treatment to a variety of patients depending on the combination.

This basic regimen of hypertension treatment was only possible with the advent of ARBs. In the hypertension treatment guideline revised in 2009, the range of patients for whom ARB therapy is applicable is wider than those of the other hypotensive drugs, and its administration for left ventricular hypertrophy, heart failure, auricular fibrillation (prevention), post-myocardial infarction, proteinuria, renal failure, late-stage cerebrovascular disorders, diabetes mellitus, metabolic syndrome, and the elderly is possible. Based on broad indications, it is presumed that the addition of ARBs to the lineup of antihypertensive drugs has greatly expanded the scope of combination therapy. In this sense, ARBs, such as candesartan, have made important contributions to the treatment of hypertension.

12.13   Global Competition for ARB R&D

Since the publication of the discovery of losartan by DuPont, there has been competition to develop ARBs involving major drug companies around the world. Between 1989 and 1995, more than 600 patent applications were filed by 60 companies, and tens of thousands of compounds were synthesized (Naka 2001). Except for successors and generics, there are six ARBs that have been launched to date. Table 12.1 lists the product names, generic names, and the company origins of the six ARBs. In 2012, azilsartan (Azilva) was marketed by Takeda, but this drug was not included in the analysis of ARB competition here because the drug is regarded as a successor to candesartan.

At the time when CV-2198 was discovered in 1976, competitors other than Takeda did not conduct R&D on nonpeptidic ARBs. This is attributable to the global trends in antihypertensive drugs towards ACE inhibitors and calcium channel blockers at that time. Therefore, it is believed that there was no other competitor at the stage when Takeda was conducting basic research on ARBs.

Table 12.1 Marketed ARBs (excluding generic products)

Generic name	Product name	Origin
Candesartan	Blopress, Atacand	Takeda Pharmaceutical
Varsartan	Diovan	Ciba-Geigy (Novartis Pharma)
Losartan	Nu-lotan	DuPont (MSD)
Olemesartan	Olmetec	Sankyo (Daiichi-Sankyo)
Telmisartan	Micardis	Dr. Karl Thomae (Boehringer Ingelheim)
Irbesartan	Avapro	Sanofi Research (Sanofi-Aventis)

Source Compiled from each company's website and drug interview forms

In 1982, the US patent for CV-2973 was issued to Takeda, and DuPont researchers focused on the development of losartan, which improved upon Takeda's CV-2961. Once publicized, the global competition of R&D for nonpeptidic ARBs began. When DuPont published the losartan discovery in 1989, drug companies around the world focused their attention on ARBs, and about 50–60 companies entered the ARB development competition. Oki, who led the chemical synthesis group for candesartan, spoke about candesartan's basic patent application:

> Until our patent application became publicly available, we made a daily check of the patent bulletin with great anxiety over the possibility that another company applied for the patent on the same compound earlier. (Nikkei Business 1997, p. 62)

12.14   Over $20 Billion Markets Worldwide

Takeda promoted the clinical development of candesartan simultaneously in Japan, the USA, and Europe, and candesartan was launched first in the UK in 1997. Candesartan became the third ARB in the world after losartan and valsartan, and soon afterward, with irbesartan and telmisartan launched, they took the antihypertensive market by storm. Figure 12.4 shows the sales data of various ARBs for the years 2000 to 2012. Sales data are based on the generic name of the drug.

Of note, the ARB market grew dramatically, reaching over 20 billion USD in 2008. Sales of ARBs dropped in 2012, which was probably caused by generic drug launches following expiration of the patents for the first ARBs. Once the generic version of an ARB is launched, it is inevitable that even the still-patented ARBs will

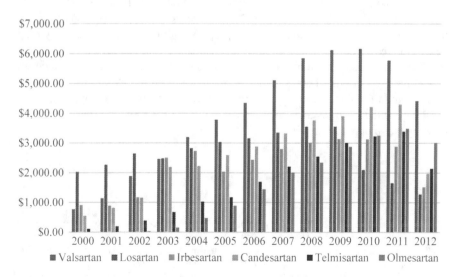

Fig. 12.4   Annual sales of ARBs (unit 1 M USD). *Source* Pharma Future (Cegedim Strategic Data)

be exposed to generic competition. The market size of ARBs on a sales basis will likely continue to decline because of the advent of generics.

12.15   Conclusion

Candesartan was created by Takeda based on the combination of basic research by Takeda, and the knowledge and information provided by academia and other companies. Takeda successfully invented CV-2198, which is the basic structure of nonpeptidic ARBs (Angiotensin-II Receptor Blockers), by starting exploratory research on diuretics from Alinamin, a star product at that time. Takeda made full use of SHRs as a pathological animal model provided by academia, and captured serendipitous events brought about by using a cutting-edge chemical synthesis method (gas-phase pyrolysis). After a lack of efficacy in clinical trials, the project was suspended until Takeda became aware of DuPont's discovery of losartan. Takeda responded immediately and synthesized candesartan in only a few months after resuming the research.

The R&D projects leading to the creation of candesartan were also innovative in that they resulted in an invention that became the common base for nonpeptidic ARBs, and had significant spillover effects on the global pharmaceutical industry, including the DuPont's invention of losartan. These achievements would have been impossible without the unique capabilities and the overall organizational strength that Takeda continued to accumulate, such as the use of SHRs, the utilization of gas-phase pyrolysis reactions, the ability to synthesize derivatives, and the ability to generate prodrugs.

It should be remembered, however, that Takeda, despite its foundational discovery, was delayed in the synthesis of the final compound, and candesartan was not a first-in-class drug. Certainly, it is difficult to make go/no-go decisions in R&D projects. In some cases, continuing R&D projects without a judicious strategy may lead to worsening performance. On the other hand, interruption of R&D projects can result in significant opportunity losses. The candesartan case is a good illustration of a company's conflict between short-term and long-term outcomes, and it may also illustrate the importance of flexible allocation of resources to cope with such conflicts and the importance of an organizational resource base for an early response to competitive opportunities.

References

Abe, K., & Tsunoda, K. (2002). Kidney domain 100 years: renal hypertension. *Journal of the Japanese Society of Internal Medicine, 91*(5), 1393–1400.

Arakawa, K. (2006). Looking back on renin-angiotensin system research in Japan (from ACE inhibitors to ARBs). Horiuchi, M. (Ed.), *THE ARB*. Medical Reviews, pp. 3–11.

Arakawa, K., Nakatani, M., Minohara, A., & Nakamura, M. (1967). Isolation and amino acid composition of human angiotensin I. *Biochemical Journal, 104,* 900–906.

Bhardwaj, G. (2006). How the antihypertensive losartan was discovered. *Expert Opinion on Drug Discovery, 1*(6), 609–18.

Bumpus, F. M., Schwarz, H., & Page, I. H. (1957). Synthesis and pharmacology of the octapeptide angiotonin. *Science, 125*(3253), 886–887.

Fukamizu, A. (2011). Chapter 5: The life sciences of mysteries-from accidental discoveries. *The Life Sciences-Image of Tsukuba Researchers*. http://www.science-academy.jp/rensai/pdf/rensai_5.pdf. Accessed in February 20, 2014.

Goldblatt, H., Lynch, J., Hanzal, R. F., & Summerville, W. W. (1934). Studies on experimental hypertension I. The production of persistent elevation of systolic blood pressure by means of renal ischemia. *The Journal of experimental medicine, 59*(3), 347–379.

Gross, F. (1958). Renin und Hypertensin, physiologische oder pathologische Wirkstoffe? *Klinische Wochenschrift, 36*(15), 693–706.

Imai, T., Miyazaki, H., Hirose, S., Hori, H., Hayashi, T., Kageyama, R., et al. (1983). Cloning and sequence analysis of cDNA for human renin precursor. *Proceedings of the National Academy of Sciences, 80*(24), 7405–7409.

Imura, Y. (2010). SHR and drug discovery research. *News Letter of the Joint Society for Models of Diseases, including SHR, 35*, 1–2.

Inada, Y., & Naka, T. (2000). Pharmacological properties and organ protective effects of a new generation of angiotensin-II receptor antagonists of candesartan cilexetil. *Journal of Science, Japanese Drug, 115*(3), 151–160.

Kagami, S. (2010). History of the Renin-Angiotensin System (RAS): From bench to bed: Why basic studies are important for clinicians. *Japanese Journal of Nephrology, 23*(2), 179–182.

Kambayashi, Y., Bardhan, S., Takahashi, K., Tsuzuki, S., Inui, H., Hamakubo, T., et al. (1993). Molecular cloning of a novel angiotensin II receptor isoform involved in phosphotyrosine phosphatase inhibition. *Journal of Biological Chemistry, 268*(33), 24543–24546.

Khairallah, P. A., Toth, A., & Bumpus, F. M. (1970). Analogs of angiotensin II. II. Mechanism of receptor interaction. *Journal of medicinal chemistry, 13*(2), 181–184.

Kokubu, T., Ueda, E., Fujimoto, S., Hiwada, K., Kato, A., Akutsu, H., et al. (1968). Peptide inhibitors of the renin–angiotensin system. *Nature, 217*, 456–457.

Kubo, K. (2006). Creation of the antihypertensive drug candesartan cilexetil. *Chem & Education, 54*(3), 138–141.

Kubo, K., Kohara, Y., Yoshimura, Y., Inada, Y., Shibouta, Y., Furukawa, Y., et al. (1993). Nonpeptide angiotensin II receptor antagonists: Synthesis and biological activity of potential prodrugs of benzimidazole-7-carboxylic acids. *Journal of Medicinal Chemistry, 36*(16), 2343–2349.

Kuno, Y., & Sato, K. (2011). *How is a drug discovery science introduction made?* Ohm.

Marks, L. S., & Maxwell, M. H. (1979). Tigerstedt and the discovery of renin: An historical note. *Hypertension, 1*(4), 384–388.

Morita, K. (2000). *New drugs are developed in this way: A development secret by the president of the researcher*. Nihon Keizai Shimbunsya.

Morrison, A. R., Nishikawa, K., & Needleman, P. (1977). Unmasking of thromboxane A2 synthesis by ureteral obstruction in the rabbit kidney. *Nature, 267*, 259–260.

Nagaoka, A. (1987). Application to drug development in spontaneously hypertensive rats (SHR). *Record of the Japanese Society of Model Animal Animals, 3*, 30–35.

Naka, T. (2001). Angiotensin II receptor antagonist: candesartan cilexetil (Blopress) (a drug developed in Japan). *Journal of Specialist Physicians of the Japanese Society of Cardiology, 9*(2), 313–318.

Naka, T., Kubo, K., & Furukawa, Y. (1995). Research and development of non-peptide angiotensin II receptor antagonists. *Journal of the Society of Synthetic Organic Chemistry, 53*(9), 802–810.

Nikkei Business. (1997). New drugs with fewer side effects that suppress the action of vasoconstrictors in the treatment of hypertension. Issue December 8, pp. 61–64.

Nishikawa, K. (2007). Development of ARBs [Basic Edition]—Finding lead compounds from research on diuretics. *Blood Pressure, 14*(5), 71–83.

Ogihara, T., & Kumahara, Y. (1982). Renin-angiotensin system inhibitor. *Medicina, 19*(11), 1998–1999.

Page, I. H. (1990). Hypertension research. A memoir 1920–1960. *Hypertension, 16*(2), 199–200.

Shimamoto, K. (2011). Obesity and hypertension. *Journal of the Japanese Society of Internal Medicine, 100*(4), 945–949.

Skeggs, L. T., Marsh, W. H., Kahn, J. R., & Shumway, N. P. (1954). The existence of two forms of hypertensin. *The Journal of Experimental Medicine, 99*(3), 275–282.

Skeggs, L. T., Kahn, J. R., & Shumway, N. P. (1956). The preparation and function of the hypertensin-converting enzyme. *The Journal of Experimental Medicine, 103*(3), 295–299.

Yanagisawa, H. (2006). History of renin-angiotensin system inhibitors angiotensin receptor antagonist. In Horiuchi, M. (Ed.), *THE ARB*. Medical Reviews, pp. 252–264.

Chapter 13
Tocilizumab (Actemra, RoActemra)

First Antibody Drug Developed in Japan

Yasushi Hara, Yoshiyuki Ohsugi and Sadao Nagaoka

Abstract Tocilizumab (Actemra) is a drug used to treat autoimmune diseases, including rheumatoid arthritis, through a novel mechanism of action discovered in Japan. It is also the first antibody drug developed in Japan. Chugai Pharmaceuticals started the basic research on autoimmune disease early, when neither a target molecule nor a disease mechanism was known. The research and development program was long and uncertain, taking 24 years to acquire the approval as a rheumatoid arthritis drug. Industry and university collaborations played critical roles in this innovation. Collaborations with the Kishimoto laboratory of Osaka University, which discovered IL-6 as well as its applicability to the autoimmune disease, were essential. Osaka University also contributed to the early implementation of a clinical trial. The discovery by Kawano at Hiroshima University of IL-6's role in multiple myeloma was also important for keeping the project alive. Collaboration with the Medical Research Council of the UK was instrumental for applying humanization technology to the antibody development as a drug. Tocilizumab in turn significantly contributed to the scientific understanding of the role of IL-6 for autoimmune diseases. It was launched significantly later than infliximab (Remicade), even though they were created at roughly the same time. This delay was largely caused by concern that drug price regulation in Japan might not allow a sufficiently high price for a new drug for rheumatoid arthritis for which low price conventional drugs were available.

Y. Hara (✉)
CEAFJP/EHESS, Paris, France
e-mail: yasushi.hara@r.hit-u.ac.jp

Faculty of Economics, Hitotsubashi University, Tokyo, Japan

Y. Ohsugi
Ohsugi BioPharma Consulting Co., Ltd, Tokyo, Japan
e-mail: ohsugi@theia.ocn.ne.jp

S. Nagaoka
Tokyo Keizai University, Tokyo, Japan
e-mail: sadao.nagaoka@nifty.com

© Springer Nature Singapore Pte Ltd. 2019
S. Nagaoka (ed.), *Drug Discovery in Japan*,
https://doi.org/10.1007/978-981-13-8906-1_13

13.1   Introduction

Rheumatoid arthritis is a common disease, affecting 72 out of every 10,000 adults aged over 18 years in the United States, and 60–100 out of every 10,000 in Japan. Rheumatoid arthritis is also known to have a high proportion of female patients, with a male to female ratio of approximately 1:4. For example, the incidence of rheumatoid arthritis in women in France is 133 out of every 10,000 aged over 40 years.

Tocilizumab (Actemra) is a drug used to treat autoimmune diseases, including rheumatoid arthritis. When introduced, it featured a new mechanism of action, namely blockade of the interleukin-6 (IL-6) receptor. It has high treatment efficacy for treatments of rheumatoid arthritis, systemic juvenile idiopathic arthritis, and Castleman's disease, which could not be provided by conventional drugs (nonbiologics). It was the first antibody drug developed in Japan and was marketed in over 130 countries in the world as of 2014. It is marketed by Chugai Pharmaceutical Co., Ltd. (Chugai) in Japan, Korea. and Taiwan, and by F. Hoffmann-La Roche, Ltd. (Roche) in other countries, including Europe, the United States, and China. Tocilizumab has the following three major characteristics: first it was the world's first IL-6 inhibitor; it was developed in Japan, and then expanded worldwide. Second, it is a therapeutic agent that fundamentally improves abnormal immune response, as opposed to other conventional treatments for rheumatoid arthritis. Third, joint research with universities and research institutes in Japan, including Osaka University, played an important role in the research and development (R&D) process, and because of that, both initial clinical trials as well as the first market launch took place in Japan.

13.2   Timeline of the R&D Process of Tocilizumab

Dr. Yoshiyuki Ohsugi from Chugai Pharmaceutical played a central role in the R&D of tocilizumab. The idea for the project came from the experimental studies he conducted in the late 1970s in the laboratory of Prof. M. Eric Gershwin at the University of California, Davis. There, he noticed that abnormal activation of B cells played an important role in the development of autoimmunity, independent of T cell involvement. Thus, he hypothesized that a drug that inhibits the activation of B cells could become a fundamental therapeutic agent in the treatment of autoimmune disease. After returning from his research stay in the United States, Ohsugi conducted a joint study with the University of Tokyo in an effort to elucidate the causes of B cell activation. This study was carried out until 1985, when it ended prematurely. After this study, Prof. Tadamitsu Kishimoto of Osaka University and his colleagues reported that IL-6, which induces B cell differentiation, may cause autoimmune diseases. With these findings, Chugai and Osaka University initiated a joint research agreement for a drug discovery program targeting IL-6. Through the course of this research, the group found it difficult to inhibit IL-6 with low molecular weight compounds, so they decided to develop a monoclonal antibody that inhibited IL-6. In 1990, they started a joint research program with the Medical Research Council (MRC) of the UK, which had developed antibody engineering technology (CDR-grafting technology), to humanize a mouse monoclonal antibody. Nine months later, they succeeded in producing a humanized antibody, which was able to inhibit IL-6 function at doses equivalent to the original mouse antibody. In vitro and in vivo animal tests utilizing these humanized antibodies were started around 1990 and completed around 1996. Thereafter, clinical studies were conducted in Japan, Europe, and the United States beginning in 1997. In Japan, the drug was approved as a treatment for Castleman's disease in 2005, and in 2008 as a treatment for rheumatoid arthritis. The drug was approved by the European Medical Agency in 2009, and the US Food and Drug Administration (FDA) in 2010.

The history of R&D from the start of discovery research to the launch is summarized as follows:

1984: Chugai Pharmaceutical began a drug discovery project for B cell inhibitors.

1985: Chugai published an article in collaboration with the University of Tokyo that identified the B cell differentiation inducer activity in the lymph nodes of mice with autoimmune diseases.

1986: Osaka University published a paper on gene cloning for IL-6. Joint research between Chugai Pharmaceutical and Osaka University began.

1988: Osaka University published a paper on the gene cloning of the IL-6 receptor, and created mouse anti-IL-6 receptor antibodies.

1989: Osaka University published a paper on gp130, and established IL-6 transgenic mice. Results were published showing that Castleman's disease is a disorder of IL-6 hyper-production.

1990: A humanized antibody was produced by collaboration between MRC and Chugai. Chugai decided to develop it as a drug for the treatment of multiple myeloma and named the drug MRA (Myeloma Receptor Antibody).

1991: Chugai initiated a preclinical study of MRA.

1992: Chugai produced a mouse version of tocilizumab, MR-16, which was a rat monoclonal antibody against the mouse IL-6 receptor, and presented the findings at a conference.

1993: Chugai published an article on humanized anti-IL-6 receptor antibody. The research team evaluated its effect in vitro using multiple myeloma cells from patients at the Kyoto Prefectural University of Medicine.

1995: IL-6 knockout mice were established.

1996: Stimulated by the introduction of infliximab (Remicade) by Tanabe Seiyaku (now Mitsubishi Tanabe Pharma Corp.), tocilizumab's target diseases expanded to cover autoimmune diseases, such as rheumatoid arthritis.

1997: Chugai began a Phase I study in healthy volunteers in Japan.

1998: Chugai started a UK Phase I study in rheumatoid arthritis patients. Chugai announced that IL-6 played an important role in the development of murine collagen-induced arthritis, and that an anti-IL-6 receptor antibody was effective for the treatment of rheumatoid arthritis.

2001: Phase II clinical trials for rheumatoid arthritis began in Japan and the UK.

2002: Roche acquired half of the stocks of Chugai.

2003: Chugai began a Phase III clinical trial in Japan.

2005: Chugai received approval in Japan for Castleman's disease. Phase III clinical trials for rheumatoid arthritis began in Europe and the United States.

2008: Approval was obtained in Japan for rheumatoid arthritis.

2010: FDA approved tocilizumab for rheumatoid arthritis.

13.3   The Mechanism of Action: Inhibition of IL-6 Signaling

In rheumatoid arthritis, synovial cells in the joints begin to proliferate for unknown reason. This is accompanied by infiltration of inflammatory cells such as macrophages and lymphocytes, and by the formation of new blood vessels. The supply of oxygen and nutrients by neovascularization in turn amplifies the infiltration of inflammatory cells, which stimulates the secretion of proteolytic enzymes by the infiltrating cells, destroying cartilage tissue in the process. When the disease state progresses, such destruction progresses to the bone tissue, causing joint stiffness and decreased joint mobility. The process by which rheumatoid arthritis arises and the progress of the disease has been investigated, but the fundamental etiology has not yet been identified.

Tocilizumab is a humanized antibody that targets a receptor for the cytokine IL-6, and blocks the action of IL-6 by blocking the binding of IL-6 to its receptor. IL-6 has been shown to exert a variety of effects and acts on many cell types, including hematopoietic stem cells, megakaryocytes (platelet precursors), mesangial cells in the kidney, and liver cells. The effects of IL-6 on T cell activation, angiogenesis, and osteoclast activation are closely related to rheumatoid arthritis (Fig. 13.1).

After binding to its cognate receptor on the cell membrane, the IL-6–IL-6 receptor complex associates with a second receptor, glycoprotein 130 (gp130). The two heterotrimers further associate with each other to form a hexamer. As a result, two molecules of gp130 fuse together in close proximity to each other in the cytoplasmic domain, activating a phosphokinase that triggers an intracellular signaling cascade, ultimately transmitting the IL-6 signal to the nucleus.

IL-6 receptor is also found in fluid compartments, such as serum and joint fluid. In this case, soluble IL-6 receptors, which are referred to as soluble receptors to differentiate them from membrane-bound receptors, are present and form complexes with IL-6. This complex binds to gp130 on cell surfaces to initiate IL-6 signaling. Thus, IL-6 can exert its effects through gp130 even on cells that do not express the IL-6 receptor. This is called "trans-signaling."

The mechanism of action of tocilizumab is to bind to the IL-6 receptor and suppress the diverse biological activities of IL-6, thereby suppressing excessive immune responses and ameliorating the symptoms of autoimmune diseases (Fig. 13.2).

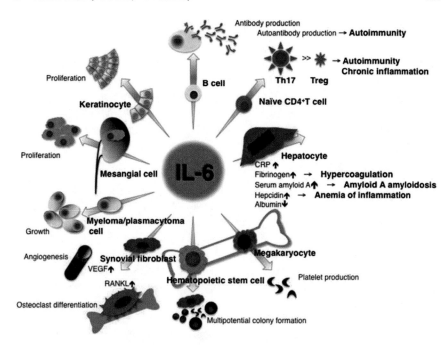

Fig. 13.1 Diverse effects of IL-6. *Source* Tanaka and Kishimoto (2012)

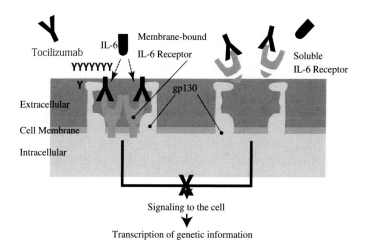

Fig. 13.2 Inhibition of IL-6 signaling by tocilizumab. *Source* Shimizu (2008)

13.4   Scientific Basis for Exploratory Research

Dr. Yoshiyuki Ohsugi, who played a central role in the tocilizumab R&D process, began his research at the University of California, Davis, in 1978 and started a collaborative research project with Prof. M. Eric Gershwin on immune cells. Professor Gershwin was a rheumatologist who had previous experience in immunologic pathology of systemic lupus erythematosus, and proposed the suppressor T cell function degeneracy theory.

Professor Gershwin instructed Ohsugi to investigate the function of B cells in New Zealand mice using B cell colony formation assay, rather than engaging in T cell research. The colony-forming technique had been published in 1975 by Donald Metcalf from the Walter and Eliza Hall Institute of Medical Research. Gershwin and Ohsugi applied this methodology to form B cell colonies by adding lipopolysaccharide (LPS). By using this technique, it became possible to analyze B cell activity in immune tissue independent of T cells. The results of the study, which were published in the *Journal of Immunology*, suggested that B cell activation is responsible for the development of autoimmune diseases (Ohsugi and Gershwin 1979; Gershwin et al. 1980; Ohsugi et al. 1982).

In 1983, during the development of a second-generation Carfenil, an immunomodulatory antirheumatic drug developed by Chugai that was launched in 1986, Chugai launched a project team to investigate and develop a superior antirheumatic drug. By 1984, a search for an inhibitor targeting B cells had officially begun (Takahashi and Ohsugi 2014), with the objective of developing a polyclonal B cell activation modulator (PBM). At the time, the team consisted of only two members: Ohsugi, and a researcher that was new to the company.

After the launch of the PBM project, the study began by investigating the causes of B cell abnormalities using MRL/lpr mice, which are systemic lupus erythematosus (SLE)-prone mice. A joint research effort with the University of Tokyo was carried out to elucidate the mechanisms of B cell activation. Lymph node cells from MRL/lpr mice were homogenized to give a soluble fraction. After several purification steps, a fraction that demonstrated B cell stimulating activity and induced autoantibody production was found, but this remained crudely purified. Unfortunately, the research team was unable to identify the molecular structure.

Subsequently, Prof. Tadamitsu Kishimoto and his colleagues at Osaka University discovered IL-6, and suggested that IL-6 was responsible for various symptoms of autoimmune diseases. This opened new avenues for discovery research into B cell inhibitors.

Box 13.1 The discovery of IL-6

Immunology research became active in Japan after Prof. Kimishige Ishizaka of John Hopkins University discovered IgE as a causative agent of allergy in 1970. At the time, Prof. Tadamitsu Kishimoto of Osaka University, who would later be involved in the tocilizumab joint research project, was studying in Prof. Ishizaka's laboratory as a postdoctoral fellow. Professor Kishimoto was engaged in research to elucidate the mechanism of antibody production using rabbits. Afterward, Kishimoto returned to Osaka University (Third Department of Internal Medicine) in 1975. Professor Kazuyuki Yoshizaki, who had been studying human lymphocytes in the same department of internal medicine, was recommended by Kishimoto to analyze the mechanism of antibody production by human B cells. This suggestion aimed at elevating the research on animal immunology to research on human immunology.

Professor Yoshizaki was given the task to analyze chronic lymphocytic leukemia (CLL) cells, which are B cell leukemia cells rather than a cell line, with the intent of constructing a simple system with a single physiological characteristic. The cancer cells were used as they are normally monoclonal and express a uniform antibody-idiotype on the cell surface. If anti-idiotype antibodies were produced and stimulated, all of CLL would be activated.

Subsequently, Yoshizaki began to purify and separate the active fraction from the supernatant of the T cell culture. In 1982, Yoshizaki published a paper on the mechanism by which B cells proliferate and differentiate into antibody-producing cells by adding T cell-derived humoral factor to B cells stimulated by anti-Ig antibody (Yoshizaki et al. 1982). In this study, it was discovered that the factor that proliferates B cells (BCGF) and the factor that promotes the differentiation of B cells and their conversion to antibody-producing cells (BCDF) are different molecules. Even if B cells divided and proliferated following stimulation with only the proliferating factor, they did not become antibody-producing cells; antibodies were produced only when growth factors were added and the cells were stimulated with a differentiation-inducing factor. Moreover, the addition of a differentiation-inducing factor without the action of a mitogen did not produce antibodies. The discovery of this differentiation-inducing factor was epoch-making, because at the time it was thought that once the B cell divides, it differentiates into antibody-producing cells. The reported article has been cited a total of 101 times and was referred to 38 times (about 30% of the total) in the 3 years immediately after the publication year, based on the data of Web of Science by Clarivate Analytics, Inc.

Professor Toshio Hirano, who had just entered the Osaka University at the same time, discovered the B cell differentiation-inducing factors independently during his secondment at Osaka Prefectural Habikino Hospital. The supervisor in the hospital informed Hirano that a large number of T-lymphocytes were contained in the pleural effusion of tuberculous pleurisy patients. Moreover, when these cells were stimulated with *Mycobacterium tuberculosis* components, strong antibody production activity was observed in the culture supernatant (Hirano et al. 1981). Hirano then moved to the laboratory of Prof. Kaoru

Onoue of Kumamoto University and spent 4 years there purifying and analyzing this soluble factor. In 1982, Hirano successfully isolated the TRF-like agent/BCDF (Teranishi et al. 1982).

In April 1982, Prof. Tetsuya Taga joined Kishimoto's laboratory as a graduate student with an interest in the mechanism of B cell differentiation. Professor Taga started studying the mechanism of action of BCDF on the antibody production of cell lines CESS and SKW6-CL4 in the Second Laboratory of the Third Department of Internal Medicine, Osaka University Hospital. A fraction with a molecular weight of approximately 20,000 based on the report of Muraguchi and Kishimoto was used as BCDF. Instead of the conventional method, Taga switched to the new method of measuring IgG produced by CESS cells, with use of the Enzyme-linked Immunosorbent Assay (ELISA). This method increased the accuracy of the activity measuring method of the BCDF, and the refining method narrowed the fraction that contained the BCDF (Muraguchi et al. 1981).

In 1984, Prof. Hirano joined the Kishimoto laboratory amidst the fierce competition to purify the B cell differentiation-inducing factor. In December of the same year, with the help of Prof. Nobuhiro Kurosawa at the Institute of Protein at Osaka University, Hirano sequenced the N-terminal partial peptide. But Hirano, in collaboration with their colleagues, decided to refine them from 100-liter cultures again because of fear that the amino acid sequence might have been incorrect. Finally, IL-6 was successfully cloned on 25 May 1986, and appeared in the 6 November 1986 issue of *Nature* journal (Hirano et al. 1986). This was the same timing of the first successful report of IL-5 gene cloning by Prof. Kiyoshi Takatsu of Kumamoto University and Prof. Tasuku Honjo of Kyoto University, and the discovery of the same molecule was reported twice in the same year under different names in different journals. It was clear that the research to discover IL-6 was extremely competitive. It is noteworthy that Japanese researchers were leading the international competition in the discovery of cytokines.

Note: This topic box is based on interviews with Kazuyuki Yoshizaki, Toshio Hirano, and Tetsuya Taga, and from the book chapter entitled "Interleukin-6" in "Cytokine Hunting-Leading Japanese Researchers." (Hirano et al. 2010). The titles of individuals in this topic box are current title.

13.5 Joint Research with Osaka University

In 1986, Prof. Kishimoto and Prof. Hirano of Osaka University presented their research at an academic conference that suggested IL-6 was responsible for autoimmune disease. After hearing this presentation, Ohsugi proposed a research collaboration with Kishimoto, and, in the same year, a drug discovery research program targeting IL-6 was launched as a collaboration between Chugai Pharmaceutical and Osaka University. Prior to starting the collaborative research with the university, the Chugai research team had independently investigated B cell inhibitors. This is considered to be one of the main reasons that the presentations by Osaka University led to such industry–academia collaboration. At that time, no pharmaceutical companies showed any interest in collaboration with Osaka University, except for Chugai (Nakajima and Kishimoto 2009).

Shortly after the start of the joint research between Osaka University and Chugai Pharmaceutical, Tosoh, Inc. joined the joint research program. Gene cloning of IL-6 was carried out mainly by Hirano of Osaka University, but a number of young researchers contributed to the project, including those sent not only by Chugai but also from Tosoh. Professor Kishimoto felt that Tosoh's participation was necessary for the progress of the project.[1] The proposed three-way joint research project would have been welcomed by Chugai, because the risk was shared among the stakeholders. Tosoh's participation began around 1987, and it continued until 1990. They co-authored a paper published shortly thereafter.

The main points of the collaborative research agreement between Prof. Kishimoto's laboratory and Chugai were: (1) the company would fund some of Prof. Kishimoto's research expenses; (2) any substance patents related to the IL-6 receptor, and any patents as the outcome of future collaborative research results would be licensed to the company for commercialization; (3) Chugai had an obligation to pay a part of its profit as patent royalty to Prof. Kishimoto, and (4) Chugai would need to obtain Kishimoto's consent to use any results obtained from this collaborative effort with any third party (Sumikura 2013).

In 1988, the Kishimoto laboratory shared the DNA sequences of the receptor protein with Chugai and Tosoh. The laboratory had been able to produce soluble IL-6 receptors of various lengths and had tested their ability to bind IL-6. While the results showed that the soluble receptors bound IL-6, these receptors were unable to inhibit the action of IL-6. They would later discover that the soluble receptors act as agonists (Fig. 13.2). Based on these results, the joint research team abandoned the development of a soluble IL-6 receptor.

In the meantime, the search for small molecule compounds that could inhibit the action of IL-6 was investigated, but no promising lead compounds were found, and the search for these compounds was abandoned. The R&D strategy was then shifted toward the development of antibody drugs. This strategy change was found to be wise years later after elucidation of hexameric crystallographic structures of IL-6, IL-6 receptor, and gp130. Small molecule drugs act by blocking certain receptor

---

[1] From interview with Kiyoshi Hogawa in 2013.

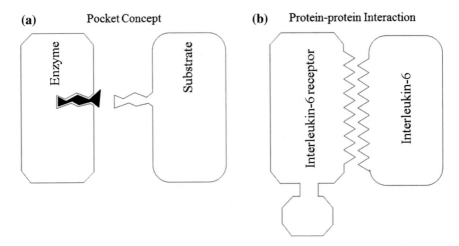

Fig. 13.3 Binding models. A low molecular weight compound can inhibit the interaction in (**a**), but not a protein–protein interaction such as that between IL-6 and the IL-6 receptor. *Source* Ohsugi (2013)

binding sites, or "keyholes" (Fig. 13.3a). However, to inhibit the action of IL-6, the binding area must be several times larger than a small molecule because IL-6 and the IL-6 receptor bind in a planar fashion over a wide area through a protein–protein interaction (Fig. 13.3b).

13.6 Humanized Antibodies

In 1985, Dr. Greg Winter and colleagues from MRC, developed and filed patents on a method of CDR grafting in which only the CDRs (complementarity-determining regions) of murine antibodies are transplanted onto human antibody genes; the patent was published in Japan at the end of 1987 (Japanese Patent No: 2912618). The presence of this patent led to collaborative research with MRC. In addition, Chugai favored the following conditions: (1) expenses for contract research were relatively low, and (2) MRC accepted visiting researchers from the company. The collaborative research with MRC started in 1990, and two researchers were dispatched to the UK from Chugai. These visiting researchers learned monoclonal antibody technology and succeeded in producing the humanized antibody in their first 9 months by improving upon the existing technology. This humanized antibody was the first antibody to be produced in the world that retained 100% of the activity of the mouse antibody.

13.7 Efficacy and Safety of Tocilizumab Were Confirmed

Using the newly synthesized humanized antibody, preclinical animal studies were conducted in mice and monkeys. Dr. Sachiko Suematsu of the Kishimoto laboratory succeeded in the preparation of mice transfected with the human IL-6 gene in 1989 (Suematsu et al. 1989). Technology was also developed that permitted researchers to genetically engineer mice. As such, a mouse line was developed in which the IL-6 gene was artificially deleted, knocked out, by Prof. Manfred Kopf at the Max Planck Institute in Germany; the study was published in 1994 (Kopf et al. 1994). These animal models were able to predict the role of IL-6 in vivo and the side effects of tocilizumab.

To conduct simulations in mouse models, the Chugai research team produced a rat monoclonal antibody, MR16-1, against the mouse IL-6 receptor, and presented the results at the annual meetings of the Japanese Society for Oncology and the Japanese Society for Immunology. Experiments with mice revealed that: (1) overproduction of IL-6 induces various lesions in mice, (2) MR16-1 can inhibit the effects of IL-6, (3) MR16-1 can suppress collagen-induced arthritis in mice, and (4) autoantibody production in New Zealand mice can be stopped, which strongly suppresses the development of SLE nephritis.

Subsequently, experiments with monkeys were conducted. Ikuo Imazeki and Hirofumi Shinkura, members of the Chugai research team, found that tocilizumab bound to some T cells in their experiments with lymphocytes in the peripheral blood of monkeys. Because tocilizumab was found to cross-react with monkey IL-6 receptors, it was possible to investigate the effects of tocilizumab in monkeys. After repeated injections of IL-6 into monkeys, C-reactive protein (CRP) levels and platelet counts increased in the blood. When tocilizumab was injected at the same time as IL-6, it was confirmed that the effect of IL-6 was almost completely blocked. In addition, it was confirmed that the use of tocilizumab suppressed the onset of monkey collagen-induced arthritis, and inhibited joint destruction.

13.8 Mass Production of Tocilizumab

To establish a reliable mass production method for tocilizumab, the preparation of a master cell bank (MCB), which is a seed cell, was required. In the process of making MCBs, a gene called dihydroxy folate reductase (DHFR) was conjugated to the upstream part of the gene for tocilizumab. The conjugated gene was introduced into Chinese hamster ovary (CHO) cells and cultivated in cell culture medium supplemented with methotrexate, a DHFR inhibitor. Cells that cannot produce DHFR do not proliferate. However, cells synthesizing DHFR, that is cells producing tocilizumab, grow more. Gradually, the concentration of methotrexate is increased, intensifying the stress. By repeating this process, cells that synthesize large amounts of tocilizumab can be selected. However, this selection process for antibody-producing cells poses a risk of gene sequence mutation, and it is not easy to establish a stable cell line. In addition, bovine spongiform encephalopathy created cell culture issues, because fetal bovine serum is a key ingredient of cell culture media. The use of peptides degraded by skipjack solved problems in culturing.

In the case of tocilizumab, it was 6 years between the successful creation of the humanized antibody and the start of clinical trials. One of the primary reasons for this delay was that it took researchers a long time to establish an optimal cell line and to produce clinical samples.

13.9   Tocilizumab for Multiple Myeloma

Because the production costs associated with an antibody drug were high, there was concern over Chugai's ability to set a sufficiently high price on tocilizumab to cover these costs, considering the restrictions set by the drug price system in Japan. Given that relatively cheap conventional drugs were already available for the treatment of rheumatoid arthritis, even though they were not very effective, the company would experience greater price restrictions if tocilizumab were developed for such disease. As a result, development of tocilizumab as a drug for autoimmune diseases was suspended.

Instead, preclinical tests for multiple myeloma began in 1991 on the basis that IL-6 was a growth factor for multiple myeloma, which is a life-threatening disease without an effective therapy. This was discovered by Prof. Michio Kawano of Hiroshima University's Department of Atomic Bomb Radiation Medicine (Kawano et al. 1988). Professor Kawano, who obtained an antibody, MH166, that bound to IL-6 and neutralized IL-6 activity from Prof. Hirano of Osaka University, discovered that the antibody suppressed the growth of myeloma cells collected from multiple myeloma patients. This study proved that myeloma cells secreted IL-6, which in turn stimulated myeloma cells to proliferate. Based on these results, the Chugai team proposed the development of a monoclonal antibody against the IL-6 receptor as a cancer drug based on a new mechanism of action, and named this antibody drug MRA.

To develop tocilizumab as a drug for multiple myeloma, it was necessary to determine the percentage of multiple myeloma patients wherein IL-6 plays a role as a growth or survival factor. This information was needed to estimate the number of patients for which an IL-6 inhibitor treatment was needed, which in turn could be used to estimate drug sales. In collaboration with Prof. Hideo Gotoh of Kyoto Prefectural College of Medicine, Prof. Chihiro Shimazaki investigated whether the external addition of IL-6 increased the growth of cancer cells ex vivo. Accordingly, the researchers found that IL-6 stimulated proliferation in approximately half of the patient samples (Shimazaki and Gotoh 1997). However, the extent of increase was not very large, at most two or three times, and was most often an increase of only a few percentage points. Because it was a disease with unmet medical needs, it was thought that this effect would be satisfactory if it contributed to treatment for at least

10% of patients. Thus, it was a reasonable experimental result. However, the clinical effectiveness remained unclear.

The potential of developing tocilizumab as a multiple myeloma drug was also examined in foreign countries. British oncologist, Prof. Gabriel Panayi, who is a key opinion leader in this field was so interested in tocilizumab that he came to Osaka to meet Prof. Kishimoto for a half-day meeting. Chugai was interested in the use of tocilizumab in patients with multiple myeloma in the UK under the compassionate use category. The purpose of this system is to provide potentially life-saving therapy to patients who have lethal disease and no other therapeutic options. In contrast, however, the research proposal from Prof. Panayi was to examine the pharmacokinetics of tocilizumab in monkeys. Their plan was to use tocilizumab in humans only after results were available from the monkey experiments, which were designed to determine whether tocilizumab could efficiently reach tumor tissues and whether it induced side effects at other tissue sites. This research plan was abandoned because Chugai needed to produce radiolabeled tocilizumab, and to prepare a large amount of research expenses.

Since the discovery of the IL-6 gene, Prof. Kishimoto was frequently invited to present at academic conferences or to deliver guest lectures across the globe. Researchers from the University of Arkansas in the United States, who had listened to a speech at an academic conference, said they wanted to conduct a clinical trial in patients with myeloma. In Europe, the study was initiated by a group of French researchers. However, in all of these studies, the development was discontinued because the reduction of tumor size expected by the clinician was not observed. Moreover, the amount of M protein in the blood was not halved, and the drug was considered ineffective (M protein is used as an indicator of the effect of a therapeutic agent for multiple myeloma). Considering the mechanism of action of tocilizumab, it is now possible to understand why the amount of M protein in the blood did not decrease.

Notably, tocilizumab inhibits the proliferation and differentiation of immature myeloma cells to mature myeloma cells, but because mature myeloma cells survive and continue to secrete M protein, the blood M protein level does not change within a short period of time. Thus, the drug was considered ineffective given that efficacy of the drug is judged by the reduction in M protein levels to half or less, and in most patients receiving tocilizumab the M protein concentrations were not decreased. Subsequently, there has been no further progress in the clinical development of tocilizumab for multiple myeloma, partly because the interest of clinical researchers switched to low molecular weight drugs with a different mechanism of action.

13.10   Return to Autoimmune Diseases as the Disease Target

Around 1996, the clinical development of tocilizumab returned to autoimmune diseases. The start of infliximab (Remicade) clinical trials in Japan is considered to have had a direct impact on this strategy change because infliximab was a competing antibody drug. Furthermore, the number of patients with rheumatoid arthritis was overwhelmingly larger than that of multiple myeloma, and the difference in the size of the social impact was also a large factor. These factors, combined with the poor outcome of clinical multiple myeloma studies, fostered a return to autoimmune diseases as the drug target.

However, the fact that tocilizumab was initially developed not as a therapeutic agent for autoimmune diseases, but as a therapeutic agent for multiple myeloma, had moved the attentions of Chugai and Osaka University researchers away from the development of cutting-edge research on autoimmune diseases. In fact, studies on the mechanism of action of infliximab and other tumor necrosis factor alpha (TNFα) inhibitors had been reported at academic conferences since 1993. However, these trends were only noticed by researchers at Chugai and Osaka University in 1995 when infliximab clinical studies began in Japan. The research team recognized the efficacy of infliximab in clinical research and changed the drug target of tocilizumab to autoimmune diseases, including rheumatoid arthritis, in response. In 1998, the remarkable clinical efficacy of tocilizumab for autoimmune diseases was reported (Yoshizaki et al. 1998).

In 1997, a Phase I study of tocilizumab began in Japan, which was completed in healthy volunteers. In Japan, only trials for cancer are allowed to skip Phase I. In the UK, which does not require a Phase I trial in healthy people, a Phase I/II clinical trial[2] among patients with rheumatoid arthritis began in 1998. One year later, a Phase II clinical trial in rheumatoid arthritis patients was initiated in Japan. Chugai asked Prof. Gabriel Panayi in Guy's, King's, and St. Thomas' School of Medicine, King's College, London, United Kingdom, to conduct a Phase I/II clinical trial of tocilizumab, and the results of the clinical trial were published in 2002 (Choy et al. 2002). Prof. Ravinder N. Maini was appointed as the principal investigator for Phase II clinical trials in Europe, which began in 2001.

The European Phase I trial took 2.5 years to complete because of the difficulty in recruiting patients. Notably, the relationship between IL-6 and rheumatoid arthritis was not well recognized among foreign investigators when compared with the Japanese immune community. Unlike Japan, where Osaka University took the initiative and conducted research in immunology, the rheumatoid arthritis community in Europe and the United States has focused on inhibitors targeting TNFα and was indifferent to the role of IL-6 in rheumatoid arthritis.

In Japan, a Phase II study in patients with rheumatoid arthritis was initiated in 2001, and a Phase III study followed in 2003. A Phase II European trial was launched

---

[2]Phase I clinical trials investigate the safety, distribution in the body, and excretion of drugs. Phase II clinical trials confirm safety and determine efficacy and dosage. A Phase I/II trial combines these tests.

in the same year. Afterward, tocilizumab was approved in Japan in 2005 for the treatment of Castleman's disease[3] and in 2008 for the treatment of rheumatoid arthritis. The FDA approved the drug for rheumatoid arthritis in 2010.

13.11 Drug Prices Determined by the Cost Calculation Method

In 2005, when tocilizumab was approved for Castleman's disease in Japan, it was judged that there was "no similar drug with similar indications and pharmacological effects" and a cost-based price determination system was adopted.[4] The total product cost of 372 USD (41,332 yen), operating income of 88.4 USD (9822 yen, 19.2% of the price), distribution costs of 52.9 USD (5874 yen; set to 10.3% of the price based on a survey by the Health Policy Bureau of the Ministry of Health, Labour and Welfare), and consumption tax of 25.7 USD (2851 yen) were summed to give the drug price ceiling of 538.9 USD (59,879 Yen; 77 USD per day) for a 200 mg/10 ml bottle.[5] The predicted patient number in the first year was estimated to be 80 patients, expanding to 120 patients in the second year.

In the drug price listing for rheumatoid arthritis in 2008, tocilizumab 200 mg/10 ml, which was listed in 2005, was referred to as a similar drug, and the National Health Insurance drug prices for the doses of 80 mg/4 ml and 400 mg/20 ml were formulated based on the inter-specification adjustment method.[6] The new calculation price for an 80 mg/4 ml bottle was 216.9 USD (24,101 yen) and a 400 mg/20 ml bottle was 1057 USD (117,459 yen); these were set using a 98.4103% standard adjustment between price and quantity.[7] The number of patients treated in the first year was estimated to be 3000 patients, with an estimated 38,000 patients at its peak.

Most drug prices in Japan are adjusted downward every 2 years, reflecting the actual transaction prices agreed between the parties under the constraints of ceiling by listed prices. The price of tocilizumab was maintained under the policy measure of promoting new drug discovery and applications for approval of new use of existing

---

[3]In lymphoproliferative disorders, IL-6 secreted by enlarged lymph nodes induces a variety of clinical manifestations. In many cases, the patient dies. It is a rare disease, and in Japan, there are approximately 200 patients. Actemra was approved as the world's first treatment for Castleman's disease.

[4]Reporting of Drug Pricing Organization. http://www.mhlw.go.jp/shingi/2005/05/dl/s0525-7a2.pdf (accessed 21 October 2014).

[5]Drug price formation of new drug: http://www.mhlw.go.jp/shingi/2005/05/dl/s0525-7a1.pdf http://www.mhlw.go.jp/shingi/2005/05/dl/s0525-7a2.pdf (accessed 21 October 2014).

[6]Rituxan 10 mg/mL, which was marketed as a drug for malignant lymphoma in 2001, and applied to rheumatoid arthritis treatment in foreign countries, was used for deriving the interspecification ratio, which was decided to be 98%. Drug price formation of new drug: http://www.mhlw.go.jp/shingi/2008/06/dl/s0604-5b.pdf (accessed 21 October 2014).

[7]Drug price formation of new drug: http://www.mhlw.go.jp/shingi/2008/06/dl/s0604-5b.pdf (accessed 21 October 2014).

drugs in 2010,[8] and it was reduced in 2012 after being designated as a drug with unexpected market expansion.[9] Interestingly, in the process of drug price setting, the prices of TNFα inhibitors such as infliximab, an antibody drug that has similar efficacy and was already available in Japan for rheumatoid arthritis, were not directly referred to when determining the initial price of tocilizumab.

13.12   Scientific Sources of Tocilizumab

The scientific sources for developing tocilizumab included: (1) the discovery by Prof. Gershwin and Dr. Ohsugi that autoimmune diseases were caused by B cell abnormalities, and the results of joint research with the University of Tokyo that suggested the presence of B cell activators in the lymph nodes of animal models of autoimmune diseases; (2) the findings by Prof. Kishimoto's laboratory at Osaka University, which discovered IL-6 and suggested that IL-6 was responsible for symptoms of autoimmune diseases; (3) the development of the technology needed to humanize the mouse antibody; (4) the discovery by Prof. Kawano that IL-6 is a growth factor for multiple myeloma; and (5) that the TNFα inhibitor infliximab improved the symptoms of rheumatoid arthritis. Notably, IL-6 had been shown to induce TNFα. The details of each are described below.

(1)   B cell activation was responsible for autoimmune disease.

The source of tocilizumab R&D goes back to the discovery that the pathogenesis of autoimmune diseases is the abnormal activation of B cells. This was Dr. Ohsugi's findings during his research stay at the University of California, Davis. After returning to Japan, Ohsugi, in collaboration with Prof. Katagiri at the University of Tokyo, found that B cell activating factors were secreted in the lymph nodes of mouse models of autoimmune diseases and determined that these factors were responsible for autoimmune diseases.

(2)   The discovery of IL-6 and its role in autoimmune disease.

Following the discovery of IL-6, the gene was cloned by Prof. Kishimoto and his laboratory members at Osaka University. In addition, the publication of clinical data suggesting that IL-6 was a causative factor in autoimmune diseases focused the discovery program for B cell inhibitors on IL-6 as a target molecule. Interestingly, this research was triggered by a patient with a benign tumor called an intra-atrial myxoma. The patient complained of an exothermic reaction, arthralgia, and fatigue, and received medical attention at Osaka Police Hospital. Blood tests revealed increases in the CRP and erythrocyte sedimentation rate (ESR), as well as the typical symptoms

---

[8]Results of promotion for new drug creation in FY 2010 Medical Price Revision. http://www.mhlw.go.jp/shingi/2010/06/dl/s0623-2c.pdf (accessed 23 October 2014).

[9]Market expansion recalculated item in H24.1.25. http://www.mhlw.go.jp/stf/shingi/2r98520000020zbe-att/2r98520000020zfi.pdf (accessed 22 October 2014).

of an immunoinflammatory disease, such as autoantibody production and hypergam-maglobulinemia. When the tumor was removed, all of the symptoms completely dis-appeared. IL-6 was suspected of involvement, and culture of the excised tumor cells indeed revealed that a large amount of IL-6 was secreted into the culture medium. These results suggested that IL-6 is deeply involved in the development of various immunoinflammatory diseases.

(3) The development of revolutionary genetic engineering technique, CDR grafting.

The discovery program for IL-6 inhibitors was unsuccessful for a long time, and the PBM project was in trouble. Antibody technology helped the research team out of the crisis. CDR grafting technology enabled the humanization of mouse antibodies, and the development of tocilizumab shifted to the development of humanized antibody drug. It had long been known that the specificity of an antibody when it binds to its corresponding antigen is determined by a region of the antibody molecule called the CDR. In 1985, Dr. Winter from MRC used genetic engineering to establish a revolutionary technique for transplanting CDRs. In the development of tocilizumab, murine CDRs were transplanted into human antibodies. Through use of this technol-ogy, side effects were drastically reduced by modifying the mouse antibody, which itself could not be clinically applied to humans. Thus, the application of an antibody drug became possible.

(4) IL-6 acts as a growth factor for multiple myeloma cells.

The discovery of the ability of IL-6 to act as a growth factor in multiple myeloma played an important role in sustaining the project by adding multiple myeloma to the potential target diseases of tocilizumab. Multiple myeloma is an intractable and lethal disease, so there was a possibility of establishing a higher drug price that would cover the cost of production. No excellent therapeutic drug exists for multiple myeloma, so it was considered to be an appropriate target disease for tocilizumab, an antibody drug with a completely new mechanism of action.

(5) TNFα inhibitor infliximab improved the symptoms of rheumatoid arthritis.

The start of the Japanese clinical trial of the antibody medicine infliximab, which targeted TNFα, played an important role in returning tocilizumab to the target disease of autoimmune diseases. The impact of TNFα-targeted antibody drugs on the advent of tocilizumab can be summarized as follows. (a) First, blocking TNFα was found to markedly improve the symptoms of rheumatoid arthritis. (b) This fact clarified that the antibody demonstrated the effectiveness for immune inflammatory diseases, such as rheumatoid arthritis, and the application of the antibody to chronic inflam-matory diseases could be carried out. (c) Finally, advances in basic research strongly suggested that TNFα and IL-6 are related to each other in the body's immune inflam-matory system for rheumatoid arthritis.

The scientific evidence for the efficacy of tocilizumab in rheumatoid arthritis is explained as follows. Stimulation of inflammatory cells with TNFα results in the pro-duction of large amounts of IL-6. In animal studies, injection of LPS, a pathogenic endotoxin, can induce septic shock, which can be suppressed by administration of anti-TNFα antibodies. However, similar suppression of shock could be observed with anti-IL-6 antibodies. The ultimate effector of septic shock is not TNFα but IL-6 induced by TNFα, as has been reported in the literature. Based on this scientific evidence, it is possible that some of the clinical effects of infliximab may be exerted through inhibition of IL-6 production. This suggested that tocilizumab, another antibody drug, was also effective for rheumatoid arthritis.

13.13 Tocilizumab Development Through Successful Academia–Industry Alliances

As a major feature of tocilizumab's R&D, academia–industry cooperation played an important role in the successful development of the drug. A summary of the collaborations throughout the R&D process is presented in Table 13.1.

The main members of the research teams from Chugai, Osaka University, and MRC are as follows. Most researchers from Chugai had secondment experience to either a university or a national research institute, indicating strong investment by Chugai on the absorptive capability of their researchers.

Table 13.1 Industry–academia collaborations throughout the tocilizumab R&D process

	Year	Collaboration partners	R&D theme	Joint research method
Basic research	1978–1981	University of California, Davis Prof. Gershwin	B cell research	Implementation of joint research by research stay of Dr. Ohsugi
	1984–1985	University of Tokyo Prof. Katagiri	Study of B cell differentiation factors	Main purpose was to provide experimental equipment for drug discovery. No official goal of the collaborative research
Applied research	1986–	Osaka University Prof. Kishimoto and his laboratory	Discovery of IL-6 inhibitors	Chugai researchers on secondment to the university. Regular meetings
	1990	Medical Research Council (UK)	Production of humanized antibodies	Chugai researchers on secondment to the laboratory. Developed royalty contracts

- Chugai Pharmaceutical

  - Dr. Yoshiyuki Ohsugi: A central role in tocilizumab R&D
    Research stay at the University of California, Davis, where he obtained core research ideas for developing tocilizumab. Consistently responsible for the R&D process from the early stages of development.
  - Hiroyasu Fukui: R&D member
    Researched anticancer drugs in the Immunology Laboratory at Tokai University, and was placed on secondment to the National Cancer Institute of National Institutes of Health (NIH) .
  - Yasuo Koishihara: R&D member
    Placed on secondment to Kumamoto University.
  - Yuichi Hirata: Cloned IL-6 receptor genes
    Placed on secondment to Prof. Kishimoto's laboratory.
    In charge of genetic cloning of G-CSF in collaboration with Tsuchiya at the Institute of Medical Sciences, University of Tokyo.
  - Masayuki Tsuchiya: Humanized anti-IL-6 receptor antibodies
    Placed on secondment to MRC.
    In charge of genetic cloning of G-CSF in collaboration with Hirata at the Institute of Medical Sciences, University of Tokyo.
  - Koh Sato: Humanized anti-IL-6 receptor antibodies
    Placed on secondment to MRC.

- Osaka University

  - Prof. Tadamitsu Kishimoto: Principal investigator of IL-6 study at Osaka University
  - Prof. Toshio Hirano: IL-6 study member
  - Prof. Kazuyuki Yoshizaki: IL-6 study member
  - Prof. Tetsuya Taga: IL-6 study member
    "Although the Kishimoto group was a group of basic researchers, a large number of clinicians from the department of internal medicine of Osaka University were enrolled. The group also supervised the clinician's immunological research, and information exchange with clinicians of related medical institutions was carried out continuously."

- MRC

  - Mary Margaret Bendig, Steven Tarran Jones, José William Saldanha
    Humanization of anti-IL-6 receptor antibodies (co-inventors).

Professor Gershwin (University of California, Davis) had made many achievements in the study of the immunologic pathology in SLE, and had proposed the theory of depressed suppressor T cell function. Professor Gershwin was also a rheumatologist. Throughout his 2.5 years of study in the Gershwin laboratory, Ohsugi obtained significant knowledge and experience on B cells.

Dr. Ohsugi, who played a central role in the R&D process of the antibody drug after returning to Japan, developed the ideas that led to tocilizumab from his basic research at the University of California, Davis. Ohsugi led the team until the later stages of R&D. Ohsugi also took the initiative to start joint research with Osaka University. One member of the research team, Dr. Fukui, had conducted research on anticancer drugs in the Immunology Laboratory of Tokai University, and had also been Placed on secondment to the National Cancer Institute of the National Institutes of Health. Similarly, Koishihara was placed on secondment to Kumamoto University, and Hirata, who was in charge of cloning the gene for G-CSF spent time at the Institute of Medical Sciences, University of Tokyo, and was then seconded to Prof. Kishimoto's laboratory where he worked to clone the IL-6 receptor gene. Tsuchiya, who was in charge of gene cloning of G-CSF in collaboration with Hirata at the Institute of Medical Sciences of the University of Tokyo, was seconded to MRC, where he and Sato were in charge of humanizing anti-IL-6 receptor antibodies. Thus most of research members of Chugai spent some time at universities and public research institutes to acquire specialist knowledge. These pre-tocilizumab R&D experiences were certainly important in implementing cutting-edge R&D, including academia–industry collaborations.

Chugai had developed biopharmaceuticals before the development of tocilizumab. These drugs included Epogin, a recombinant human erythropoietin preparation, which was approved in Japan in 1990, and Neutrogin, a recombinant human G-CSF preparation, which was approved in Japan in 1991.[10]

During collaborative research with Katagiri during the basic research stage in the 1980s, the concrete roles of each party were not described beforehand. Instead Chugai lent the devices required for research to the University of Tokyo, hoping for deeper understanding of B cell differentiation factors, which was aimed at creating opportunities for drug development. No contract was developed with regard to patent rights.

In collaboration with Prof. Kishimoto's laboratory at Osaka University, which began as a result of the discovery of IL-6 in 1986, Chugai placed researchers in the Kishimoto laboratory to acquire the skills necessary to find inhibitors. The Osaka University stood to benefit because they saved on research labor costs and Chugai benefited through the improved capability of its corporate scientists, as well as through the development of networking with scientists in the university.

Regular meetings between Osaka University and Chugai on the progress of the joint research were held once a month, and information was closely exchanged. The research results of Osaka University were handed over to Chugai, and Chugai took charge of the patent applications. The availability of cutting-edge information from Osaka University, one of the world's leading immunology institutions, was a great benefit for the company. In doing so, the corporate scientists of Chugai improved their scientific knowledge to understand the basic research carried out in Osaka University (Nagaoka and Akaike 2013). Thus, clear division of labor between industry

---

[10]Subsequently, Neutrogin was launched as Granocyte in Korea and Ireland in 1993. It was then launched in Germany, the UK, Australia, Thailand, and France in 1994; in Belgium and Luxembourg in 1995; and in Taiwan in 1997. From Drug Interview Form: Recombinant human G CSF Preparation Neutrogin Inj. 50 µg, Neutrogin Inj. 100 µg, Neutrogin Inj. 250 µg.

and academia, close both-way communications, and a high absorptive capability of Chugai for scientific advancement significantly contributed to the R&D process of tocilizumab.

> **Box 13.2 Contribution of drug discovery to science: the role of IL-6 revealed by tocilizumab.**
>
> During the tocilizumab R&D process, the role of IL-6 in the pathogenesis and progression of autoimmune diseases, especially rheumatoid arthritis, was clarified. First, MR16-1, a murine version of tocilizumab was created by Chugai scientists for preclinical studies, and this played a key role in elucidating the contribution of IL-6 to immunoinflammatory diseases. In doing so, IL-6 was found to be an essential factor in the development of many models of arthritis, including murine collagen arthritis. Similarly, IL-6 was found to be essential for the development of autoimmune diseases such as SLE and MS. Furthermore, through administration of anti-IL-6 receptor antibodies to mice, researchers increased their understanding of the molecule's function because only IL-6 was inhibited.
>
> Much scientific progress was also produced during the clinical studies. Previously, no one knew what role IL-6 played in the body's immune system in rheumatoid arthritis. However, tocilizumab was found to significantly improve joint swelling, joint pain, anorexia, general malaise, anemia, and fever, demonstrating the magnitude of the role played by IL-6. In addition, acute phase

Table 13.2 Clinical manifestations of rheumatoid arthritis associated with IL-6, elucidated by tocilizumab

IL-6 Effect	Clinical signs of rheumatoid arthritis
B cell activation	Production of rheumatoid factor, hypergammaglobulinemia, and cytokines
T cell activation	Induction of arthritis by inducing differentiation of Th17 cells
Induction of adhesion molecules	Increased lymphocytic infiltration into joint sites
Production of acute-phase proteins of the liver	Enhanced CRP production, organ deposition of amyloid
Inhibition of hepatic albumin synthesis	Hypoalbuminemia, increased ESR
Hepatic production of hepcidin	Anemia
VEGF expression in fibroblasts	Neovascularization and proliferation of articular synoviocytes
RANKL expression in fibroblasts	Activation of osteoclasts to promote bone resorption and joint destruction
Potentiation of leptin action	Anorexia
Pain threshold decreased	Pain

proteins such as CRP, SAA, and fibrinogen in the blood, and inflammatory markers such as ESR, as well as hypergammaglobulinemia and autoantibodies, were found to improve to normal levels, demonstrating that various symptoms observed in patients are induced by IL-6. Tocilizumab was also shown to improve blood levels of MMP-3 and VEGF to normal levels and to improve bone resorption and osteogenesis markers in patients.

The results of these clinical trials were reflected in the basic research, which showed that IL-6 induces the production of VEGF on synovial fibroblasts, and that IL-6 induces the expression of RANKL, a molecule essential for inducing osteoclasts. The mechanism of action of tocilizumab for the prevention of bone destruction was thus elucidated, contributing to many scientific developments. The clinical manifestations of rheumatoid arthritis associated with IL-6 elucidated by tocilizumab are summarized in Table 13.2.

For these contributions to science, pioneering efforts of both industry and academia were indispensable. Professor Yoshizaki recalls in his own book chapter as follows (Japanese Interferon-Cytokine Society 2010). A research team at Osaka University filed an application to the Ethics Committee for a license to administer tocilizumab to patients with intractable diseases of Castleman's disease and multiple myeloma. However, the Ethics Committee emphasized the existence of clinical trials overseas as a condition for approving the application. Facing such response, Prof. Kishimoto said that "As long as such emphasis on the existence of clinical trials abroad is placed on the implementation of domestic trials, it is impossible to develop new medicines in Japan. Therefore, the clinical research papers by Japanese scientists are not accepted by top clinical journals such as the *New England Journal of Medicine* or *Lancet*. As long as the research of a Japanese clinical researcher remains someone's follow-up, or the follow-up of the follow-up, it cannot hope to become the top of the world forever in the clinical research field." As a result, the applications for the treatments were finally approved.

Thus, the challenge of novelty and creativity has led to a virtuous circle between the creation of innovative drugs and creative research outcomes, respectively.

13.14   Tocilizumab Sales Exceeded 2 Billion USD

The sales of Actemra in Japan, Europe, the United States, and other countries are shown in Fig. 13.4. In Japan, the sales figure counts only sales for treatment of Castleman's disease from 2005 to 2007; thereafter, sales increased sharply as the drug was approved for rheumatoid arthritis in Japan in 2008. As a treatment for rheumatoid arthritis, sales began in Europe in 2009 and in the United States in 2010.

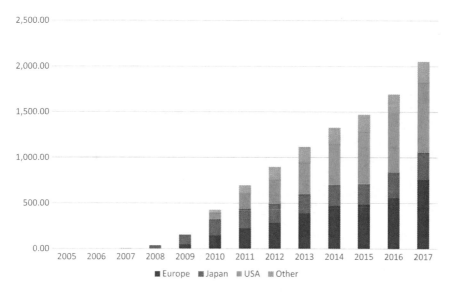

Fig. 13.4 Trends in Actemra sales (unit: 1 M USD). *Source* Roche and Chugai financial reports

In 2017, worldwide sales exceeded 2 billion USD. While sales in Europe and the United States steadily increased, sales in Japan were more stagnant at around 0.2 billion USD. This could be attributed to a 25% reduction in drug prices as a result of a market expansion recalculation in 2012.[11]

The clinical effects of tocilizumab include that it stops the progress of joint destruction, improves the functionality of joints, and improves the patient's quality of life unlike any conventional rheumatoid arthritis therapy that focuses on relieving the pain (Jones et al. 2010). Furthermore, not only are symptoms improved in the joints; improvements are also noted in systemic manifestations such as antipyresis, appetite, and fatigue. These effects are also major features of tocilizumab.

Figure 13.5 shows the results of Actemra's Phase III study, SATORI. Approximately half of the patients in this study achieved clinical remission as judged by the clinician within 6 months of treatment. The vertical axis in Fig. 13.5a is the disease activity score (DAS28),[12] and the horizontal axis shows the elapsed time (weeks) after tocilizumab or methotrexate (MTX) was administered once every 4 weeks. DAS28 levels remained largely unchanged in the control group (MTX) but significantly declined in the tocilizumab-treated group, indicating improved health conditions. Statistically significant differences relative to the control group were observed at all time points after week 4.

---

[11] Market expansion recalculated item, http://www.mhlw.go.jp/bunya/iryouhoken/iryouhoken15/dl/gaiyou_yakka_5.pdf (accessed 24 August 2014).

[12] Score based on the European League Against Rheumatism calculation and is used as a measure of disease activity. The sum of the number of swollen and painful joints, the erythrocyte sedimentation rate, and the general symptoms of a given 28 joints is given by multiplying each of them by a factor. If the DAS28 score value is less than 2.6, it is evaluated as remission.

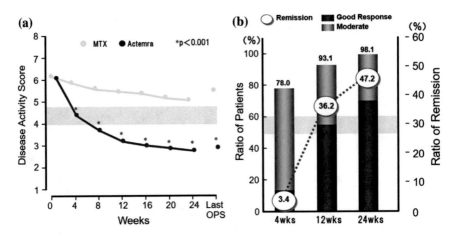

Fig. 13.5 Effect of tocilizumab on rheumatoid arthritis. **a** Trends in mean values in the respective groups with significant difference from methotrexate (MTX) control group ($p < 0.001$: paired t-test) (Nishimoto et al. 2009). **b** Level of response over 24 weeks (Ohsugi and Kishimoto 2011; modified)

In addition, the left vertical axis of Fig. 13.5b shows the percentage of patients achieving symptomatic improvements (complete and effective), and the right vertical axis shows the DAS28 response rate or ratio of remission. As the dose was repeated, the proportion of patients showing improvement increased, confirming that after 24 weeks, six doses, the proportion of patients achieving a "good response" and "moderate response" reached 98%. Numbers in the circle indicate the DAS28 remission rate; after 24 weeks, 47% achieved remission.

These results indicate that tocilizumab is effective in relieving symptoms in patients who do not respond adequately to methotrexate, a standard treatment for rheumatoid arthritis. Furthermore, in a comparative study with adalimumab (Humira) without concomitant methotrexate, higher therapeutic effects were reported (Gabay et al. 2013). Tocilizumab, unlike other rheumatic drugs, acts directly on IL-6, which plays a fundamental role in the immune response.

13.15 Competition Among Antibody Drugs

Table 13.3 lists the major rheumatoid arthritis antibody drugs currently marketed in Japan. Actemra (launched in 2005) is the only inhibitor targeting IL-6, while Remicade (launched in 1998), Enbrel (launched in 1998), Humira (launched in 2003), and Simponi (launched in 2009) are antibody drugs targeting TNFα. Of note, the discovery project for Actemra started the earliest in 1986.

Figure 13.6 shows trends in the global sales, including indications other than rheumatoid arthritis, for each drug. While Actemra has increased its share year by year, its sales are about one-tenth of the two leading antibody drugs (Humira and

Table 13.3 Antibody drugs used to treat rheumatoid arthritis

Product name	Drug type	Year when discovery project launched	Year of launch	Discovery firm	Sales/clinical developers
Remicade	Anti-TNFα	1988	1998	Centcor	Johnson & Johnson/Mitsubishi Tanabe Pharma
Enbrel	Anti-TNFα	1991	1998	Immunex	Pfizer, Takeda Chemical Industries
Actemra	IL-6 receptor inhibitor	1986	2005 2008[a]	Chugai	Chugai, Pharmaceutical Roche
Humira	Anti-TNFα	1992	2003	BASF (BBC)	Abbott Japan, Eisai
Orencia	T cell-selective co-stimulation modulator		2010	Bristol-Myers	Bristol-Myers
Simponi	Anti-TNFα		2009	Centcor	Janssen Pharma, Mitsui Tanabe Pharma

Source Medtrack

[a] Approved for rheumatoid arthritis

Remicade). Notably, Remicade and Enbrel, which were put on the market at a relatively early stage, have maintained a high market share. In addition, Humira's sales base share, which increased sharply from 6.7% in 2003 to 29.3% in 2012, overtook Remicade's sales share (27.6% in 2012). In 2017, Humira's sales share was 39.9% while the sales share of Remicade decreased (16.4%); shares of Actemra slightly increased, but remained at 4.4% in 2017.

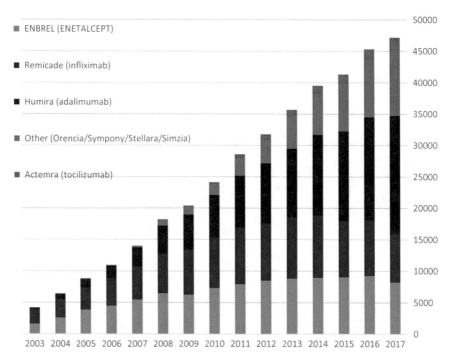

Fig. 13.6 Trends in rheumatoid arthritis drug sales from 2002 to 2017 (unit: 1 M USD) (Enbrel sales are the sum of Amgen, Pfizer (non-North America) and Takeda Chemical Industries (Japan). Sales of Remicade include sales and export in USA by Johnson & Johnson, sales in Europe by Merck, and sales in Japan by Mitsubishi Tanabe Pharma. Sales of Humula include sales of Abbott Laboratories and Eisai (Japan). Actemra sales include Roche and Chugai Pharmaceuticals sales. Orencia sales are based on Bristol-Myers Squibb, Stellara sales on Johnson & Johnson, and Cimzia sales on UCB. Simponi is the sum of sales by Johnson & Johnson, Merck, which has sales rights in Europe, and Mitsubishi Tanabe Pharma, which has sales rights in Japan.). *Sources* Pharma Future (Uto Brain Division, Cegedim Strategic Data) and Financial Reports. Sales data include sales for indications other than rheumatoid arthritis

13.16 Comparison with R&D Process of Remicade (Infliximab)

Infliximab is a chimeric anti-TNFα monoclonal antibody[13] that was developed by Centocor Inc. (Janssen Biotech, a subsidiary of Johnson & Johnson). Like tocilizumab, its main target disease is rheumatoid arthritis, but it is also used to

---

[13] Antibodies in which the constant region has been replaced with a human gene, leaving the variable region from the mouse.

treat a variety of other diseases, including Crohn's disease,[14] ulcerative colitis,[15] ankylosing arthritis,[16] psoriatic arthritis,[17] and Bechet's disease.[18] The main difference from tocilizumab is its combined use with MTX, which is indispensable for the administration to rheumatoid arthritis patients to prevent the appearance of human anti-chimeric antibody, because it is the chimeric monoclonal antibody.

The target molecule of infliximab is TNFα, one type of so-called inflammatory cytokine, which induces inflammation. It was discovered as a factor inducing hemorrhagic necrosis of tumors by killing vascular endothelial cells; thus, it is called tumor necrosis factor. Infliximab improves the symptoms of autoimmune diseases by neutralizing TNFα and injuring cells that produce TNFα. Similar to tocilizumab, infliximab was produced by industry–university cooperation.

Centcor was a start-up company established in the United States in 1979 with experience in reagent and drug development prior to the development of infliximab. In the early 1980s, Centcor was responsible for the production of monoclonal antibodies to be provided to major pharmaceutical companies and research institutes and provided reagents to over 80 organizations (Marks 2009). Then in the latter half of the 1980s, Centcor made humanized antibodies and accumulated cooperative relationships with many research institutes and universities.

Centcor then began its own drug development. Centoxin, a treatment for septic shock,[19] was approved in 1991. However, infringement of another company's patent was found, and the effectiveness of the drug itself came under a cloud and the approval was finally withdrawn (Marks 2012). In 1994, with support from Eli Lilly, Centcor gained approval for ReoPro, a cardiovascular treatment developed in collaboration with the State University of New York. Thus, prior to the development of infliximab, Centcor had accumulated experience in antibody drug production and academia–industry collaborations, including a failure experience. Failures in Centoxin led Centcor to set very high standards for later bio-tech drug developments.

A joint research agreement between New York University and Centcor, which later led to the development of infliximab, was signed in 1984. Initially, the main

---

[14] A type of inflammatory bowel disease in which the mucosa of the large and small intestines is chronically inflamed or ulcerated for no known reason. (http://www.nanbyou.or.jp/entry/81 accessed 9 October 2013).

[15] An inflammatory disease of the large intestine in which erosions or ulcers form in the lining of the large intestine. (Centers for Intractable Diseases: http://www.nanbyou.or.jp/entry/62 accessed 9 October 2013).

[16] Inflammation of the hip or shoulder joints. Inflammation called enthesitis occurs at the heel and on other parts of the tendon that attach to the bone. (http://www.rheuma-net.or.jp/rheuma/rm120/kouza/kyochoku.html accessed 9 October 2013).

[17] Psoriasis, a skin disorder, is a complication of swollen, painful arthritis. (http://www.rheuma.net.or.jp/ rheuma/rm120/kouza/kansen.html accessed 9 October 2013).

[18] Chronic recurrent systemic inflammatory disease with four main symptoms: aphthous ulcers of the oral mucosa, vulvar ulcers, cutaneous symptoms, and ocular symptoms. (http://www.nanbyou.or.jp/entry/187 accessed 9 October 2013).

[19] An inflammatory condition that results from a systemic bacterial infection, with a significant reduction in tissue streaming. (Merck Manual: http://merckmanual.jp/mmpej/sec06/ch068/ch068a.html accessed 9 October 2013).

objective was to develop an enzyme-linked antibody assay for human IFN-gamma (Vilcek 2009). New York University obtained TNF samples cloned from Genentech in 1985. New York University then produced a mouse anti-TNFα monoclonal antibody in 1988. In 1991, the clinical efficacy of mouse anti-TNFα antibody was investigated in patients with septic shock, and the results were reported. However, the results were disappointing. It was also proven that the human antibody for the mouse antibody appeared in the patient. From 1989 to 1991, Centcor created chimeric antibodies to attenuate the antigenicity of mouse antibodies and make them easier to use in humans. Infliximab was originally developed as a treatment for septic shock. It can be presumed that this was intended to be a successor drug for Centoxin, but there were no articles reporting the results of clinical trials. The development of infliximab as a drug for septic shock was apparently not successful.

The change of targeted diseases from septic shock to rheumatoid arthritis for infliximab was made through collaboration with Prof. Ravinder N. Maini and Prof. Marc Feldmann of the Kennedy Institute of Rheumatology at Oxford University, both prominent researchers in the field of rheumatoid arthritis. Dr. James Woody, who was a research director at Centocor at that time, introduced infliximab to Prof. Maini and Prof. Feldman. Woody had previously served as a visiting researcher in the laboratory of Feldmann and co-authored an article with them (Maoz et al. 1977). Woody, who subsequently moved to Centcor, agreed to provide infliximab to Maini and Feldmann, and it was studied in rheumatoid arthritis patients in 1992 (Feldman and Maini 2003). Maini and Feldmann found that hamster antibodies to TNFα/β showed preventive and therapeutic effects against arthritis in murine collagenous arthritis models and published the research result in the same year (Williams et al. 1992). Thereafter, clinical development of infliximab was conducted at a remarkable rate.

Chimerization of TNFα antibodies was published in 1993 by Dr. Knight of Centcor (Knight et al. 1993). The results of a Phase I clinical trial for patients with rheumatoid arthritis were also published in the same year (Elliot et al. 1993). The preparation of the MCB was started in 1991.[20] The Kennedy Institute of Rheumatology at the University of Oxford held a method-of-use patent for infliximab for rheumatoid arthritis, and Centcor concluded licensing agreements and had rights for commercial use (Maini et al. 2004). The substance patent for infliximab was co-filed by New York University and Centcor in the United States in 1994 (priority date: 1991). The Phase II clinical trial was started in 1993, and the clinical trial in Japan was started by Mitsubishi Tanabe Pharma in 1995. The initiation of infliximab's clinical trial in Japan had a great impact on the R&D of tocilizumab. Thereafter, in 1997, a Phase III clinical trial was initiated and, in 1999, Johnson & Johnson acquired Centcor. In the same year, FDA approved the drug for the treatment of rheumatoid arthritis, followed by Japanese approval in 2003.

The advent of infliximab and its rapid introduction into the market had a huge impact on the tocilizumab R&D strategy and market sales. The market for tocilizumab was already dominated by the TNFα inhibitors because of the late entry

---

[20]FDA Remicade Product Review. p. 14.

of tocilizumab to the market compared with infliximab, etanercept, and adalimumab. Given that tocilizumab had a different target molecule, and was a whole new drug, the information on efficacy and safety was overwhelmingly lacking relative to TNFα inhibitors. For that reason, the FDA approved tocilizumab as a second-line biological drug. Therefore, tocilizumab was approved as an antibody drug that can be used only for patients who cannot control their condition with the preceding biological drugs. Even in Japan, there was initially a tendency for tocilizumab to be administered only when the preceding biological preparation was ineffective.[21]

When comparing tocilizumab and Remicade's R&D process and its network, the active and sustained involvement of universities is a common point of the R&D for both antibody drugs. The Osaka University immune research team in the case of tocilizumab, and the New York University and Cambridge University research teams in the case of infliximab, were deeply involved in the R&D process. Such an academia–industry alliance system played an important role in appropriately grasping the potential efficacy of highly novel antibody drugs. It is also noteworthy that the companies responsible for development were not specialized in the development of low molecular weight compounds, but had experience in the development of bio-pharmaceuticals.

The time spent producing the candidate antibodies was almost the same for both drugs, and the antibodies were developed at almost the same time; tocilizumab in 1990, infliximab in 1991. However, for infliximab, the preclinical trial was completed in about 1 year, while tocilizumab required a period of about 6 years for the completion of preclinical trials, although preclinical studies were initiated in the same year. The two main reasons for these differences were: (1) tocilizumab was temporarily targeted to non-autoimmune diseases, whereas infliximab was able to switch the main target diseases from septicemia to rheumatoid arthritis after the participation of Professors Maini and Feldmann; and (2) tocilizumab took significantly more time for preclinical trials than infliximab.

13.17   Conclusion

In the R&D of tocilizumab, an entrepreneurial corporate scientist started the research on autoimmune disease early when the target molecule and disease mechanism were still unknown. He also strongly promoted the collaborations with university and research institutes, specifically that with Osaka University, a leading research university in the field of immunology. As a result, Chugai succeeded in the development of a highly innovative antibody drug with a new target and a new mechanism of action.

Research and development took 24 years from the start of basic research in 1984 to its approval as a rheumatoid arthritis drug in 2008. In the meantime, the tocilizumab R&D team encountered at least two crises that could have discontinued the R&D

---

[21] Chugai Pharmaceutical. IL-6 Discovery. http://ra-online.jp/pr/bone/act/il6/il6/002.html (accessed 5 October 2013).

program. First, basic research was initiated in the search for B cell inhibitors, but it was not possible to identify the target molecules. This situation was resolved by the discovery of IL-6 and its applicability to autoimmune diseases. Next, it was not possible to find a lead compound from soluble receptors or low molecular weight compounds. The crisis was solved by a discovery by Prof. Kawano at Hiroshima University of the applicability of antibodies to multiple myeloma, which is a life-threatening disease with no effective therapy. This disease status meant that the high production cost of tocilizumab could be accommodated. In both cases, the crisis of discontinuation was overcome by incorporating scientific progress into drug discovery.

Furthermore, an industry–university collaboration with Osaka University established a mutually beneficial relationship so that science and innovation contributed mutually to progress, rather than a unilateral relationship of knowledge transfer from a university to a company for its commercialization. The development of mouse versions of tocilizumab and the drug itself greatly advanced the scientific understanding of IL-6. At the same time, Osaka University, which had a profound understanding of the mechanism of action of tocilizumab, actively promoted clinical trials in Japan and tocilizumab was approved first in Japan. Notably, there was limited interest in IL-6-targeting drugs among foreign researchers who were more focused on TNFα inhibitors.

In contrast to tocilizumab, the TNFα inhibitors targeting rheumatoid arthritis were developed in Europe and the United States, and were all discovered by bio-start-up companies. Infliximab was developed by Centcor, adalimumab by BASF's BBC, and etanercept by Immunex. However, in Japan, Chugai, a pharmaceutical company, began to develop tocilizumab, which eventually reached the market. The presence of in-house scientists capable of absorbing cutting-edge science, who also had extensive experience in the field of biotechnology, and the managerial understandings of biopharmaceuticals led to success of the tocilizumab R&D program.

From the viewpoint of global R&D competition, the scientists in Japan such as at Osaka University led the world in the field of immunology, so Osaka University provided the seeds for tocilizumab to Chugai very early. Chugai itself started its basic research early, so that the discovery project for tocilizumab started the earliest among the major competing drugs. Furthermore, drug candidate development took place at almost at the same time (about 1990) as that for infliximab. However, tocilizumab was launched 9 years later as a rheumatoid arthritis drug. One of the reasons for such a delay was the concern that tocilizumab would be unable to set a sufficiently high price to cover the production cost for a rheumatoid arthritis drug, partly because of the restrictions of the drug price system in Japan. Therefore, the company switched the drug target to multiple myeloma, although clinical trials for this application failed. As a result, development as an agent for treating autoimmune diseases was suspended. Afterward, it returned to the development for autoimmune diseases, and clinical development in Japan advanced smoothly, while clinical development in Europe and the United States was slower. Roche acquired half of the stocks of Chugai on the way to the clinical trial of tocilizumab for autoimmune diseases, which realized the worldwide development of tocilizumab.

References

Choy, E. H. S., Isenberg, D. A., Garrood, T., Farrow, S., Ioannou, Y., Bird, H., et al. (2002). Therapeutic benefit of blocking interleukin-6 activity with an anti–interleukin-6 receptor monoclonal antibody in rheumatoid arthritis: A randomized, double-blind, placebo-controlled, dose-escalation trial. *Arthritis and Rheumatology, 46*(12), 3143–3150.

Elliott, M. J., Maini, R. N., Feldmann, M., Longfox, A., Charles, P., Katsikis, P., et al. (1993). Treatment of rheumatoid-arthritis with chimeric monoclonal-antibodies to tumor-necrosis-factor-alpha. *Arthritis and Rheumatism, 36*(12), 1681–1690.

Feldmann, M., & Maini, R. N. (2003). TNF defined as a therapeutic target for rheumatoid arthritis and other autoimmune disease. *Nature Medicine, 9*, 1245–1250.

Gabay, C., Emery, P., van Vollenhoven, R., Dikranian, A., Alten, R., Pavelka, K., Klearman, M., Musselman, D., Agarwal, S., Green, J., Kavanaugh, A., ADACTA Study Investigators. (2013). Tocilizumab monotherapy versus adalimumab monotherapy for treatment of rheumatoid arthritis (ADACTA): A randomised, double-blind, controlled phase 4 trial. *Lancet, 381*(9877), 1541–1550.

Gershwin, M. E., Ohsugi, Y., Ahmed, A., Castles, J. J., Scibienski, R., & Ikeda, R. M. (1980). Studies of congenitally immunologically mutant New-Zealand mice. IV. Development of autoimmunity in congenitally athymic (nude) New-Zealand black x white F1-hybrid mice. *Journal of Immunology, 125*(3), 1189–1195.

Hirano, T., Teranishi, T., Toba, H., Sakaguchi, N., Fukukawa, T., & Tsuyuguchi, I. (1981). Human helper T cell factor(s) (ThF). I. Partial purification and characterization. *Journal of Immunology, 1266*, 517–522.

Hirano, T., Yasukawa, K., Harada, H., Taga, T., Watanabe, Y., Matsuda, T., et al. (1986). Complementary-DNA for a novel human interleukin (BSF-2) that induces lymphocytes-B to produce immunoglobulin. *Nature, 324*(6092), 73–76.

Hirano, T., Yoshizaki, K., & Kishimoto, T. (2010). Interleukin-6 in *Cytokine Hunting, Japanese Society of Interferon and Cytokine Research*, 157–194.

Jones, G., Sebba, A., Gu, J., Lowenstein, M. B., Calvo, A., Gomez-Reino, J. J., et al. (2010). Comparison of tocilizumab monotherapy versus methotrexate monotherapy in patients with moderate to severe rheumatoid arthritis: The AMBITION study. *Annals of the Rheumatic Diseases, 69*(1), 88–96.

Kawano, M., Hirano, T., Matsuda, T., Taga, T., Horii, Y., Iwato, K., et al. (1988). Autocrine generation and requirement of BSF-2/IL-6 for human multiple myelomas. *Nature, 332*(6159), 83–85.

Knight, D. M., Trinh, H., Le, J. M., Siegel, S., Shealy, D., Mcdonough, M., et al. (1993). Construction and initial characterization of a mouse-human chimeric anti-TNF antibody. *Molecular Immunology, 30*(16), 1443–1453.

Kopf, M., Baumann, H., Freer, G., Freudenberg, M., Lamers, M., Kishimoto, T., et al. (1994). Impaired immune and acute-phase responses in interleukin-6-deficient mice. *Nature, 368*, 339–342.

Maini, R. N., Breedveld, F. C., Kalden, J. R., Smolen, J. S., Furst, D., St Weisman, M. H., et al. (2004). Sustained improvement over two years in physical function, structural damage, and signs and symptoms among patients with rheumatoid arthritis treated with infliximab and methotrexate. *Arthritis and Rheumatism, 50*(4), 1051–1065.

Maoz, A., Woody, J., & Feldman, M. (1977). Purification of antigen-specific helper and suppressor lymphocytes. *Israel Journal of Medical Science, 13*(10), 1057.

Marks, L. V. (2009). Collaboration—A competitor's tool: The story of Centocor, an entrepreneurial biotechnology company. *Business History, 51*(4), 529–546.

Marks, L. V. (2012). The birth pangs of monoclonal antibody therapeutics: The failure and legacy of Centoxin. *mAbs, 4*, 403–412.

Muraguchi, A., Kishimoto, T., Miki, Y., Kuritani, T., Kaieda, T., Yoshizaki, K., et al. (1981). T cell-replacing factor- (TRF) induced IgG secretion in a human B blastoid cell line and demonstration of acceptors for TRF. *Journal of Immunology, 127*, 412–416.

Nagaoka, S., & Akaike, S. (2013). Management forum; Aiming for a successful industry-academia collaboration: Recognition of university's original existence should be recognized (Interview with Prof. Tadaamitsu Kishimoto: Project Professor, Frontier Research Center for Immunology, Osaka University) (in Japanese). *Hitotsubashi Business Review, 61*(3), 170–178.

Nakajima, A., & Kishimoto, T. (2009). Surprises of the new modern immunology stories: Antibody drugs and autoimmune (in Japanese). Kodansha.

Nishimoto, N., Miyasaka, N., Yamamoto, K., Kawai, S., Takeuchi, T., Azuma, J., et al. (2009). Study of active controlled tocilizumab monotherapy for rheumatoid arthritis patients with an inadequate response to methotrexate (SATORI): Significant reduction in disease activity and serum vascular endothelial growth factor by IL-6 receptor inhibition therapy. *Modern Rheumatology, 19,* 12–19.

Ohsugi, Y. (2013). Shinyaku Actemra No Tanjo - Kokusan Hatsu No Koutai Iyakuhin (in Japanese). Iwanami Kagaku Library, Iwanami Shoten.

Ohsugi, Y., Gershwin, M. E., Ahmed, A., Skelly, R. R., & Milich, D. R. (1982). Studies of congenitally immunological mutant New Zealand mice. VI. Spontaneous and induced autoantibodies to red-cells and DNA occur in New Zealand x-linked immunodeficient (xid) mice without phenotypic alterations of the xid gene or generalized polyclonal B-cell activation. *Journal of Immunology, 128*(5), 2220–2227.

Ohsugi, Y., & Gershwin, M. E. (1979). Studies of congenitally immunological mutant New-Zealand mice. III. Growth of lymphocyte-B clones in congenitally athymic (nude) and hereditarily asplenic (dh- +) NZB mice—Primary B-cell defect. *Journal of immunology, 123*(3), 1260–1265.

Ohsugi, Y., & Kishimoto, T. (2011). IL-6 signaling and its blockade with a humanized anti-interleukin-6 receptor antibody in rheumatoid arthritis: Advent of a new and innovative therapeutic drug, Tocilizumab. *Current Rheumatology Reviews, 7*(4), 288–300.

Shimazaki, C., & Gotoh, H. (1997). Antitumor effects of human myeloma models and anti-human IL-6 receptor antibodies. *Clinical Blood, 38*(4), 281–284.

Shimizu, T., Chugai Pharmaceutical Co., Ltd Product Manager. (2008). http://www.chugai-pharm. co.jp/html/meeting/pdf/080522jShimizu.pdf. Accessed October 27, 2014.

Suematsu, S., Matsuda, T., Aozasa, K., Akira, S., Nakano, N., Ohno, S., et al. (1989). IgG1 plasmacytosis in interleukin 6 transgenic mice. *Proceedings of the National Academy of Sciences, 86*(19), 7547–7551.

Sumikura, K. (2013). Contribution of university basic research to the creation of new drugs: Case-based analysis (in Japanese). *Japan Intellectual Property Society.*

Takahashi, Y., & Ohsugi, Y. (2014). Third secretary for DMARDs and development in Japan: From Lobenzarit to Anti-IL-6 Inhibitors in Japan (in Japanese). *Japan Rheumatism Foundation News, 122*(5). http://www.rheuma-net.or.jp/rheuma/rm220/pdf/news122.pdf. Accessed June 24, 2014.

Tanaka, T., & Kishimoto, T. (2012). Targeting interleukin-6: All the way to treat autoimmune and inflammatory diseases. *International Journal of Biological Sciences, 8*(9), 1227–1236. https:// doi.org/10.7150/ijbs.4666.

Teranishi, T., Hirano, T., Arima, N., & Onoue, K. (1982). Human helper T cell factor(s) (ThF). II. Induction of IgG production in B lymphoblastoid cell lines and identification of T cell-replacing factor- (TRF) like factor(s). *Journal of Immunology, 128,* 1903–1908.

Vilcek, J. (2009). From INF to TNF: A journey into realms of lore. *Nature, 10*(6), 555–557.

Williams, R. O., Feldmann, M., & Maini, R. N. (1992). Antitumor necrosis factor ameliorates joint disease in murine collagen-induced arthritis. *Proceedings of the National Academy of Sciences of the United States of America, 89*(20), 9784–9788.

Yoshizaki, K., Nakagawa, T., Kaieda, T., Muraguchi, A., Yamamura, Y., & Kishimoto, T. (1982). Induction of proliferation and Ig production in human B leukemic cells by Anti-immunoglobulins and T cell factors. *Journal of Immunology, 128,* 1296–1301.

Yoshizaki, K., Nishimoto, N., Mihara, M., & Kishimoto, T. (1998). Therapy of rheumatoid arthritis by blocking IL-6 signal transduction with a humanized anti-IL-6 receptor antibody *20*(1–2), 247–259.

Chapter 14
Nivolumab (Opdivo)

Science-Based Antibody Drug, Which Opened a New Category of Cancer Treatments

Yasushi Hara and Sadao Nagaoka

Abstract Nivolumab is a humanized anti-PD-1 monoclonal antibody drug against cancer, based on immune checkpoint blockade as an entirely new mechanism of action. The new gene (PD-1), its function, and the possibility that its suppression can be used to block cancer growth were identified in the laboratory of Prof. Tasuku Honjo at Kyoto University, starting from a serendipitous basic science discovery. Despite concrete demonstrations by Honjo's team of the potential of PD-1 signal inhibition to treat cancer, subsequent investment in drug development did not readily occur. Strong skepticism against cancer immunotherapy prevailed, so that Ono Pharmaceutical, a Japanese pharmaceutical company that joined in the development process of nivolumab from an early stage, could not find a collaborating firm, either domestic or international, that had antibody technology. However, an American startup company appreciated the method-of-use patent and the discovery underlying the patent and participated in the project. This case demonstrates how pure basic research aimed at deepening understanding of fundamental questions led unexpectedly to a major discovery of great practical value. The subsequent use-oriented basic research by the university as well as the use of diverse capability and multiple views helped exploit a path-breaking scientific discovery for innovation.

Y. Hara (✉)
CEAFJP/EHESS, Paris, France
e-mail: yasushi.hara@r.hit-u.ac.jp

Faculty of Economics, Hitotsubashi University, Tokyo, Japan

S. Nagaoka
Tokyo Keizai University, Tokyo, Japan

© Springer Nature Singapore Pte Ltd. 2019
S. Nagaoka (ed.), *Drug Discovery in Japan*,
https://doi.org/10.1007/978-981-13-8906-1_14

14.1 Introduction

Nivolumab (Opdivo; development numbers ONO-4538, MDX-1106, and BMS-936558) is a humanized anti-PD-1 monoclonal antibody for the treatment of malignant tumors. It was developed through basic research in Prof. Tasuku Honjo's Laboratory at Kyoto University, through its alliance with Ono Pharmaceutical, and the participation by Medarex Inc. (Bristol-Myers Squibb). The drug was based on an entirely new mechanism of action (blockade of immune system checkpoints), and was the first humanized IgG4 monoclonal antibody against human PD-1 (programmed cell death protein 1). In contrast to ipilimumab (Yervoy), which is marketed by Bristol-Myers Squibb and was another pioneer drug based on immune checkpoint blockade against CTLA-4 (cytotoxic T-lymphocyte-associated protein 4), nivolumab directly blocks the immune escape pathway of cancer cells, in addition to increasing the activity of activated T cells. This work has stimulated the research and development (R&D) of new drugs based on similar mechanisms of action and has changed cancer therapeutics in a fundamental way. Professor Tasuku Honjo was awarded the 2018 Nobel Prize in Physiology or Medicine jointly with Prof. James P. Allison, who discovered that blocking of CTLA-4 strengthens the T cell anti-tumor response, for their discovery of cancer therapy by inhibition of negative immune regulation. They opened a new category of cancer treatments.

Conventional cancer treatments include surgery, radiotherapy, and anticancer drugs, including molecular targeted drugs. Surgery and radiotherapy are used for patients with early stage disease, while anticancer drugs have been further combined for patients with late-stage disease or metastatic disease. In the meantime, there have been attempts to use cancer immune cell therapy to utilize or enhance the immune function that a human body naturally possesses. In the 1980s, activated autologous lymphocyte therapy to attack cancer cells was performed using lymphocytes with cytotoxic activity but it was not able to show a significant effect (Egawa 2009). In such therapy, lymphocytes were extracted from the patient's body, and then cultured in vitro; interleukin 2 (IL-2) was added and administered to patients, such as lymphokine-activated killer (LAK) and tumor-infiltrating lymphocyte (TIL) therapy.

The fundamental reason that these cancer therapies focusing on the internal immune mechanism did not deliver an effective result in the past is that they did not sufficiently control the immune escape mechanism or the immune-suppressive mechanism of cancer cells. Immune checkpoint inhibitors are unique in directly addressing this problem. Immune checkpoint inhibitors, such as nivolumab and ipilimumab, are characterized by: (1) instead of directly attacking cancer cells, these drugs attack cancer cells by activating immune responses that are suppressed; (2) because of the ability of the immune system to respond to any mutation of cancer cells, the problem of drug resistance is unlikely to occur, so that the improvement of long-term survival of a patient can be expected; and (3) the ability of the immune system to attack all cancer cells, in principle as all cancer cells express non-self antigens; thus, the new drugs are applicable to a wide variety of cancers because of the mechanism of action (Tamada 2012).

This chapter provides detailed accounts of how such a path-breaking drug that is changing cancer therapeutics in a fundamental way was developed through basic research in a university, and then through academia–industry alliances involving both pharmaceutical companies and bio-start-ups. A brief comparison of the drug discovery process between nivolumab and ipilimumab is also provided (Box 14.1).

14.2 Overview of Nivolumab R&D

In 1992, Prof. Tasuku Honjo and his laboratory members in the Faculty of Medicine at Kyoto University, reported that programmed cell death 1 (PD-1; CD279) had been discovered as a gene whose expression was enhanced upon induction of programmed death of T cells. PD-1 belongs to the CD28/B7 family, a group of molecules that regulate T cell activation, expressed on activated T and B cells (Ishida et al. 1992) (Fig. 14.1).

In 1998, it was clarified that immune function was enhanced in vivo by deficiency of PD-1 in the PD-1 knockout mouse produced in the Honjo laboratory, and it was indicated that PD-1 negatively regulated the immune response (Nishimura et al. 1998). Following this, the search for the ligands of PD-1 was carried out; PD-L1 (B7-H1, CD274) was reported in 2000, and PD-L2 (B7-DC, CD273) in 2001. Furthermore, in 2002, it was confirmed that tumorigenesis was suppressed when the mouse cancer cell line (J558L) was transplanted into PD-1-deficient mice. In the case when the same mouse cell line was transplanted into a wild-type mouse, the tumor thrived. The group also showed that PD-L1 antibody blocked tumor growth in mice (Iwai et al. 2002). Based on these results, a method-of-use patent application was

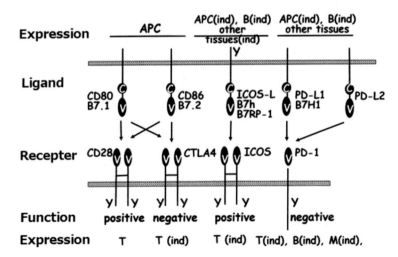

Fig. 14.1 Immune response regulatory molecules. *Source* Prof. Tasuku Honjo

filed in collaboration between Prof. Tasuku Honjo and Ono Pharmaceutical, which was intended to stimulate corporate investment in drug development for a drug that treats cancer by inhibiting PD-1-mediated immunosuppressive signals. In 2003, a PCT application was filed (WO 2004/004771), and in January 2004 the patent was made public. Afterward, the American bio-startup Medarex, which had antibody preparation technology, participated as a joint R&D partner of Ono Pharmaceutical to develop the antibody. The clinical trial was started in the following year. In 2009, Medarex was acquired by Bristol-Myers Squibb, and they expanded the scope of clinical trials of nivolumab. As a result of these clinical trials, the drug was approved and launched in Japan and the United States in 2014 and in Korea and Europe in 2015, first for the treatment of melanoma, but approvals for the treatment of other cancers have since been granted.

We can identify the following four major features of this R&D process:

(1) The discovery of PD-1 was a serendipitous discovery, which was not anticipated from the research on cell selection in the thymus, the original aim of the research of the Honjo laboratory.

(2) In addition to the discovery of PD-1, the Honjo laboratory used PD-1-deficient mice to elucidate the functions of PD-1 and found the possibility of negative regulation of immune responses. Following these findings, they also clarified the mechanism by which PD-1 is regulated by PD-1 ligands through international collaborative research. The group then clarified that PD-L1 antibodies can be used to block cancer growth in a mouse model, and concretely demonstrated the possibility of drug development by inhibiting PD-1 signals. On the basis of these findings, a method-of-use patent for PD-1 signal inhibition in cancer therapy was filed jointly with Ono Pharmaceutical to encourage corporate investment in drug development.

(3) Basic research support from the government fund, such as KAKEN, played an important role for the long-term basic research, which took almost 10 years from the identification of PD-1 in 1992 to the concrete demonstration of the possibility of a cancer drug in 2002.

(4) Despite concrete demonstrations of the potential of PD-1 signal inhibition to treat cancer, investment in drug development did not readily occur. Ono Pharmaceutical did not possess its own technology for antibody drugs, so they requested cooperation from a dozen domestic and foreign pharmaceutical companies with antibody pharmaceutical technology. However, almost all firms refused based on their skepticism of cancer immunotherapy. Fortunately, Medarex, an American bio-startup company, recognized the potential, as well as the value of the method-of-use patent, and showed its intention to participate. Medarex contributed to the production and clinical development of antibody drugs in collaboration with Ono Pharmaceutical.

Nivolumab was approved for the treatment of refractory malignant melanoma in 2014. As its mechanism of action suggests, it is active against a variety of cancer types and has been approved for the treatment of lung cancer, renal cell carcinoma, Hodgkin's lymphoma, and head and neck cancer in Japan, the USA, and Europe

Table 14.1 Research and development status of nivolumab by disease

Development status	Japan	Abroad
Phase I clinical study	• Cases of hepatocellular carcinoma • Carcinoma of the biliary tract • Solid tumors (in combination with mogamulizumab)	• Cases of hepatocellular carcinoma • Hematologic cancers (T-cell lymphoma, multiple myeloma, chronic leukemia; Western countries) • Chronic myelogenous leukemia • Hepatitis C • Solid tumors (in combination with mogamulizumab)
Phase II clinical study	• Virus-positive and virus-negative solid tumors • Glioblastoma • Ovarian cancer • Urothelial cancer	• Solid cancers (triple-negative breast cancer, gastric cancer, spleen cancer, small cell lung cancer, bladder cancer; Western countries) • Virus-positive and virus-negative solid tumors • Colorectal cancer • Diffuse large B cell lymphoma • Follicular lymphoma • Urothelial cancer
Phase III clinical trial		• Gastric cancer (Korea and Taiwan) • Glioblastoma • Esophageal cancer • Small cell carcinoma

(continued)

(Table 14.1). More types of carcinoma are under clinical trials, especially considering the drug has become increasingly available in many other countries.

Of some importance, the mechanism of action described above suggests the likelihood of autoimmune disease as an adverse effect of nivolumab. In the Phase II clinical study in Japan, some side effects were observed in 30 of 35 patients evaluated for safety. Therefore, the pharmaceutical labelling has indicated a list of infusion reactions (Ono Pharmaceutical and Bristol Myers 2015a). In September 2015, warnings were issued to prepare for adverse reactions caused by excessive immune reactions (Ono Pharmaceutical and Bristol Myers 2015b).

14.3 Mechanisms of Action and Characteristics of Nivolumab

Cancer cells emerge in the body of healthy individuals but are usually eliminated before proliferating as cancer cells by the immune system. However, cancer cells

Table 14.1 (continued)

Development status	Japan	Abroad
Under application		• Non-small cell lung cancer (approved in Taiwan, in Phase III clinical trials in Korea)
Launched	• Malignant melanoma (July 2014) • Non-small cell lung cancer (December 2015) • Renal cell carcinoma (August 2016) • Hodgkin's lymphoma (December 2016) • Head and neck cancer (March 2017) • Gastric cancer (September 2017)	• Malignant melanoma (USA, in 2014; Europe; approved in Korea, and pending in Taiwan) • Non-small cell lung cancer (Europe) • Advanced lung cancer (USA, 2015) • Metastatic renal cell carcinoma (USA, 2015; Europe) • Hodgkin's lymphoma (USA, 2016; Europe) • Non-small cell lung cancer (excluding non-squamous cell carcinoma; Europe and Taiwan) • Head and neck cancer (USA, 2016; Europe) • Previously treated locally advanced or metastatic urothelial carcinoma (USA, 2016; Europe) • Metastatic colorectal cancer (USA, 2017) • Hepatocellular carcinoma (USA, 2017) • Small cell lung cancer (USA, 2018)

Source Ono Pharmaceutical (2018), Opdivo Approval History in Drugs.com (2019). Opdivo: EPAR—Product Information (2019)

can inactivate the immune system and grow. Cancer cell inactivation of the immune system is mediated by the binding of T cell-associated PD-1 to PD-L1 expressed on a cancer cell, which transmits an inhibitory signal to lymphocytes, suppressing T cell activation and causing T cells to stop attacking cancer cells. It has been reported that there is an inverse correlation between PD-L1 expression in resected tumor tissues and postoperative survival time in a variety of human carcinomas, including ovarian, esophageal, renal cell, spleen, and urothelial carcinomas (Taube et al. 2012). However, the relationship between the efficacy of antibody treatment with nivolumab and PD-L1 expression in cancer cells has not been conclusively established.

Nivolumab binds to the extracellular domain of PD-1 (PD-1 ligand binding domain), which inhibits the binding of PD-L1 and PD-L2 to PD-1 between antigen-presenting cells and activated T cells. This causes the activation of cancer antigen-specific T cells to be enhanced, as illustrated in Fig. 14.2. In addition, by inhibiting the binding of PD-L1 to PD-1 between T cells and cancer cells, the ability of T

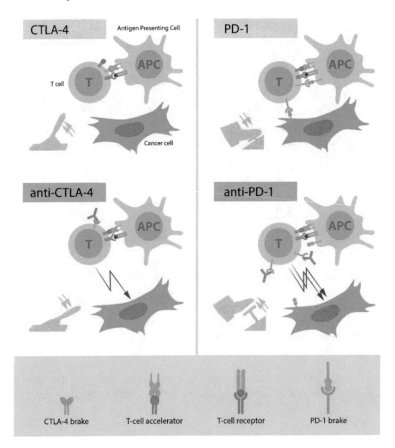

Fig. 14.2 CTLA-4 and PD-1 signaling pathway. *Source* Nobel Media AB (2018), © The Nobel Committee for Physiology or Medicine. Illustrator: Mattias Karlén

cells to attack cancer cells is restored. It is presumed that nivolumab plays these two roles. In contrast, the CTLA-4 antibody drug ipilimumab is primarily targeted at proliferating activated T cells and may be ineffective if the PD-1 PD-L1 pathway is established.

Nivolumab shows high response rates even in refractory cancers for which conventional therapy was ineffective, and it also has a sustained tumor shrinkage effect. Figure 14.2 shows the changes over time in the target lesions from Phase II clinical trials in patients with refractory malignant melanoma who were administered nivolumab. Even if refractory, the response rate is found to be high. In addition, a comparative study with dacarbazine (DTIC), which has been used for conventional treatment, demonstrated that the survival rate after 1 year was 72.9% with nivolumab and 42.1% with dacarbazine, for untreated patients with BRAF (v-raf murine sarcoma viral oncogene homolog B1) wild-type unresectable or metastatic malignant

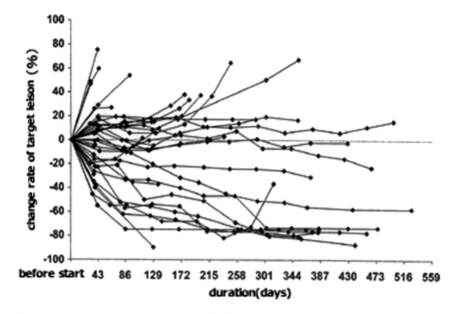

Fig. 14.3 Percentage change in target lesion in the Japanese phase II clinical study in patients with refractory malignant melanoma. *Source* Yoshida et al. (2015)

melanoma. Furthermore, in the case of nivolumab, the effect on the survival rate continued thereafter, but the effect decreased for dacarbazine (Fig. 14.3).

14.4 Timeline from the Start of Basic Research to the Market Launch

Notably, it took more than 20 years from the serendipitous discovery of PD-1 to the first market launch of nivolumab. The following is the timeline of the major events of nivolumab development.

Basic research and drug discovery

1992: PD-1 was identified in Prof. Tasuku Honjo's laboratory in the Faculty of Medicine, Kyoto University.
1994: Gene patent applications for human PD-1 were filed.
1996: Honjo laboratory succeeded in generating PD-1 knockout mice and demonstrated that PD-1 deficiency enhances immune function in vivo, suggesting that PD-1 negatively regulates immune responses.
2000: PD-L1 was identified by the collaboration of Kyoto University, the Genetics Institute (GI), and Harvard University.

2001: PD-L2 was identified by collaborative research between Kyoto University, GI, and Harvard University.

2002: It was confirmed that tumor growth was suppressed when the tumor was transplanted into PD-1-deficient mice and that cancer growth could be blocked using a PD-L1 antibody in mice.

July 2002: Based on the above findings, a method-of-use patent for activating an immune response and treating cancer and infectious diseases by inhibiting the immunosuppressive action mediated by PD-1 was filed jointly by Prof. Tasuku Honjo of Kyoto University and Ono Pharmaceutical.

2002: Ono Pharmaceutical asked over ten major pharmaceutical companies in Japan and abroad with offices in Japan about joint development of cancer immunotherapy drugs, but could not find a collaborating firm. However, Medarex appreciated the method-of-use patents and the underlying technology, and subsequently entered into joint development with Ono Pharmaceutical.

May 2005: A patent for a fully human monoclonal IgG4 antibody (nivolumab, development code: ONO-4538/MDX-1106/BMS-936558) was filed by Ono Pharmaceutical and Medarex. It was entitled "Method for cancer treatment with human monoclonal antibody and anti-PD-1 antibody alone or in combination with other immunotherapy against Programmed Death 1 (PD-1)."

Clinical Studies

August 2006: Investigation new drug application approved by US FDA.

2006: Phase I study of nivolumab in solid tumors was launched in the United States (CA209001 study).

2008: Phase I repeated study of nivolumab was launched in the United States (CA209003 study).

2008: Clinical studies were launched in patients with solid tumors in Japan (ONO-4538-01 study).

July 2009: Bristol-Myers Squibb acquired Medarex, as well as the rights to develop and commercialize anti-PD-1 antibodies in North America.

2011: Phase II clinical study of nivolumab was launched for malignant melanoma in Japan (ONO-4538-02 study).

June 2013: Japanese orphan drug designation (for Stage III and IV melanoma).

December 2013: Ono Pharmaceutical applied for manufacturing and marketing approval for nivolumab for malignant melanoma in Japan.

May 2014: Nivolumab was designated as a breakthrough therapy for Hodgkin's lymphoma by the FDA (thereafter, designated as breakthrough therapy for melanoma in September 2014, and for progressive renal cell carcinoma and non-squamous non-small cell lung carcinoma in September 2015).

July 2014: Marketing approval was granted in Japan for non-curable malignant melanoma.

September 2014: Drug price listing and marketing began in Japan.

December 2014: Approved in the United States for malignant melanoma that cannot be cured by surgery.

March 2015: Approval obtained for metastatic squamous non-small cell lung cancer in the United States.

March 2015: Approved in Israel for unresectable malignant melanoma.

March 2015: Approved in South Korea for non-curable malignant melanoma.

June 2015: The European Commission approved nivolumab for the treatment of patients with advanced melanoma that was previously treated.

July 2015: The European Commission approved nivolumab for the treatment of locally advanced or metastatic squamous cell lung cancer after chemotherapy.

October 2015: Approved for non-squamous, non-small cell lung cancer in the United States.

14.5 Background to Discovery Research

Before the discovery of PD-1, the laboratory of Prof. Tasuku Honjo at Kyoto University boasted many pioneering achievements in immunology, including the cloning of the interleukin-2 receptor in 1984, the cloning of interleukins 4 and 5 in 1985 and 1986, respectively, the elucidation of the mechanism of class switching, and the isolation of AID genes (Kinashi and Honjo 2010). At that time, it was a very large laboratory with about 50 scientists.[1]

Medicine Chemistry Course 1, School of Medicine Department, Kyoto University, for which Professor Tasuku Honjo was appointed, had been conducting joint research with Ono Pharmaceutical for a long time since Prof. Osamu Hayaishi was appointed as its chief. In the process of Ono Pharmaceutical's entry into R&D of prostaglandin in 1960s, Prof. Hayaishi advised Ono Pharmaceutical to send its researchers to the laboratory of E. J. Corey at Harvard University to facilitate technology transfer from the Corey laboratory to Ono Pharmaceutical (see Chap. 7: Pranlukast). Professor Honjo's joint papers with Ono Pharmaceutical date back to 1988.

14.6 Discovery and Isolation of PD-1 and Generation of PD-1 Knockout Mice

In 1992, the isolation and identification of PD-1 was performed in the Honjo laboratory. The group had identified an mRNA transcript whose expression was enhanced upon the induction of T cell programmed cell death (Ishida et al. 1992). However, this discovery was not expected from the original research objective, which was to characterize cell selection in the thymus, and it was a serendipitous discovery. On 5 June of the same year, a PD-1 gene patent (Patent Number: JP-A-5-336973, "Novel Polypeptides Related to Programmed Cell Death and DNAs Encoding the Same")

---

[1] From interview with Prof. Tasuku Honjo.

was filed jointly by Prof. Tasuku Honjo and Ono Pharmaceutical as co-applicants. To apply for the substance patent for PD-1, Prof. Honjo asked Ono Pharmaceutical for cooperation. At the time, national universities in Japan did not have the infrastructure for patent application and maintenance. Also, taking into account the expenses associated with patent application and maintenance, applications were filed jointly.[2] The same situation occurred with the subsequent application for the method-of-use patent, and Ono Pharmaceutical contributed expert knowledge about patent applications, which was lacking at the university, in addition to the expense. In 1994, the results of analysis of the human PD-1 gene structure were published (Shinohara et al. 1994).

However, the specific function of PD-1 remained unclear at this point. Therefore, Hiroyuki Nishimura, a PhD student at the time of his membership in the Honjo laboratory since 1991, began preparing PD-1 knockout mice (Kyoto University 2005) . The production technology of the knockout mouse was established in the United States in the latter half of the 1980s, and the expertise was acquired through cooperative research. PD-1 knockout mice were produced in the Honjo laboratory as a world first.

It took several years to prepare pure lines of knockout mice that researchers were able to analyze for the immune functions of PD-1, but in 1996 it was confirmed that the PD-1 deletion caused autoimmune disease. The research was first presented at a meeting for the International Dermatology Association in Cologne, Germany (May 1998). The reasons for such time-consuming research included the need to prepare pure mouse lines and the need for a long observation period, including the aging of the mice after treatment, to determine the effects of PD-1 deficiency, because PD-1 deficiency causes very slight symptoms.[3]

> It was our first experience [making] knockout mice, almost the first time in Japan, and it took about 2 years. Knockout mice usually have symptoms right away, but our mice had no symptoms right away, and they had no symptoms for 1 month or so. They are very mild symptoms and appeared after 3 to 6 months. But they were autoimmune symptoms, showing that it involves the control of the immune system.

PD-1 knockout mice show autoimmune disease-related symptoms, such as increased spleen weight and increased circulating immunoglobulins, suggesting that PD-1 negatively regulates immune responses in vivo (Nishimura et al. 1999). Studies using PD-1 knockout mice also suggested that dilated cardiomyopathy may be caused by autoimmune disease because of PD-1 deficiency (Nishimura et al. 2001). In addition, the laboratory published an earlier work on the PD-1 signaling pathway (Okazaki et al. 2001).

Shiro Shibayama from Ono Pharmaceutical was a visiting researcher at the Honjo laboratory at the time the PD-1 research project was initiated. He directly observed the advances of the research in the laboratory, such as the preparation of knockout mice and the evidence suggesting the role of PD-1 in the immune system. Therefore,

---

[2]From interview with Prof. Tasuku Honjo.
[3]From interview with Prof. Tasuku Honjo.

in 2000, Ono Pharmaceutical began researching PD-1-targeted drug development.[4] At that time, however, PD-1 was only recognized in the firm as a "totally unknown" substance, and it was not easy for Shibayama to obtain the company approval to embark upon drug development, even after the discussions with Medarex began. It took a considerable amount of time for the corporation to make a decision to invest fully in developing nivolumab (Toyo Keizai Online 2015).

14.7 International Collaboration to Discover the Ligands of PD-1

Following the elucidation of the function of PD-1, an international collaboration to discover its cognate ligand was initiated to improve the understanding of the mechanisms by which PD-1 acts. The American firm Genetics Institute (GI) had a device that could electrically measure the ligands in the assay (Biacore screening system), and this inspired Prof. Honjo to propose a collaborative study (Okazaki and Honjo 2007). Beginning in 1999, joint research was conducted between the Honjo laboratory, Ono Pharmaceutical, and GI, which led to the discovery of PD-L1 (Freeman et al. 2000). In this collaboration, although the samples were also provided by Kyoto University, GI filed PD-L1 patents in August 2000 (Number: WO2001014557 and WO2001014556) that did not confer any ownership to the Honjo laboratory. In 2001, PD-L2 was also discovered through a collaboration between Kyoto University, GI, and Harvard University (Latchman et al. 2001).

14.8 Pursuing the Applicability of Anti-PD-1 Antibodies as Cancer Therapeutic Agents

Professor Honjo decided to explore the possibility of developing cancer treatments by suppressing PD-1 and activating immunity. Professor James Allison at the University of California, Berkeley, published a paper in 1996 suggesting the possibility of treating cancers with antibodies to CTLA-4, an immunosuppressive regulatory molecule (Leach et al. 1996). However, CTLA knockout mice died of autoimmune diseases in a few weeks. Hence, there had been little progress in developing drugs that inhibit CTLA-4. Given that PD-1-deficient mice responded mildly to their autoimmune disease, Prof. Honjo believed that targeting PD-1 could lead to therapeutic agents with fewer side effects. Using PD-1-deficient mice, the Honjo laboratory showed that the loss of PD-1 function enhanced tumor immunity. Focusing on the interaction between PD-1 and PD-L1, they also conducted a mouse experiment using PD-L1 antibody that strongly suggested inhibition of the PD-1/PD-1 ligand pathway was a powerful method for cancer immunotherapy (Iwai et al. 2002).

---

[4]From interview with Ono Pharmaceutical staff.

In addition, to promote corporate R&D investment, the Honjo laboratory further investigated the potential effects of anti-PD-1 antibodies on specific cancers and obtained a number of promising findings. First, anti-PD-1 antibodies reduced melanoma metastasis (Iwai et al. 2004). Second, study of the expression patterns of PD-L1 in human ovarian cancer demonstrated that the survival rate after the surgery of ovarian cancer patients was inversely correlated with the expression rate of PD-L1 (Hamanishi et al. 2006). Furthermore, several institutions have reported that PD-L1 expression is inversely associated with prognosis of patients with kidney cancer (Thompson et al. 2004), esophageal cancer (Ohigashi et al. 2005), gastric cancer (Wu et al. 2006), urothelial cancer (Nakanishi et al. 2007), pancreatic cancer (Nomi et al. 2007), and malignant melanoma (Hino et al. 2010).

14.9 Application for Method-of-Use Patents and Creation of Nivolumab Through Collaboration with Medarex

On 3 July 2002, Prof. Tasuku Honjo and Ono Pharmaceutical jointly filed a method-of-use patent for the treatment of cancer and infectious diseases by inhibiting PD-1-mediated immunosuppressive signals based on the aforementioned results of the effects of inhibiting PD-1 signals. The patent was filed through the PCT route on 2 July of the following year and was published on 15 January 2004 (WO 2004/004771). After the patent application was filed, Ono Pharmaceutical started to examine the possibility of developing anti-PD-1 antibody as a cancer therapeutic. However, Ono Pharmaceutical did not have the in-house technology to create an antibody drug, so the company sought cooperation with more than ten pharmaceutical companies in Japan and abroad that had the necessary technology. Unfortunately, no company was willing to participate in the project, because the common perception in Japan and abroad was that cancer immunotherapy was not trustworthy, and the experimental data in mice was insufficient to overturn these prejudices (Weekly Diamond 2015).

Later, however, Medarex, an American bio-startup that valued the method-of-use patent and the technology underlying the patent, showed willingness to participate in the co-development. Medarex found the discovery based on the disclosed patent by Honjo and Ono Pharmaceutical. It was the only firm to express interest. The company had been developing a humanized monoclonal IgG1 antibody, ipilimumab, against CTLA-4. In addition, they had proprietary humanized antibody generation technology (UltiMAb), which could be applied to the generation of a humanized antibody against PD-1. Nivolumab and ipilimumab are drugs targeted at the same diseases, but their mechanisms of action differ, and the combination can enhance effectiveness.

In 2005, Medarex and Ono Pharmaceutical successfully synthesized an anti-PD-1 antibody nivolumab, later branded as Opdivo, using UltiMAb technology. The IgG4 subclass, which has no antibody-dependent cell cytotoxicity (ADCC) or complement-dependent cytotoxicity (CDC) effects, was selected to prevent damage to the bound activated T cells, and structural stabilization by amino acid replacement (S228P) of the IgG4H hinge region was implemented (Yoshida et al. 2015). Nivolumab was produced using Chinese hamster ovary (CHO) cells by gene recombination technology (Oyama et al. 2014). On 9 May 2005, an antibody-based substance patent ("Human Monoclonal Antibodies to Programmed Death1 (PD-1) and Methods for Treating Cancers Combined with Anti PD-1 Antibodies alone or with Other Immunotherapies") was filed jointly by Ono Pharmaceutical and Medarex (Patent No. 44361545).

Following this, preclinical pharmacological tests were carried out. Based on a joint research contract between Medarex and Ono Pharmaceutical, the validation experiments using anti-PD-1 antibody were conducted, and it was determined that the growth of the tumor was suppressed in prophylactic and therapeutic administration of anti-PD-1 antibody to murine tumor models (Korman et al. 2007). Nivolumab was shown to exhibit antitumor effects by enhancing antigen-specific T cell activation through inhibition of the binding of PD-1 to PD-L1 and PD-L2 and by enhancing the immune response against tumor cells (Wong et al. 2007). Studies also showed that the new antibody demonstrated neither ADCC nor CDC effects on activated human T cells expressing PD-1 (Wang et al. 2014). Finally, the pharmacological studies showed enhanced activation of cancer antigen-specific T cells and of cytotoxic activity of T cells against cancer cells, which is the mechanism of action of nivolumab.

14.10   Conducting Phase I Clinical Studies

In 2006, a Phase I study (CA209001/MDX-1106-01 study) was initiated in the United States to determine the safety, tolerability, and pharmacokinetics of nivolumab in human subjects with solid tumors. In the study, nivolumab was administered intravenously to 39 patients with advanced or recurrent malignancies (non-small cell lung cancer, melanoma, renal cell cancer, prostate cancer, or colorectal cancer) that had failed to respond to the existing treatments. One patient with colorectal cancer receiving 3 mg/kg, one with malignant melanoma receiving 10 mg/kg, and one with renal cell carcinoma showed antitumor effects (Brahmer et al. 2010). A Japanese Phase I clinical trial for solid tumors (ONO-4538-01 study) and an American Phase I repeated-dose clinical trial (CA209003 study) were both initiated in 2008. Within these trials, the drug was found to be effective against several carcinomas, including malignant melanoma, renal cell carcinoma, and non-small cell lung cancer (Oyama et al. 2014). At the beginning of these clinical trials, the prevailing distrust of cancer immunotherapy led to low priorities being given to nivolumab at medical institutions performing clinical trials, so recruiting patients was difficult. Despite these circumstances, clinicians were gradually shifting to more favorable postures as the patients treated with nivolumab improved. However, despite the low priority given by medical institutions, some physicians energetically helped Ono Pharmaceutical, which had no experience in clinical development of oncology drugs.[5]

In July 2009, Medarex was acquired by Bristol-Myers Squibb (Bristol-Myers 2009). It was a buyout that valued Medarex's therapeutic antibody pipeline, including nivolumab, and the technologies to produce therapeutic antibody drugs. This acquisition also led to changes in the division of labor between Ono Pharmaceutical and the partner, reflecting the high clinical development capabilities of Bristol-Myers Squibb. In 2005, the joint R&D agreement between Ono Pharmaceutical and Medarex gave Ono the right to develop nivolumab in all regions other than North America. However, the 2011 agreement with Bristol-Myers restricted Ono Pharmaceutical's rights to development and commercialization of nivolumab only in Japan, South Korea, and Taiwan (Ono Pharmaceutical 2005).

14.11 Late-Stage Clinical Studies and Approval

In 2011, a Phase II clinical study (ONO-4538-02 study) was initiated in Japan for the treatment of malignant melanoma. Based on the results of this study, Ono Pharmaceutical filed an application in December 2013 for the approval of manufacture and sale of nivolumab, and, in July 2014, obtained approval for the indication of malignant melanoma in Japan. This was the first approval for nivolumab in the world. It was launched in September 2014.[6]

In 2012, the results of clinical trials with PD-1 antibodies were presented at the 48th American Society of Clinical Oncology. In June of the same year, the results of the clinical trial were published in the *New England Journal of Medicine*, which showed that PD-1 inhibitors had sustained effects (Topalian et al. 2012) (Fig. 14.4). With the publication of the results of these clinical studies, nivolumab began to attract the interest of research institutes and pharmaceutical companies around the world (Awada 2015).

Because of the efficacy of nivolumab in malignant melanoma, renal cell carcinoma, and non-small cell lung cancer in a Phase II study, Bristol-Myers Squibb initiated a Phase III clinical trial. In a Phase III study (CheckMate 066) of untreated, unresectable melanoma in the absence of a BRAF mutation, nivolumab ($n = 210$) was compared with dacarbazine ($n = 218$) for 18 months and was found to significantly prolong survival (Robert et al. 2015) (Fig. 14.5). In December 2014, Bristol-Myers Squibb received FDA approval of nivolumab for the treatment of unresectable or metastatic melanoma that had progressed after treatment with the anti-CTLA-4 drug ipilimumab and a BRAFV600 mutation-positive BRAF inhibitor (Ono Pharmaceuticals 2014). In March 2015, nivolumab received further FDA approval for the treatment of patients with metastatic squamous cell lung cancer that had progressed

---

[5]From interview with Ono Pharmaceutical staff.
[6]Drug Interview Form: Opdivo.

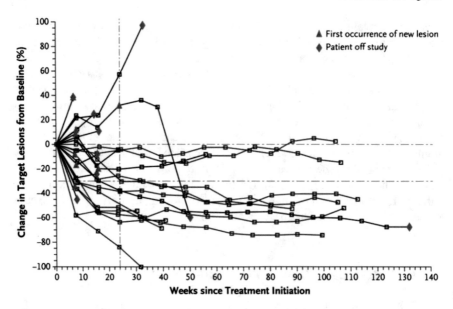

Fig. 14.4 Changes in target lesions after nivolumab treatment for malignant melanoma. *Source* Topalian et al. (2012)

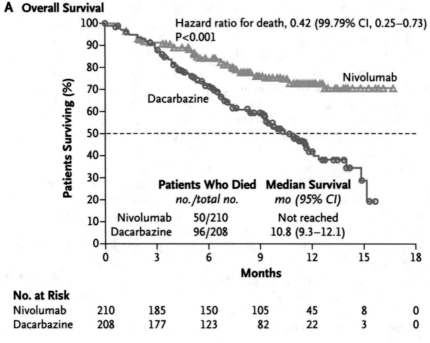

Fig. 14.5 Comparison of survival in the nivolumab-treated and dacarbazine-treated groups. *Source* Robert et al. (2015)

during or after treatment with platinum-based chemotherapy (Ono Pharmaceuticals 2015c).

In July 2015, the European Medicines Agency approved nivolumab for the treatment of locally advanced, and of metastatic, squamous cell lung cancer after chemotherapy (Ono Pharmaceuticals 2015d). In November 2015, an application was accepted by the European Medicines Agency for approval for an additional indication for nivolumab in patients with advanced renal cell carcinoma (Ono Pharmaceuticals 2015e, 2015f).

14.12   Scientific Basis for the Discovery of Nivolumab

As scientific sources used in nivolumab development, the following three findings had a direct impact:

(1) Discovery of PD-1 and the elucidation of its function by Prof. Tasuku Honjo's laboratory at Kyoto University.

The discovery of PD-1 by the Honjo laboratory in 1992 and its isolation for genetic identification was the beginning of the drug discovery process leading to nivolumab. In addition to discovering new molecules, PD-1 knockout mice were developed by introducing technologies established in the United States in the latter half of the 1980s. Careful and long-term research using these knockout mice suggested that PD-1 negatively regulates the immune response and created possibilities of further research leading to drug development.

(2) Experimental confirmation in mice that blocking PD-1 signaling is a powerful tool for cancer immunotherapy.

Professor Honjo's laboratory discovered that tumors cannot be engrafted in PD-1 knockout mice, and they also convincingly showed by the experiment of injecting mice with an anti-PD-L1 antibody that blocking PD-1 signaling could become a powerful cancer treatment.

(3) Use of Medarex humanized antibody technology; UltiMAb.

Using Medarex's proprietary humanized antibody technology, UltiMAb, nivolumab was created as a humanized anti-PD-1 monoclonal antibody. Subclass selection and structural stabilization were implemented to prevent T cells from inhibiting their binding.

14.13   Use of Public Research Grants, KAKEN, and Other Grants

A Grant-in-Aid for Scientific Research (KAKEN) supported Prof. Honjo's long-term research related to PD-1. KAKEN is the main grant-in-aid in Japan for researchers at universities and gives considerable freedom and flexibility to the Japanese scientists pursuing curiosity-driven basic research. First, PD-1 was discovered while engaging in another basic research project (cloning interleukin-2, 4, and 5, and elucidating the mechanism of class switching) supported by KAKEN. The discovery was serendipitous, and was not intended objective of the original research project.[7]

Second, since the discovery of PD-1 in 1992, PD-1 research was promoted by Prof. Honjo who regularly obtained KAKEN grants, including those for the formation of a Center and Excellence (COE). Professor Honjo obtained the following grants related to PD-1 research: (1) specially promoted research of KAKEN from 1992 to 1996 on "Research on the mechanism of lymphocyte differentiation: molecular mechanism of gene rearrangement and selective cell death by antigen," including isolation of the PD-1 gene. (2) From 1995 to 1999, the scientific research fund for COE formation on "Research on the functional control of higher order biological systems," including the development of the PD-1 gene-deficient mouse. (3) KAKEN Scientific Research (B) from 2000 to 2001 on "Isolation of ligands for PD-1 receptor in autoimmune diseases and their applications as immunotherapies." (4) Specially promoted research of KAKEN (COE) from 2000 to 2004 on "Construction and response of acquired biological information and research of abnormalities and conditions thereof," including clarification of the effect of the deficiency of PD-1 promoting the development of autoimmune diseases. (5) KAKEN Scientific Research (B) from 2002 to 2004 on "Investigation of the involvement of PD-1/PD ligands in models of rheumatoid arthritis and murine arthritis."

In addition, "Research support project for innovative research on drugs and medical devices" of the National Institutes of Biomedical Innovation, Health, and Nutrition supported the research project "Development of a new cancer therapy by inhibiting PD-1 immunosuppressive receptor signaling" from 2005 to 2009, which included the research on PD-1 inhibition therapy for cancer and joint research with Kyoto University Hospital. In this study, joint clinical research with Kyoto University Hospital revealed: (1) that PD-L1 expression was negatively associated with survival in ovarian cancer and malignant melanoma, (2) that the PD-1 and PD-L1 co-crystals were structurally analyzed to reveal the binding mode, and (3) that the binding of PD-1 and its ligand rendered T cells unresponsive to the same antigen that the T cells encountered earlier in experiments with PD-1 knockout mice (The National Institute of Biomedical Innovation 2005).

---

[7] From interview with Prof. Tasuku Honjo.

14.14 Sales of Opdivo

Opdivo recorded sales of 25 million USD (2.5 billion yen) in the first post-launch fiscal year (2014), 209 million USD in 2015, and 1026 million USD in 2016 in Japan, as shown in Fig. 14.6. The increase in sales was driven by an increase in the number of patients receiving the drug for approved diseases, as well as the inclusion of more target diseases. The sales in 2017, however, dropped to 0.9 billion USD, which will be explained later. In addition, Ono Pharmaceutical obtained 81 million USD (8.2 billion Yen) in 2015, 264 million USD in 2016, and 552 million USD in 2017, as patent royalty payment for Opdivo from Bristol-Myers Squibb.[8] Bristol-Myers Squibb has reported significantly increasing worldwide sales of Opdivo: 0.9 billion USD in 2015, 3.8 billion USD in 2016, and 4.9 billion USD in 2017.[9] Thus, Opdivo became a global blockbuster drug soon after its launch.

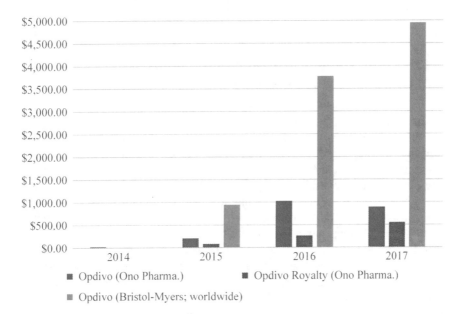

Fig. 14.6 Annual sales of Opdivo and royalty payments from Bristol-Myers Squibb to Ono Pharmaceutical (unit: 1 M USD). Sales data in Japan based on data published by Ono Pharmaceutical and converted to USD using OECD PPP exchange rate. *Source* Ono Pharmaceutical, Bristol-Myers Squibb company financial data

---

[8]Ono Pharmaceutical Co., Ltd. (2015, 2016, 2017, 2018), Annual report. https://www.ono.co.jp/jpnw/ir/ir_library.html#j_houkoku. Accessed 7 Aug 2018.

[9]Bristol-Myers Squibb (2016, 2017, 2018) Annual report. https://s21.q4cdn.com/104148044/files/doc_financials/quarterly_reports/2017/q4/Q42017-Earnings-Release.pdf. https://s21.q4cdn.com/104148044/files/doc_financials/quarterly_reports/2016/Q42016-Earnings-Release.pdf. Accessed 7 Aug 2018

In Japan, the drug price is regulated. If there is a similar drug for the target disease of the new drug, the price of the new drug is set with the prevailing price of the existing drug as a reference: the same price if there is no significant improvement of utility and a premium added if significant improvement exists. If there is no similar drug, the price is set based on the cost of the drug, taking into account its novelty. The latter rule applied to nivolumab when it was approved for the treatment of malignant melanoma. Because the number of applicable patients was small, the approved drug price was relatively high at 6584.86 USD (729,849 yen) for a 100 mg/10 mL bottle (MHLW 2016a). Because the approval in Japan preceded that in foreign countries, there was no adjustment based on its foreign price. Subsequently, the expected market size for Opdivo rapidly expanded and concerns were expressed on the fiscal burden of its high price. These concerns led to the decision of the Central Social Insurance Medical Council of the Ministry of Health, Labor and Welfare (MHLW) to cut the drug price of Opdivo by 50% in February 2017 as an exceptional response (MHLW 2016b; Japan Central Social Insurance Medical Council 2016).

14.15 Major Immune Checkpoint Inhibitors and Combination Therapy

The successful demonstration of immune checkpoint inhibitors as anticancer drugs by nivolumab and ipilimumab stimulated many of the world's leading pharmaceutical companies to enter the R&D competition for a new drug based on immune checkpoint inhibition as the mechanism of action. As a result, the types of molecular targets have also expanded considerably. A list of drugs is shown in Table 14.2. There are six cancer immunity checkpoint inhibitors that were on the market as of 2018: ipilimumab (Yervoy, Bristol-Myers Squibb) launched in 2011, nivolumab (Opdivo, Bristol-Myers Squibb, Ono Pharmaceutical) and pembrolizumab (Keytruda, Merck) both launched in 2014,[10] atezolizumab (Tecentriq, Genentech) launched in 2016, and avelumab (Bavencio, Pfizer) and durvalumab (Imfinzi, AstraZeneca) launched in 2017. All of these antibody drugs, expect for ipilimumab, target the PD-1 pathway.

Ipilimumab is primarily specialized in proliferating activated T cells and may be ineffective if the PD-1–PD-L1 pathway is established to attack cancerous cells, while nivolumab inhibits the binding of the PD-1–PD-L1 pathway by which activated T cells attack cancer cells, although it also increases activated T cells (Fig. 14.2). Thus, combination of the two drugs may provide a better balance between increasing the number of activated T cells and enhancing their ability to attack, and clinical studies have supported such conjecture. As a result, on 1 October 2015, the FDA approved a combination therapy of nivolumab and ipilimumab for malignant melanoma (Ono Pharmaceutical and Bristol-Myers 2015). In July 2014, Ono Pharmaceutical and

---

[10]In terms of anti-PD-1 antibodies, nivolumab's approval in Japan was the earliest in the world in July, followed by the US approval in September for pembrolizumab, and the US approval in December for nivolumab.

Table 14.2 Approved target diseases of immune checkpoint inhibitors other than nivolumab (as of August 2018)

Product name (generic name if not marketed)	Molecular targets	Targeted disease	Firm that developed the drug
Tencentriq (atezolizumab; MPDL3280A)	PD-L1 (CD-274)	Locally advanced or metastatic urothelial carcinoma	Genentech
		Metastatic non-small cell lung cancer	(Sales: Roche)
Bavencio	PD-L1 (CD-274)	Metastatic Merkel cell carcinoma (MCC)	Pfizer
(avelumab)		Advanced or metastatic urothelial carcinoma (UC)	
Imfinzi (durvalumab; MEDI4736)	PD-L1	Locally advanced or metastatic urothelial carcinoma	Medi-mune (Sales: AstraZeneca)
		Unresectable, stage III non-small cell lung cancer (NSCLC)	
Keytruda (Pembrolizumab)	PD-1	Melanoma	Merck
		Non-small cell lung cancer	
		Head and neck squamous cell cancer	
		Classical Hodgkin lymphoma	
		Primary mediastinal large B cell lymphoma	
		Urothelial carcinoma	
		Microsatellite instability-High cancer	
		Gastric cancer	
		Cervical Cancer	
Yervoy (Ipilimumab)	CTLA-4	Unresectable or metastatic melanoma	Medarex, Bristol-Myers Squibb
		Adjuvant melanoma	
		Advanced renal cell carcinoma	
		Microsatellite instability-high (MSI-H) or mismatch repair deficient (dMMR) metastatic colorectal cancer	

Sources Prepared by Clinicaltrials.gov, and made from prescription information of antibody drugs and clinical trials information such as (Bavencio 2017; Keytruda 2014; Pfizer 2018; Roche 2018; Tencentriq 2018)

Bristol-Myers Squibb entered a strategic alliance in response to the demonstration of the effectiveness of combination therapy in clinical development as well as for the joint implementation of globally integrated clinical development.[11]

There exists a significant share of cancer patients for each disease in whom the approved immune checkpoint inhibitors are not effective. As a result, the companies developing these drugs are also making partnerships with firms with other types of cancer drugs for combination therapies. In the case of Bristol-Myers Squibb and Ono Pharmaceutical, collaborations with Kyowa Hakko Kirin (Poteligeo), Eli Lilly (Galunisertib), Novartis (Zykadia, Capmatinib, and EGF816), Insight (INCB24360), Seattle Genetics (Adcetris), and Celgene (Abraxane) are in progress.

Box 14.1 Exploiting frontier science for drug development: a comparison with ipilimumab (Yervoy)
There are two major similarities between nivolumab and ipilimumab. Like nivolumab, ipilimumab is directly based on the scientific discovery from basic research at a university (UC Berkeley Cancer Research Laboratory) on the mechanisms of T cell activation and suppression. Furthermore, a bio-startup in the United States (in fact, the same firm) played an important role in the process of exploiting the scientific finding for drug development, because of the limited interest of major pharmaceutical firms in development of the drug based on its entirely new mechanism of action: inhibition of immune checkpoint.

Ipilimumab (Yervoy) is a recombinant, humanized monoclonal antibody that binds to the cytotoxic T-lymphocyte-associated antigen 4 (CTLA-4). It suppresses tumor proliferation by enhancing tumor antigen-specific T cell proliferation, its activation, and its cytotoxicity by blocking the binding of CTLA-4 and its ligands B7.1 (CD80) and B7.2 (CD86) on antigen-presenting cells (see Fig. 14.2 for an illustration). It also exhibits antitumor effects by enhancing tumor immune responses caused by the decrease of the function of regulatory T cells (Treg) and its numbers in tumor tissues. While the first clinical trial of ipilimumab was launched early in 2001 that covered several cancers (melanoma, prostate cancer, and lymphoma), it was successfully approved only for unresectable or metastatic melanoma in 2011, a decade later. As of May 2018, ipilimumab has been approved in more than 50 countries and regions worldwide (Bristol Myers and Ono Pharmaceutical 2018). Like nivolumab,

---

[11] According to press releases by Ono Pharmaceutical in 2014, which indicated that (1) the two companies jointly develop and commercialize nivolumab (Opdivo), ipilimumab (Yervoy), and the three tumor immune-related compounds at an early stage of development as single agents and in combination therapy; (2) Ono Pharmaceutical has the right to jointly develop and commercialize the four compounds of Bristol-Myers Squibb; (3) Bristol-Myers Squibb has the right to jointly develop and commercialize nivolumab in Japan, South Korea, and Taiwan, and to expand its leadership in tumor immunology; (4) under this partnership, patients in Japan, South Korea, and Taiwan will be integrated into global clinical trials to conduct efficient clinical research.

adverse reactions associated with autoimmune diseases have been reported for ipilimumab.

CTLA-4 was discovered as a new superfamily of immunoglobulins in 1987, followed by the identification of the human CTLA-4 gene in 1988 (Dariavach et al. 1988). In the mid-1990s, Prof. James Allison, the director of the UC Berkeley Cancer Research Laboratory, began CTLA-4 research as a part of his basic research on the mechanisms of T cell activation and suppression, together with a postdoctoral fellow Cynthia A. Chambers and a graduate student Matthew F. Krummel (UC Berkley Cancer Research Laboratory 2015). To elucidate the function of CTLA-4, Krummel and Allison generated antibodies to CTLA-4 and demonstrated that CD28 and CTLA-4 exerted opposite effects on lymphokine production and growth and that T cell activation was controlled by the signals generated by these two molecules (Krummel and Allison 1995). The researchers then published a paper in *Science* in 1996 that showed that blocking of the binding of CTLA-4 to T cells by the antibodies was able to suppress the growth of cancers (Leach et al. 1996). In July 1995, they filed a patent application for a CTLA-4 inhibitor with the applicant as The Regents of the University of California, and inventors as Leach, Krummel, and Allison (US5811097, "Blockade of T lymphocyte down-regulation associated with CTLA-4 signaling"); the patent was granted in 1998.

One of the unique characteristics of the commercialization of this CTLA-4 inhibitor is that it attracted no interest from major pharmaceutical firms and furthermore it went through three transfers of licenses to different bio-startup firms until full-scale development by Medarex. In May 1998, Nexter Pharmaceuticals, a United States bio-startup, acquired exclusive options for developing drugs that inhibit CTLA-4 from UC Berkley (Nexter Pharmaceuticals 1998). Its main objective was to utilize CTLA-4 in their drug discovery pipelines. In March 1999, Nexter was acquired and merged with Gilead Sciences (Gilead Sciences 1999). Options for CTLA-4 were also passed on to Gilead at this stage.

In August 1999, Medarex obtained exclusive sublicense for developing an CTLA-4 inhibitor from Gilead Sciences (Medarex 1999). According to the press release, Medarex intended to exploit its own antibody-making technology to create a fully humanized antibody drug that was later called Yervoy; it also had the right to develop a small molecule drug to block CTLA-4, and a part of the patent royalties from commercialization would be allocated to UC Berkeley and Gilead (ADIS R&D Profile 2010). Clinical trials were subsequently initiated, and in November 2001, a Phase II clinical trial was initiated in patients with melanoma and prostate cancer. The Phase II clinical trial for lymphoma was started in September 2002.

In January 2005, Medarex and Bristol-Myers signed a collaborative agreement to develop and commercialize ipilimumab and gp100 (MDX-1379) peptide vaccines. As previously mentioned, Medarex was acquired by Bristol-Myers Squibb in July 2009 (Bristol Myers 2009).

In March 2011, ipilimumab was approved in the United States for unresectable or metastatic melanoma. In Europe, the drug was approved in July 2011 for advanced-stage melanoma (unresectable or metastatic) in previously treated adults. In October 2013, the drug was expanded to include untreated advanced-stage melanoma. In Japan, the drug was designated as an orphan drug in March 2013. It received marketing approval in July 2015 and was launched on 31 August 2015. As of May 2018, ipilimumab has been approved in more than 50 countries and regions worldwide (Bristol Myers and Ono Pharmaceuticals 2018).

14.16 Conclusion

The development of nivolumab started with a serendipitous discovery from basic research in the laboratory of Prof. Tasuku Honjo at Kyoto University, which had been conducting cutting-edge basic science research on the immune mechanism for many years. However, PD-1 was discovered in an unexpected and serendipitous manner when conducting basic research on cell selection in the thymus. The development of ipilimumab also started from pure basic research by Prof. James Allison on the mechanism of activation and suppression of T cells. These two cases are excellent examples of how basic research aimed at deepening the understanding of fundamental questions, rather than research for specific applications or problem-solving, can become a direct source of a breakthrough innovation. The two professors shared the 2018 Nobel Prize in Physiology or Medicine.

In addition to discovering and isolating PD-1, the research in Prof. Honjo's laboratory played the central role in elucidating the effects and the mechanism of action of inhibiting PD-1. That is, in addition to pure basic research, the laboratory also engaged in use-oriented basic research, with the intention of promoting corporate investment in drug development based on its scientific discovery. It generated PD-1 knockout mice, clarified the function of PD-1 by such mice, and demonstrated that the inhibition of the binding between PD-1 and its ligand by an antibody works as a treatment of cancer. Patent applications were also actively pursued. In the process of conducting these "pure" and use-oriented basic researches, grants-in-aid for scientific research played an important role. With these grants, graduate students and postdoctoral fellows worked on unique research themes that required a long-term investment, including the creation and experimentation of PD-1 knockout mice, which was a time-consuming process. Thus, the case of nivolumab (Opdivo) also demonstrates the contribution of use-oriented basic research by the Honjo's laboratory to making a bridge between science and innovation, especially given the general reluctance of major pharmaceutical companies to invest in completely unknown technology.

In the implementation of the research, the knowledge and capability of the world were combined so that state-of-the-art research methods were utilized. The technology to produce a knockout mouse model was acquired from a university in the United States early on, and, in the search for PD-1 ligands, joint research with overseas research institutes, such as GI and Harvard University, was implemented. The formation of such international collaboration networks seems to have been made possible by Prof. Tasuku Honjo's intellectual leadership in the academic world.

Despite the concrete demonstration of the potential of PD-1 signal inhibition for cancer therapy in mouse models, major pharmaceutical companies in Japan and abroad showed no interest to invest in R&D because of their high distrust of cancer immunotherapy. Ono Pharmaceutical, however, had a long-term relationship with Prof. Honjo's laboratory and was in a unique position to observe the progress of the research of PD-1 directly. Thus, Prof. Honjo and Ono Pharmaceutical applied for a joint method-of-use patent. However, the approval for investing in drug development was not easily obtained even in Ono Pharmaceutical, and it took considerable time to make the decision.

Under these circumstances, the participation of the bio-startup Medarex in the joint development became an important driving force to enable Ono Pharmaceutical to prepare and clinically develop antibody drugs. Medarex had experience in developing an antibody drug, ipilimumab, based on its proprietary antibody-making technique. Thus, this case also demonstrates the importance of exploiting diverse capabilities and diverse views to develop a ground-breaking new medicine. In addition, the method-of-use patent by Ono Pharmaceutical and Prof. Honjo played an important role in promoting the participation of Medarex in terms of information provision and incentive.

Poor performance of past cancer immunotherapies has led to many clinical physicians holding skeptical views of cancer immunotherapy. However, many clinicians quickly changed their attitude after learning of the results of clinical trials, leading to the eventual launch of nivolumab. The 2009 acquisition of Medarex by Bristol-Myers Squibb created a global clinical-development framework for nivolumab, including combination therapy.

References

ADIS R&D Profile. (2010). Ipilimumab. *Drugs R D. 10*(2), 97–110.

Awada, H. (2015). The development of Nivolumab, a humanized anti-human PD-1 monoclonal antibody. *Regulatory Science of Pharmaceutical and Medical Devices, 46*(3), 126–130.

Bavencio. (2017). *Bavencio, highlights of prescribing information.* https://www.emdserono.com/content/dam/web/corporate/non-images/country-specifics/us/pi/bavencio-pi.pdf. Accessed August 11, 2018.

Brahmer, J. R., Drake, C. G., Wollner, I., Powderly, J. D., Picus, J., Sharfman, W. H., et al. (2010). Phase I study of single-agent anti-programmed death-1 (MDX-1106) in refractory solid tumors: safety, clinical activity, pharmacodynamics, and immunologic correlates. *Journal of Clinical Oncology, 28*(19), 3167–3175.

Bristol-Myers. (2009). *Bristol-Myers Squibb to acquire Medarex.* http://news.bms.com/press-release/partnering-news/bristol-myers-squibb-acquire-medarex. Accessed August 3, 2015.

Bristol-Myers and Ono Pharmaceutical Co., Ltd. (2018). *Pharmaceutical interview form/human-type anti-human CTLA-4 monoclonal antibody drug Yervoy intravenous infusion 50 mg, highlights of prescribing information.* https://packageinserts.bms.com/pi/pi_yervoy.pdf. Accessed August 11, 2018.

Dariavach, P., Mattéi, M. G., Golstein, P., & Lefranc, M. P. (1988). Human Ig superfamily CTLA-4 gene: chromosomal localization and identity of protein sequence between murine and human CTLA-4 cytoplasmic domains. *European Journal of Immunology, 18*(12), 1901–1905.

Drug Interview Form Opdivo Intravenous Drip Infusion 20 mg/Opdivo Intravenous Drip Infusion 100 mg (2014).

Egawa, S. (2009). *Trends toward cancer therapeutic good healthcare.* Kawase Shobo Shinsha Co., Ltd.

Freeman, G. J., Long, A. J., Iwai, Y., et al. (2000). Engagement of the PD-1 immunoinhibitory receptor by a novel B7 family member leads to negative regulation of lymphocyte activation. *Journal of Experimental Medicine, 192*, 1027.

Gilead Sciences. (1999). *Gilead sciences and Nexstar pharmaceuticals to merge.* https://web.archive.org/web/19990903023959/, http://www.nexstar.com/web/NeXnews.nsf/f9bd633013d67127872565f6007eb58f/cd672da2852df1c2872567270043d3e6?. Accessed January 1, 2016.

Hamanishi, J., Mandai, M., Iwasaki, M., Okazaki, T., Tanaka, Y., Yamaguchi, K., et al. (2006). Programmed cell death 1 ligand 1 and tumor-infiltrating CD8 + T lymphocytes are prognostic factors of human ovarian cancer. *PNAS, 104*(9), 3360–3365.

Hino, R., Kabashima, K., Kato, Y., Yagi, H., Nakamura, M., Honjo, T., et al. (2010). Tumor cell expression of programmed cell death-1 ligand 1 is a prognostic factor for malignant melanoma. *Cancer, 116*(7), 1757–1766.

Imfinzi. *Highlights of Prescribing Information.* https://www.azpicentral.com/imfinzi/imfinzi.pdf#page=1. Accessed August 11, 2018.

Ishida, Y., Agata, Y., Shibahara, K., & Honjo, T. (1992). Induced expression of PD-1, a novel member of the immunoglobulin gene superfamily, upon programmed cell death. *The EMBO Journal, 11*(11), 3887–3895.

Iwai, Y., Ishida, M., Tanaka, Y., Okazaki, T., Honjo, T., & Minato, N. (2002, September 17). Involvement of PD-L1 on tumor cells in the escape from host immune system and tumor immunotherapy by PD-L1 blockade. *PNAS 99*(19), 12293–12297.

Iwai, Y., Terawaki, S., & Honjo, T. (2004). PD-1 blockage inhibits hematogenous spread of poorly immunogenic tumor cells by enhanced recruitment of effector T cells. *International Immunology, 17*(2), 133–144.

Japan Central Social Insurance Medical Council. (2016, July 27). *The central social insurance medical council general assembly 334th minutes.* https://www.mhlw.go.jp/stf/shingi2/0000137208.html. Accessed August 9, 2018.

Keytruda. (2014). *Keytruda, highlights of prescribing information.* https://www.merck.com/product/usa/pi_circulars/k/keytruda/keytruda_pi.pdf. Accessed August 11, 2018.

Kinashi, T., & Honjo, T. (2010). *(IL)-4, cloning IL-5. Cytokine hunting* (pp. 98–102). Kyoto University Scientific Press.

Korman, A., Chen, B., Wang, C., Wu, L., Cardarelli, P., & Selby, M. (2007). Activity of Anti-PD-1 in murine tumor models: Role of "Host" PD-L1 and synergistic effect of Anti-PD-1 and Anti-CTLA-4. *Journal of Immunology, 178*, 48–37.

Krummel, M. F., & Allison, J. P. (1995). CD28 and CTLA-4 have opposing effects on the response of T cells to stimulation. *The Journal of Experimental Medicine, 182*(2), 459–465.

Kyoto University, Department of Molecular Biology, Graduate School of Medicine. (2005). "Chaos" Prof. Honjo's Laboratory History: Prof. Tasuku Honjo Professor Retirement Memorial Album (pp. 118–119).

Latchman, Y., Wood, C. R., Chernova, T., et al. (2001). PD-L2 is a second ligand for PD-1 and inhibits T cell activation. *Nature Immunology, 2*, 261.

Leach, D. R., Krummel, M. F., & Allison, J. P. (1996). Enhancement of antitumor immunity by CTLA-4 blockage. *Science, 271*(5256), 1734–1736.

Medarex. (1999). *Medarex signs development agreement with Gilead for Anti-CTLA-4 therapeutics.* https://web.archive.org/web/20061016120604/, http://www.medarex.com/cgi-local/item.pl/19990831-555415. Accessed January 1, 2016.

Ministry of Health, Labour and Welfare. (2016a). *Drug price calculation for new drugs.* https://www.mhlw.go.jp/file/05-Shingikai-12404000-Hokenkyoku-Iryouka/0000131478.pdf. Accessed August 9, 2018.

Ministry of Health, Labour and Welfare. (2016b). *About dealing with expensive drugs (draft).* https://www.mhlw.go.jp/file/05-Shingikai-12404000-Hokenkyoku-Iryouka/0000131476.pdf. Accessed August 9, 2018.

Nakanishi, J., Wada, Y., Matsumoto, K., Azuma, M., Kikuchi, K., & Ueda, S. (2007). Overexpression of B7-H1 (PD-L1) significantly associates with tumor grade and postoperative prognosis in human urothelial cancers. *Cancer Immunology, Immunotherapy, 56*(8), 1173–1182.

National Institute of Biomedical Innovation. (2005). "Project to Promote Basic Research in the Fiscal Year of Health and Medical Sciences" (1) Research PD-1 on the development of new cancer therapies by inhibiting immunosuppressive receptor signals, aiming at diagnosing, treating, and preventing drugs in areas where therapeutic methods are not available or existing therapeutics are insufficiently developed (e.g. lifestyle-related diseases, which are increasing with aging). http://www.nibio.go.jp/shinko/kisoken/2005report/pdf/report07-09.pdf. Accessed August 24, 2015.

Nexter Pharmaceuticals. (1998). *NeXstar obtains exclusive option to CTLA-4 blockade technology.* http://www.thefreelibrary.com/NeXstar+Obtains+Exclusive+Option+to+CTLA-4+Blockade+Technology-a020612341. Accessed January 1, 2016.

Nishimura, H., Minato, N., Nakano, T., & Honjo, T. (1998). Immunological studies on PD-1 deficient mice: implication of PD-1 as a negative regulator for B cell responses. *International Immunology, 10*, 1563.

Nishimura, H., Nose, M., Hiai, H., Minato, N., & Honjo, T. (1999). Development of lupus-like autoimmune diseases by disruption of the PD-1 gene encoding an ITIM motif-carrying immunoreceptor. *Immunity, 11*, 141–151.

Nishimura, H., Okazaki, T., Tanaka, Y., Nakatani, K., Hara, M., Matsumori, A., et al. (2001). Autoimmune dilated cardiomyopathy in PD-1 receptor-deficient mice. *Science, 291*(5502), 319–322.

Nobel Media AB. (2018). https://www.nobelprize.org/prizes/medicine/2018/press-release/. Accessed October 7, 2018.

Nomi, T., Sho, M., Akahori, T., Hamada, K., Kubo, A., Kanehiro, H., et al. (2007). Clinical significance and therapeutic potential of the programmed death-1 ligand/programmed death-1 pathway in human pancreatic cancer. *Clinical Cancer Research, 13*(7), 2151–2157.

Ohigashi, Y., Sho, M., Yamada, Y., Tsurui, Y., Hamada, K., Ikeda, N., et al. (2005). Clinical significance of programmed death-1 ligand-1 and programmed death-1 ligand-2 expression in human esophageal cancer. *Clinical Cancer Research, 11*(8), 2947–2953.

Okazaki, T., & Honjo, T. (2007). PD-1 and PD-1 ligands: From discovery to clinical application. *International Immunology, 19*(7), 813–824.

Okazaki, T., Maeda, A., Nishimura, H., Kurosaki, T., & Honjo, T. (2001). PD-1 immunoreceptor inhibits B cell receptor-mediated signaling by recruiting src homology 2-domain-containing tyrosine phosphatase 2 to phosphotyrosine. *Proceedings of the National Academy of Sciences of the United States of America, 98*(24), 13866–13871.

Ono Pharmaceutical. (2005). *Ono Pharmaceutical Co., Ltd. concludes joint research agreement on therapeutic antibodies with Medarex, Inc. U.S.A.* https://www.ono.co.jp/jpnw/news/pdf/2005/n05_0512.pdf. Accessed July 23, 2015.

Ono Pharmaceutical, Bristol-Myers Squibb. (2015). *A combination therapy of Opdivo and Yervoy for wild-type or metastatic melanoma approved by the U.S. Food and Drug Administration (FDA).* https://www.ono.jp/jpnw/n15_1005.pdf. Accessed January 2, 2016.

Ono Pharmaceutical, Bristol-Myers Squibb, Kyowa Hakko Kirin. (2015). *Concluding an agreement with Ono Pharmaceutical Co., Ltd., Bristol-Myers Squibb Co., Ltd., and Kyowa Hakko Kirin Optivo (generic name: nivolumab) on immunotherapy for advanced solid tumors with mogamulizumab.* https://www.ono.co.jp/jpnw/PDF/n14_1210_02.pdf. Accessed September 24, 2015.

Ono Pharmaceutical Co., Ltd. (2015). *Main progress in reported R&D activities and development products in the 67th Phase of the project report.* https://www.ono.co.jp/jpnw/ir/pdf/j_houkoku/67/09_10.pdf. Accessed September 17, 2015.

Ono Pharmaceutical Co., Ltd., Bristol-Myers Co., Ltd. (2015a). *Attachment-Antitumor Agent-View of the humanized anti-human PD-1 monoclonal antibody, Optivo Intravenous Infusion 20 mg/100 mg.* https://www.opdivo.jp/contents/pdf/open/tenpu.pdf. Accessed September 18, 2015.

Ono Pharmaceutical Co., Ltd. Bristol-Myers Co., Ltd. (2015b). *Revised precautions notice: View Opdivo intravenous infusion 20 mg/100 mg.* https://www.opdivo.jp/contents/pdf/open/precautions_revised.pdf. Accessed September 18, 2015.

Ono Pharmaceuticals. (2015c). *Bristol-Myers Squibb received accelerated clearance from the U.S. Food and Drug Administration (FDA) for Opdivo (generic name: nivolumab).* https://www.ono.co.jp/jpnw/PDF/n14_1226.pdf. Accessed August 12, 2015.

Ono Pharmaceuticals. (2015d). *For the first European PD-1 immune checkpoint inhibitor in which the European panel demonstrated Nivolumab (Nivolumab BMS) prolongs survival in previously approved patients with advanced squamous cell lung cancer.* https://www.ono.co.jp/jpnw/PDF/n15_0722_01.pdf. Accessed August 23, 2015.

Ono Pharmaceuticals. (2015e). *Opdivo (generic name: nivolumab) of Ono Pharmaceutical Co., Ltd. and Bristol-Myers Squibb Co., Ltd. obtained approval from the FDA for an expanded indication for patients with previously treated, unresectable, recurrent Non-Small-Cell Lung Cancer (NSCLC) and offered a life extension to more patients.* https://www.ono.co.jp/jpnw/PDF/n15_1014.pdf. Accessed January 2, 2016.

Ono Pharmaceuticals. (2015f). *Ono Pharmaceutical Co., Ltd. Bristol-Myers Squibb, European Medicines Agency accepts application for additional indication of Opdivo (Nivolumab) in patients with previously treated advanced renal cell carcinoma.* https://www.ono.co.jp/jpnw/PDF/n15_1106.pdf. Accessed January 2, 2016.

Ono Pharmaceutical, Status of Development Pipeline. https://www.ono.co.jp/eng/investor/pdf/fr/2018/0510/11.pdf. Accessed August 12, 2018.

Opdivo Approval History. https://www.drugs.com/history/opdivo.html. Accessed January 10, 2019.

Opdivo: EPAR—Product Information. https://www.ema.europa.eu/documents/product-information/opdivo-epar-product-information_en.pdf. Accessed January 10, 2019.

Open the ultimate drug for cancer: How to create the new drug, Opdivo, with the potential for a new treatment. Toyo Keizai Online. http://toyokeizai.net/articles/-/76852. Accessed September 1, 2015.

Oyama, Y., Ichido, M., & Yoshida, T. (2014). Human-type anti-human PD-1 monoclonal antibody "Nivolumab". *Medical Science Digest, 40*(8), 412–416.

Pfizer. (2018). *Pfizer Product Pipeline.* https://www.pfizer.com/science/drug-product-pipeline. Accessed August 12, 2018.

Robert, C., Long, G. V., Brady, B., Dutriaux, C., Maio, M., Mortier, L., et al. (2015). *Nivolumab in Previously Untreated Melanoma without BRAF Mutation, 372,* 320–330.

Roche. (2018). *Roche product development portfolio.* https://www.roche.com/research_and_development/who_we_are_how_we_work/pipeline.htm. Accessed August 13, 2018.

Shinohara, T., Taniwaki, M., Ishida, Y., Kawaichi, M., & Honjo, T. (1994). Structure and chromosomal localization of the human PD-1 gene (PDCD1). *Genomics, 23*(3), 704–706.

Tamada, K. (2012). Research and development of new therapies targeting immune escape mechanisms in cancer. *Yamaguchi Medical, 61*(1, 2), 5–10.

Taube, J. M., Anders, R. A., Young, G. D., Xu, H., Sharma, R., McMiller, T. L. et al. (March 28, 2012). Colocalization of inflammatory response with B7-H1 expression in human melanocytic lesions supports an adaptive resistance mechanism of immune escape. *Science Translational Medicine, 27, 4*(127), 127ra37.

Tencentriq. (2018). *Highlights of prescribing information.* https://www.gene.com/download/pdf/tecentriq_prescribing.pdf. Accessed August 11, 2018.

Thompson, R. H., Gillett, M. D., Cheville, J. C., Lohse, C. M., Dong, H., Webster, W. S., et al. (2004). Costimulatory B7-H1 in renal cell carcinoma patients: Indicator of tumor aggressiveness and potential therapeutic target. *PNAS, 101*(49), 17174–17179.

Topalian, S. L., Hodi, F. S., Brahmer, J. R., Gettinger, S. N., Smith, D. C., McDermott, D. F. … Sznol, M. (2012). Safety, activity, and immune correlates of Anti–PD-1 antibody in cancer. *The New England Journal of Medicine, 366, 26*, 2443–2454.

Toyo Keizai Online. (2015). *Cancer Therapeutics that enable new treatments - The R&D process of Opdivo (in Japanese).* https://toyokeizai.net/articles/-/76852. Accessed September 1, 2015.

UC Berkley Cancer Research Laboratory. (2015). *The Story of Yervoy (Ipilimumab) The idea of checkpoint blockade and the revitalization of immunotherapy.* http://crl.berkeley.edu/discoveries/the-story-of-yervoy-ipilimumab/. Accessed September 26, 2015.

Wang, C., Thudium, K. B., Han, M., Wang, X. T., Huang, H., Feingersh, D., et al. (2014). In vitro characterization of the anti-PD-1 antibody nivolumab, BMS-936558, and in vivo toxicology in non-human primates. *Cancer Immunology Research, 2*(9), 846–856.

Weekly Diamond, April 18, 2015.

Wong, R. M., Scotland, R. R., Lau, Roy L., Wang, C., Korman, A. J., Kast, W. M., et al. (2007). Programmed death-1 blockage enhances expansion and functional capacity of human melanoma antigen-specific CTLs. *International Immunology, 19*(10), 1223–1234.

Wu, C., Zhu, Y., Jiang, J., Zhao, J., Zhang, X. G., & Xu, N. (2006). Immunohistochemical localization of programmed death-1 ligand-1 (PD-L1) in gastric carcinoma and its clinical significance. *Acta Histochemica, 108*(1), 19–24.

Yoshida, Y., Koda, K., Nakao, S., & Naoyama. (2015). Pharmacological properties and clinical effects of a novel immune checkpoint inhibitor human PD-1 antibody, nivolumab (opsivo intravenous infusion 20 mg, 100 mg). *Pharmacology Journal, 146*, 106–114.

Chapter 15
Sources of Innovation of Drug Discovery in Japan and Its Implications

Sadao Nagaoka

Abstract This final chapter analyzes the sources of innovation by exploiting the case studies as well as a complementary large scale survey on drug discovery projects in Japan in a unified framework, and discusses their potential implications on innovation management and policy. The focal issues of discussion are: (A) knowledge sources of the discovery projects; (B) the dynamic relationship between drug discovery and scientific progress; (C) coping with uncertainty; (D) uniqueness and competition in discovery; and (E) implications on management and policy. We hope that the discussion of these topics will deepen our understandings of pharmaceutical innovations.

15.1 Knowledge Sources

15.1.1 Knowledge Sources Inspiring Discovery Projects

In over half of the twelve drug-discovery cases (12 groups of 15 drugs) studied in this volume, recent important scientific progress directly inspired the drug-discovery projects. Table 15.1 outlines the nature of innovation as well as the sources of the ideas for the discovery projects that have been described in the previous chapters. In 7 of the 12 cases, recent scientific progress achieved at a university and/or research institute was a direct trigger of the discovery project. The following four projects show particularly close relationships between the scientific discovery and the subsequent drug discovery, as indicated by the short lag between the year of scientific discovery and the year of drug discovery:

S. Nagaoka (✉)
Tokyo Keizai University, Tokyo, Japan
e-mail: sadao.nagaoka@nifty.com

© Springer Nature Singapore Pte Ltd. 2019
S. Nagaoka (ed.), *Drug Discovery in Japan*,
https://doi.org/10.1007/978-981-13-8906-1_15

Table 15.1 Nature of innovation and ideas inspiring discovery projects

Group	Generic and product names (Japan, USA, and Europe)	Year of discovery[a]	Therapeutic domain	Nature of innovation	Source of idea for discovery project (progress in science or known compound)	Scientific progress as key input to discovery project
1	Compactin (discontinued in Phase 2)	1974	High cholesterol	Novel mechanism of action that competitively inhibits HMG-CoA reductase, thereby lowers LDL cholesterol in blood by upregulating LDL receptors for the disease with no effective therapy	(1) Cholesterol synthesis mechanisms, including discovery of HMG-CoA as a rate-limiting enzyme (1966) (2) Internal cholesterol synthesis as main source for human body (1970)	○ (Yes)
	Pravastatin (Mevalotin, Pravachol)	1980		• Hydrophilic, unlike existing statins • Tissue selectivity for liver	Compactin (Mevalotin discovered from a metabolite of compactin, 1974)	
2	Rosuvastatin (Crestor)	1991		• Highly potent LDL-cholesterol lowering effect • Increasing effect on HDL cholesterol	Existing statins, especially fluvastatin (1984) which was the first statin totally synthesized	

(continued)

Table 15.1 (continued)

Group	Generic and product names (Japan, USA, and Europe)	Year of discovery[a]	Therapeutic domain	Nature of innovation	Source of idea for discovery project (progress in science or known compound)	Scientific progress as key input to discovery project
3	Leuprorelin (Leupron USA only)	1973	Prostate cancer	Treatment agents with novel mechanisms of action	(1) Discovery of LH-RH at Tulane University (1971) (2) A study of the relationship between prostate cancer progression and men's hormones (1941)	○ (Yes)
	Leuprorelin (Leuplin, Viadur)	1983		DDS-based sustained-release injectable formulations first commercialized in the world		
4	Ofloxacin (Tarivid, Floxin)	1980	Broad-spectrum antibiotic	• Broad antibacterial spectrum • Strength of antimicrobial activity and good pharmacokinetics	(1) Nalidixic acid (antibiotic licensed by Daiichi Pharmaceutical) was the lead compound (2) Norfloxacin (published in 1979)	
	Levofloxacin (Cravit, Levaquin)	1985		World's first optically active new quinolone synthetic antibiotic (high antimicrobial properties and low side effects)		

(continued)

Table 15.1 (continued)

Group	Generic and product names (Japan, USA, and Europe)	Year of discovery[a]	Therapeutic domain	Nature of innovation	Source of idea for discovery project (progress in science or known compound)	Scientific progress as key input to discovery project
5	Tamsulosin (Harnal, Flomax)	1980	Prostatic hyperplasia	Improvement of dysuria by a new mechanism of action (fast effect, fewer side effects)	(1) In-house hypertensive drug (Lowgan) was the lead compound (2) Progress in understanding the subtypes of $\alpha 1$ receptors distributed in the prostate (3) Clinical study on efficacy of α-blockers in urinary system	
6	Pranlukast (Onon)	1984	Bronchial asthma	Novel mechanism of action (CysLT receptor antagonists), highly effective oral agents	(1) Discovery of LT (1979) and chemical synthesis (2) Elucidation of causal mechanism of asthma attack	○ (Yes)
7	Tacrolimus (Prograf)	1984	Organ rejection prophylaxis	• Significantly stronger immunosuppressive effect than existing drugs (cyclosporine) • Reduction of side effects	Progress in understanding mechanism of rejection in transplantation (T lymphocyte proliferation, involvement of IL-2 in the process) in 1976/77	○ (Yes)

(continued)

Table 15.1 (continued)

Group	Generic and product names (Japan, USA, and Europe)	Year of discovery[a]	Therapeutic domain	Nature of innovation	Source of idea for discovery project (progress in science or known compound)	Scientific progress as key input to discovery project
8	Pioglitazone (Actos, Glustin)	1985	Diabetes	Novel mechanism of action (improvement of insulin resistance by thiazolidinedione compound) and significant reduction of side effects of hypoglycemia	Starting from in-house basic research, progress of understanding of diabetes disease mechanism (insulin resistance) was important	
9	Donepezil (Aricept)	1987	Alzheimer's disease	World's first successful inhibitor of Alzheimer progression (effective, low side effects)	(1) Choline hypothesis (1976) (2) Research on transport system at the blood–brain barrier	○ (Yes)
10	Candesartan (Blopress, Atacand, Amitas)	1990	High blood pressure and congestive heart failure	• Novel mechanism of action (angiotensin II receptor antagonism) • Effective even at low doses, low side effects	(1) Starting from in-house basic research, elucidation of the mechanism of renin–angiotensin system (1950s) (2) Losartan of DuPont (patent published January 1988), the research for which was in turn induced by earlier patents by Takeda	

(continued)

Table 15.1 (continued)

Group	Generic and product names (Japan, USA, and Europe)	Year of discovery[a]	Therapeutic domain	Nature of innovation	Source of idea for discovery project (progress in science or known compound)	Scientific progress as key input to discovery project
11	Tocilizumab (Actemra)	1992	Rheumatoid arthritis	Novel mechanism of action (IL-6 receptor inhibition) for significantly more fundamental treatment	(1) Basic research at the University of California Davis (recognition of cause of autoimmune diseases as B-cell activation phenomenon) (2) Osaka University discovery of IL6 (1986) and suggestion that it causes autoimmune diseases (1986)	○ (Yes)
12	Nivolumab (Opdivo)	2005	Cancer	• Novel mechanism of action (inhibition of PD-1 signals restores T cell ability to attack cancer cells) • Long-term survival effect, applicable to wide range of cancers	(1) Discovery and function of PD-1 (1998) (2) Confirmation by mouse experiments (2002) that inhibition of PD-1 signaling is a potent method of cancer immunotherapy	○ (Yes)
Summary/average		1985			Four cases in which known compounds were important sources of idea for discovery project	Seven cases

[a]Year of discovery is identified by priority year of the basic patent

- Leuprorelin (Leuplin, Leupron, Viadur) (1973)[1]: The discovery of luteinizing hormone-releasing hormone (LH-RH) at Tulane University (1971) directly stimulated the search for derivatives with stronger activity.
- Pranlukast (Onon) (1984): Leukotrienes (LTs) were discovered at the Karolinska Institute in Sweden and found to be a causative agent of asthma attacks. Immediately afterward, a method of LT synthesis was discovered at Harvard University (1980). These elements of progress directly triggered a discovery project for LT receptor antagonists.
- Tocilizumab (Actemra) (1992): The discovery of interleukin-6 (IL-6) and its receptor at Osaka University (1986 and 1988, respectively) provided direct targets for a drug-discovery project and triggered academia–industry collaboration.
- Nivolumab (Opdivo) (2005): The functional characterization of PD-1 in 1988 and the confirmation that the inhibition of PD-1 signaling is an effective tool for cancer immunotherapy in mouse models in 2002, both at Kyoto University, provided a specific target for drug discovery.

In the following three cases, scientific progress also led to the idea of the discovery project, although the year of the underlying scientific discovery and that of the drug discovery are further apart than those of the above cases.

- Compactin (1974): It was discovered in 1966 that HMG-CoA reductase was the rate-limiting enzyme for cholesterol biosynthesis, following the studies on the mechanism of the synthesis the 1950s and 60s. The discovery project with this molecule as a target started in 1971 at Sankyo.
- Tacrolimus (Prograf) (1984): In 1976/77, IL-2 was found to induce proliferation of T cells and to enhance the immune response. The discovery project for an IL-2 inhibitor was initiated in 1982, based on these findings.
- Donepezil (Aricept) (1987): Based on the acetylcholine hypothesis (1982), acetylcholinesterase was targeted by a drug-discovery project. Targeting this enzyme was not new, but none had been previously successful so that this was no more a promising recognized potential target at the time. However, the discovery project was initiated at Eisai, based on the judgment that the hypothesis had not yet received adequate investigation, and the competitors were absent.

In the following two cases, the discovery projects started from in-house basic research.[2]

- The two discovery projects which yielded pioglitazone (Actos, Glustin) (1985) and candesartan (Blopress, Atacand, Amitas) (1990) began as in-house basic research investigating diabetes mellitus in the 1960s and diuretics in the early 1970s, respectively, at Takeda. Long-term cumulative research led to the discovery of these two highly successful drugs.

---

[1]Leuprorelin, the drug substance of the drug-delivery system formulation leuprorelin, was discovered in 1973 and sold in the United States in 1985 as a self-administered daily Lupron Injection.

[2]The fact that basic research was also carried out by Chugai (Dr. Ohsugi) in the case of tocilizumab would have been important for its early recognition of the drug-discovery opportunities targeting IL-6.

Pre-existing drugs or patent disclosures were the critical source of the idea for the discovery research in the following four projects:

- In the case of tamsulosin (Harnal, Flomax) (1980), work began as a discovery project for a hypertension drug, which used a pre-existing in-house chemical compound as a lead compound.
- Ofloxacin (Tarivid, Floxin) (1980) and levofloxacin (Cravit, Levaquin) (1985), and rosuvastatin (Crestor) (1991) had pre-existing drugs with an effective mechanism of action that served as lead compounds for the discovery project.
- The project leading to the discovery of candesartan (Blopress, Atacand, Amitas) (1990) was suspended once. Ultimately, the final chemical compound was discovered by making use of DuPont's losartan patent (published January 1988), the discovery of which in turn made use of the patent publications of the suspended project of Takeda.

Although there are cases in which in-house basic research or pre-existing drugs became the source of these highly innovative drugs, scientific progress was the most frequent source of ideas for these 12 discovery projects.

15.1.2 Transfer of Knowledge Through Direct Collaboration Among Researchers

As discussed above, recent scientific advances at universities and national laboratories were often important sources to initiate a drug-discovery project. But the question is—how did Japanese companies optimize this technology transfer process? In many cases, direct collaboration between university scientists and corporate researchers played an important role for the initiation of discovery research. In the following five cases, key corporate researchers worked in university laboratories and other institutions that were conducting cutting-edge research in the field, which led to the early absorption of those ideas into pharmaceutical research and development programs.

- Leuprorelin (Leuplin, Leupron, Viadur): Dr. Fujino of Takeda Pharmaceutical engaged in research on LH-RH under Prof. Darrell Ward at the University of Texas,[3] and launched an early research and development project on LH-RH derivatives immediately after returning to Japan.
- Pranlukast (Onon): Dr. Toda, who played the central role in the synthetic research of the discovery project, studied under Prof. Corey of Harvard University, the inventor of a synthetic method for LTs.
- Tacrolimus (Prograf): Dr. Goto, then of Fujisawa Pharmaceutical, studied the transplant rejection mechanism at the National Cancer Institute, which prompted him to launch a discovery program for immune suppressants after he returned home.

---

[3]Professor Darrell Ward was a colleague of Roger Guillemin at Baylor College of Medicine at the University of Texas, who was in the race with Prof. Schally at Tulane University to isolate LH-RH.

- Tocilizumab (Actemra): Dr. Ohsugi of Chugai Pharmaceutical worked with Prof. Gershwin at the University of California, Davis, which led to the launch of a B cell-based autoimmune disease research project following his return to Japan. Immediately after he learned of the discovery of IL-6 at Osaka University at an academic conference, he recognized the potential of IL-6 as the target for a drug treating autoimmune disease and launched a joint discovery project between Chugai Pharmaceutical and Osaka University.
- Nivolumab (Opdivo): Dr. Shibayama recognized that a drug discovery targeting PD-1 was valuable while visiting Dr. Honjo's Laboratory at Kyoto University, and promoted the establishment of a research and development project by Ono Pharmaceutical, based on the results of the research advances at Kyoto University.

In the following three cases, the research experience at advanced research institutions played an important role in helping corporate scientists acquire novel research methods, and shape research strategies.

- Compactin: The US research experience by Dr. Endo of Sankyo, led to his recognition of the research competition in the USA and the importance of a unique research project, and prompted him to adopt the strategy of screening candidates from natural products. Furthermore, he discovered that arteriosclerosis was a serious health concern in the USA, which made him to recognize the value of a drug that lowers blood cholesterol.
- Leuprorelin (Leuplin, Leupron, Viadur): Dr. Okada of Daiichi Pharmaceutical, who played the central role in the development of sustained-release formulations, studied drug delivery system technology at the University of Kansas, before initiating the project. Dr. Yamamoto, who was the collaborator for Okada at Takeda, also worked at the University of Texas on a microcapsule process using polylactic acid.
- Candesartan (Blopress, Atacand, Amitas): Dr. Nishikawa, who was the leader of the pharmacology group at Takeda Pharmaceutical, was engaged in research on the renin–angiotensin system (RAS) at the University of Washington, and this experience helped him to discover the angiotensin II (AII) antagonism of the lead compound CV-2198.

These cases indicate that direct contact of corporate researchers with cutting-edge research at universities and research institutions was very important for them to understand the progress of science and to launch corporate research programs that make use of this recent progress. Studies on transfers of early stage biotechnology from universities to industry in the United States indicate that having local connections, living and working together in the same geographical region, between the university researchers who developed the new knowledge, and the biotech companies willing invest in its commercialization, is important to effectively transfer knowledge from academia to industry (Zucker et al. 1998). The case studies of Japanese drug discoveries covered in this book show that human interactions were similarly important for the transfer of knowledge, with regards to the breakthrough drugs developed in Japan. They confirm the importance of tacit knowledge embodied in human beings. However, the technology transfer was realized primarily by the movement of corporate researchers from firms to universities. This is a different mechanism of technology transfer to overcome geographical distance, compared to the start-up system in the USA.

15.1.3 Research Tools and the Discovery of the New Use of a Compound

The introduction of new research tools often made essential contributions to the success of discovery project research, as summarized in Table 15.2. In 10 of the 12 projects discussed, advances in research tools such as animal models and optical resolution played very important roles, and in most of such cases (9 cases), the new tools exploited were contributed significantly by universities or national research institutes. These cases confirm that new tools are important embodiments of scientific advances for drug discovery. Thus, the ability of firms to make use of the progress of science through exploiting and developing the advances in research tools was also of great importance.

- Compactin: By improving the screening technology developed by academic researchers, compactin was discovered after the screening of active substances from over 6000 microorganism stocks within 2 years.
- Pravastatin (Mevalotin, Pravachol): Preclinical studies were efficiently conducted using a new animal model with hyperlipidemia.
- Leuprorelin (Leuplin, Leupron, Viadur): Learning of peptide synthesis technology by a corporate researcher at the Institute for Protein Research at Osaka University played an important role in the first successful synthesis of leuprorelin.
- Levofloxacin (Cravit, Levaquin) and tamsulosin (Harnal, Flomax): The advancement of optical resolution technology allowed the researchers to acquire a new drug with maximum activity and minimal side effects.
- Pranlukast (Onon) : Ono researchers were able to learn the method for synthesizing LT developed at Harvard University early after its discovery, enabling early development of a screening system for LT receptor antagonists.
- Tacrolimus (Prograf): The mixed lymphocyte reaction, which Dr. Kino from Fujisawa had learned at Tokai University, was used for screening.
- Pioglitazone (Actos, Glustin): The presence of model mice with spontaneous diabetes (KKAy mice) was a prerequisite for initiating discovery programs.
- Donepezil (Aricept): Computer-aided drug design, which Dr. Kawakami from Eisai learned at the University of Tsukuba, contributed to improving the drug design, which had faced serious tradeoff concerns between bio-availability, efficacy, and safety.
- Candesartan (Blopress, Atacand, Amitas): The model animals SHRs (spontaneously hypertensive rats) and SHRSPs (stroke-prone SHRs) developed by universities enabled the company to carry out compound screening with high efficiency.

Table 15.2 Advances in research tools and discovery of the use of new drugs

Group	Generic and product names (Japan, USA, and Europe)	Significant advances in research tools: ○ contributed by a university or national research institute	Discovery of use: ○ contributed by a university or national research institute
1	Compactin (discontinued in Phase II)	○ Advances in experimental method for cell-free cholesterol synthesis	
	Pravastatin (Mevalotin, Pravachol)	○ Model animals with hyperlipidemia (WHHL rabbits at Kobe University)	
2	Rosuvastatin (Crestor)		
3	Leuprorelin (Leupron USA only)	○ Peptide synthesis technology from Osaka University Protein Research Institute (through researcher stay)	
	Leuprorelin (Leuplin, Viadur)		
4	Ofloxacin (Tarivid, Floxin)		
	Levofloxacin (Cravit, Levaquin)	• Advances in optical resolution technology made at Nagoya University (for optical resolution of Tarivid)	
5	Tamsulosin (Harnal, Flomax)		○ Suggestion of use of α1-blockers for treatment of urinary disorders
6	Pranlukast (Onon)	○ Early adoption of LT synthesis technology enabled early construction of screening systems	
7	Tacrolimus (Prograf)	○ Screening methods such as mixed lymphocyte reaction (learned at Tokai University)	○ Elucidation of mechanism of action of tacrolimus contributed to expansion of use for atopic dermatitis

(continued)

Table 15.2 (continued)

Group	Generic and product names (Japan, USA, and Europe)	Significant advances in research tools: ○ contributed by a university or national research institute	Discovery of use: ○ contributed by a university or national research institute
8	Pioglitazone (Actos, Glustin)	○ Model animals (KKAy mice, etc.) developed at Nagoya University, used for screening	Discovery of blood sugar-lowering (hypoglycemic) effect of a compound discovered in research project for a lipid-lowering drug
9	Donepezil (Aricept)	○ Structural design using CADD (computer-aided drug design learned at Tsukuba University), contributing to the creation of highly optimized drug • Optical resolution by using a new column (no significant difference was found)	
10	Candesartan (Blopress, Atacand, Amitas)	○ Spontaneously hypertensive rats (SHRs introduced from Kyoto University and co-developed SHRSP)	Discovery of blood pressure-lowering (hypotensive) effect of the compound found in diuretic research
11	Tocilizumab (Actemra)	○ Genetic engineering technology for antibody humanization (CDR transplantation of MRC)	○ Discovery of a new use of Actemra (for multiple myeloma cells) at Hiroshima University
12	Nivolumab (Opdivo)	• Genetic engineering technology for antibodies humanization (Medalex UltiMAb)	
Summary (Number of cases with significant advances in research tools or with discovery of use)		Ten cases (seven Japanese sources)	Five cases
Frequency of universities and national research institutes as sources		Nine cases	Three cases

- Tocilizumab (Actemra): A joint research program with the Medical Research Council (MRC) led to the successful design of a new drug based on a newly developed technology for humanized antibodies.
- Nivolumab (Opdivo): Succeeded in the generation of humanized antibodies in cooperation with Medarex, which possessed its own humanized antibody and was willing to participate in a project that many regarded as not promising.

To bring innovative drugs to market, innovations were also required for the manufacturing process of the drugs, or for their clinical trials, for one-third of the projects. New scientific or technological advancements were utilized to solve these problems.

- Pravastatin (Mevalotin, Pravachol): A two-step fermentation process was necessary; first to compactin and then to pravastatin. Industry–university cooperation was utilized for their efficient implementation.
- Leuprorelin (Leuplin, Leupron, Viadur): There was a need to develop a new, non-catalytic polymer synthesis method for microcapsules, and Eli Lilly's patent literature provided important suggestions.
- Levofloxacin (Cravit, Levaquin): The use of asymmetric synthesis technology was necessary for mass production of levofloxacin.
- Tacrolimus (Prograf): Clinical trials using cutting-edge transplantation technology were guided by Prof. Starzl at the University of Pittsburgh.

The discovered drug candidate may have different uses than originally envisioned, and the identification of these new uses increases the value of the R&D program and the ultimate probability of developing a successful drug, even if the newly found use may not be ultimately commercialized. Among the case studies, as shown in Table 15.2, there are five cases of new-use discovery, three of which eventually led to the drug launch for the newly identified indication. These three cases are examples of serendipity, in which in-house discovery research identified new uses that differed from the original objective of the discovery project. In the first three of the following five cases, the progress of clinical research in universities played an important role for discovering the new uses of the drugs.

- Tamsulosin (Harnal, Flomax): The applicability of this drug candidate to benign prostatic hyperplasia was suggested by a clinical research paper on the effectiveness of alpha-blockers. The original aim of the project was to develop antihypertensive drugs.
- Tacrolimus (Prograf): The elucidation of the mechanism of action of tacrolimus has contributed to the expansion of new applications such as the treatment of atopic dermatitis.
- Tocilizumab (Actemra): The finding by Prof. Kawano of Hiroshima University, which indicated the applicability of the drug to treating multiple myeloma, was very important for sustaining the research and development project in the company.
- Pioglitazone (Actos, Glustin): The hypoglycemic effect was found for a compound discovered in the project while searching for a lipid-lowering drug, and the purpose of the project was redefined to the antidiabetic drug.

– Candesartan (Blopress, Atacand, Amitas): In the diuretic research by Takeda, it
was found that the compound discovered possessed hypotensive action, and the
purpose of the project was redefined.

15.1.4 Survey Evidence on Knowledge Sources for Drug Inventions

The importance of science, research tools, and university–industry collaborations as
sources for drug-development projects were confirmed by a large-scale survey for
drug development of new molecular entities in Japan (Nagaoka et al. 2015), covering
234 projects. The survey included those approved for market launch from 1990 to
2011, those under clinical trials as of 2012, as well as discontinued projects that were
randomly selected and matched by field and year to those ongoing projects.[4] The
survey asked how important each knowledge source was for the discovery projects:
essential for the idea, essential for implementation, or nonessential but significant
delay if not available.

As shown in Fig. 15.1, scientific publications were essential for the idea of the
discovery project for 38% in the follow-on projects with a pre-existing drug and 33%
for novel projects with no such drugs. Combining the cases where they were essential
for implementing the discovery projects or their absence would have caused their
significant delay, they were very important for around two-thirds of both types of
projects. Research equipment and materials were more important in implementation
than in idea generation and very important for around 30% for both types of projects.
University and industry collaborations (including collaborations with researchers in
professional research organizations; PRO) were more essential for novel projects:
very important for 30% of the novel projects and 20% of the follow-on projects.
Specifically, 8% for the idea in novel projects, relative to 2% for the idea in follow-
on projects. Patent literature was significantly more important for follow-on projects:
very important for more than 30% of the follow-on projects and 10% of the novel
projects. Thus, absorbing scientific progress (disembodied in literature or embodied
in researchers and in research equipment and materials) is often essential for innovative drug discovery projects, and knowledge embodied in researchers is especially
important for novel projects, consistent with our case studies.

---

[4]Projects approved for market launches accounted for 22%, those under clinical trials accounted
for 45%, and discontinued projects accounted for 33%. The projects discontinued at the preclinical
stage were significantly under-sampled.

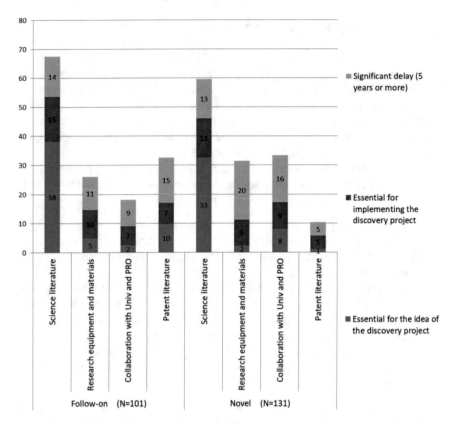

Fig. 15.1 Knowledge sources of discovery project generation and implementation. *Source* Nagaoka et al. (2015)

15.2   Dynamic Relationship Between Drug Discovery and Scientific Progress

15.2.1   Status of Science When the Discovery Research Begins

As indicated above, advances in science, including the discovery of targets, often brought drug-discovery opportunities to the forefront. At the same time, however, the case studies also show that science is quite often "incomplete" when firms initiate their discovery research. Once a new drug-discovery opportunity is recognized, even when the science remains incomplete, a firm often initiates its discovery research project, partly because early initiation is the key to the first-mover advantage in research and development competition. As our case studies demonstrated, the relevant science could be clarified as the drug discovery advanced, so that the drug-discovery process often helped in "completing" the scientific understanding.

Table 15.3 summarizes whether the target (including its function) was known at the time of starting a discovery project (or basic research program) by the pharmaceutical firm, and if the existence of a compound acting on this (or the mechanism of action) was known at that time. Notably, the target was unknown in four cases. Moreover, the mechanism of action for the target was unknown in seven cases, including three cases for which the target was known. In the following three projects, although the target was known when the discovery research began, there was no prior compound informing the mechanism of action, so that it was unknown whether any drug could be discovered for such a target.

– Compactin/pravastatin: Although HMG-CoA reductase was known to be a candidate target, it was unknown whether there were any compound that would competitively inhibit it. Furthermore, the mechanism of action of such an inhibitor remained entirely unknown.
– Leuprorelin (Leuplin, Leupron, Viadur): Although it was expected that increasing the activity of LH-RH would have some effect, it was only later discovered in the project that leuprorelin downregulates the LH-RH receptor.
– Candesartan (Blopress, Atacand, Amitas): It was unclear whether there were compounds that inhibit the angiotensin II (ATII) receptor. The ATII receptor subtype 1, to which candesartan binds, was discovered only in 1991, a year after the discovery of the drug.

Consequently, the mechanism of action of a drug to be developed was known at the start of the project in only the following five projects. In the projects of rosuvastatin (Crestor), ofloxacin (Tarivid, Floxin), and levofloxacin (Cravit, Levaquin), the lead compounds embodying the known mechanism of action existed prior to their discovery projects. In the case of donepezil (Aricept), there was a prior approved drug with the same mechanism of action (tacrine). However, it failed to serve as a lead compound, and furthermore it was withdrawn because of high toxicity. In the case of pranlukast (Onon) and nivolumab (Opdivo), both the target and the basic mechanism of action were clarified by the prior research of universities before the start of the discovery research by the enterprises.

However, in seven cases, the mechanism of action was unclear. In addition, in the case of donepezil, a prior withdrawn drug (tacrine) indicated the difficulty of finding a potent compound with minimal side effects. Thus, corporate discovery research was often (in around two-thirds of the cases) initiated at a stage of incomplete scientific understanding. This incomplete science is an important reason why drug discovery often faces unexpected difficulties, as we will see in the next section. The incompleteness of science also means that it is often important for a drug-discovery project to promote elucidating the mechanisms of action by industry–university cooperation.

Table 15.3 Completeness of science when discovery projects began

Group	Generic and product names (Japan, USA and Europe)	Mechanism of action of drug	Target (including function) known at time of initiating research by pharmaceutical firm? (○: Yes; X: No)	Existence of a compound acting on this target or mechanism of action known at time of starting discovery project? (○: Yes; X: No)
1	Compactin (discontinued in Phase II) Pravastatin (Mevalotin, Pravachol)	Competitive inhibition of HMG-CoA reductase, resulting in upregulation of LDL receptors and absorption of blood cholesterol	○	X Statins completely unknown, mechanism for regulating cholesterol metabolism through LDL receptors was discovered by Goldstein and Brown, in parallel with development of statins
2	Rosuvastatin (Crestor)		○	○
3	Leuprorelin (Leupron USA only) Leuprorelin (Leuplin, Viadur)	Downregulation of LH-RH receptors	○ Discovered in immediate past	X Downregulation was discovered in discovery project by joint research with Abbott in 1976
4	Ofloxacin (Tarivid, Floxin) Levofloxacin (Cravit, Levaquin)	Inhibition of DNA gyrase activity	○	○
5	Tamsulosin (Harnal, Flomax)	Antagonist of the α1 adrenergic receptor	X Application to urinary disorders found after discovery of compound	X Mechanism of prostatic selective action was elucidated later in the 90s in collaboration with University of Tokyo and University of Essen
6	Pranlukast (Onon)	Antagonist of CysLT receptors	○ Discovered in immediate past	○

(continued)

Table 15.3 (continued)

Group	Generic and product names (Japan, USA and Europe)	Mechanism of action of drug	Target (including function) known at time of initiating research by pharmaceutical firm? (○: Yes; X: No)	Existence of a compound acting on this target or mechanism of action known at time of starting discovery project? (○: Yes; X: No)
7	Tacrolimus (Prograf)	Selective inhibition of IL-2 production from T cells (targeted enzyme calcineurin)	X	X It was later found that complex of tacrolimus and binding protein inhibits calcineurin and suppresses IL2 production
8	Pioglitazone (Actos, Glustin)	Activation of PPAR γ	X PPAR γ target (mechanism of action) was elucidated in 1995	X Mechanism of action found later by joint research with Toyama Medical and Pharmaceutical University, Kyoto University, Kobe University, and Tokyo University
9	Donepezil (Aricept)	Inhibition of AChE	○	○ However, unknown whether drugs would have both high efficacy and low side effects
10	Candesartan (Blopress, Atacand, Amitas)	Antagonist of angiotensin II (AtII) receptor	○ ATII receptor	X Antagonist of ATII receptor was unknown, and isolation of ATII receptor subtype was found ex post in 1991
11	Tocilizumab (Actemra)	IL-6 receptor inhibition	X Unknown during basic research stage, discovered by Osaka University	X Mechanism of action was found in project by joint research with Osaka university
12	Nivolumab (Opdivo)	Inhibition of PD-1 signaling	○ Function of PD1 was elucidated in immediate past	○ Discovery of PD1 ligands had clarified a mechanism of action of antibody drug
Frequency of X (unknown target or mechanism)			Four cases	Seven cases

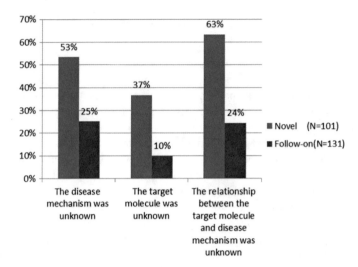

Fig. 15.2 Incomplete science when drug-discovery projects began. *Source* Nagaoka et al. (2015)

15.2.2 Survey Evidence on Incomplete Science When Drug Discovery Begins

Figure 15.2 provides complementary evidence on how "incomplete" the science was when the discovery project began, based on the large-scale survey on drug-discovery projects in Japan introduced early (see Sect. 15.1.4). We measure the incompleteness by the extent of which the disease mechanism was unknown, the target molecule was unknown, and the relationship between the target molecule and disease mechanism was unknown when the discovery project began. As expected, the novel projects with no prior drugs faced significantly higher uncertainty. The target molecule was unknown in almost 40% of the novel projects. Furthermore, the relationship between the target molecule and disease mechanism was unknown for around two-thirds of novel projects. It is also noteworthy that the disease mechanism is unknown for 25% of the projects, even if there existed prior drugs. These results indicate high uncertainty of drug-discovery projects, consistent with the findings from our case studies presented in this book.

15.2.3 Drug Discovery as Impetus to Scientific Progress

Given that drug discovery often starts while the science is incomplete, drug-discovery efforts create opportunities to provide tools for advancing science. The drug candidate discovered as a result of drug-discovery efforts often serves as an effective verification tool to determine the mechanism of how the target molecule in the body performs its function. Such contribution of a drug discovery is illustrated in the following four cases:

- Compactin: The mechanism by which statins selectively lower LDL cholesterol in the blood through LDL receptors was elucidated by Brown and Goldstein at the University of Texas, who had discovered the LDL receptor at around the same year of the discovery of compactin. Elucidation of this mechanism in turn ensured that statins were safe despite cholesterol being an essential substance in the body. This development played an important role for Merck to resume clinical trials of lovastatin, which had been discontinued following the discontinuation of the clinical trials of compactin. Statins in turn was used to clarify the feedback control mechanism mediated by the HMG-CoA enzyme.
- Pioglitazone (Actos, Glustin): Joint research between Takeda and Upjohn led to the recognition of siglitazone and pioglitazone as the new basic skeleton of a drug compound, and to the promotion of insulin resistance research. The discovery of the thiazolidinediones actual targets (PPAR gamma) was made in 1995, long after the discovery of the molecular basis of the first drug (siglitazone) in 1978.
- Tacrolimus (Prograf): The mechanism by which tacrolimus suppresses immune responses (i.e., the action of the complex of tacrolimus and FKBP) was subsequently clarified. The discovery of tacrolimus also contributed to the elucidation of the T cell signal transduction function, and it also became a trigger to the chemical biology development.
- Tocilizumab (Actemra): It was unclear what role IL-6 played in the body's immune system in rheumatoid arthritis. It was found that the inhibition of IL-6 by tocilizumab significantly improved joint swelling, joint pain, anorexia, general fatigue, and fever, demonstrating that IL-6 induces these various symptoms observed in patients (Box 13.2).

Thus, while science creates drug-discovery opportunities, drug-discovery efforts also create opportunities to advance science by providing new tools. They are in a dynamic relationship that drives each other.

15.3   Coping with Uncertainty

Successful commercialization involved coping with unexpected difficulties. Uncertainties associated with the drug-discovery process were revealed not only through the emergence of unexpected difficulties but also through cases of serendipitous discovery and unexpected good luck.

15.3.1 Unexpected Difficulties

Although all 12 cases included in this book launched drugs to the market, there were unexpected difficulties in the course of research and development in three-quarters of these cases, and half of these crises could have led to discontinuation of the projects, as shown in Table 15.4.

In the following case it was very difficult to find a firm willing to invest in the discovery project.

- Nivolumab (Opdivo): Pharmaceutical companies generally had a strong distrust of cancer immunotherapy because of their past experiences. Major domestic and foreign pharmaceutical companies that had bases in Japan showed no willingness to invest in R&D for developing a drug, at the offer from Ono Pharmaceutical. Even within Ono Pharmaceutical, it was not easy to obtain an approval for investment for developing a humanized antibody, and the decision-making took considerable time.

The following three projects were discontinued, although two of them were later resumed.

- Compactin/ Pravastatin (Mevalotin, Pravachol): During preclinical studies of compactin, the discontinuation of the project was considered twice, first in the face of inefficacy in rats and subsequently because of the suspicion of hepatotoxicity. While clinical development was launched, it was discontinued because of suspected tumors as a result of long-term toxicity studies in dogs for which extremely high doses of statins were administered.[5]
- Rosuvastatin (Crestor): Although Crestor was recognized as having a strong cholesterol-lowering effect, it was not well characterized, and was discontinued in the second phase. Later, AstraZeneca resumed the clinical development.
- Candesartan (Blopress, Atacand, Amitas): The first candidate compound, CV-2973, was found to be not sufficiently effective in clinical trials, and the project was discontinued.

In addition, the following two projects encountered serious difficulties, which could have led to discontinuations:

- Tocilizumab (Actemra): There were multiple crises faced in the development of this drug. First, a target molecule could not be identified in the search for B cell inhibitors. Second, a drug candidate could not be found from small molecule compounds in the drug-discovery research with IL-6 as a target. Finally, the choice to use an antibody as a drug candidate made its production cost very high, which might not be recovered by the regulated price with a limited markup over the low price of the existing conventional drugs in Japan.

---

[5]The discontinuation of clinical development of compactin by Sankyo made Merck also suspend the clinical trials of lovastatin. Later, however, Merck resumed clinical research and lovastatin was launched as Mevacor as the world's first commercialized statin.

Table 15.4 Unexpected difficulties, discontinuations and their solutions

Group	Generic and product names (Japan, USA and Europe)	Unexpected difficulties (○: crisis that led to or could have led to discontinuation)	Overcoming difficulties	Contribution of voluntary independent research, including *yami* research (○: Yes)
1	Compactin (discontinued in Phase II)	○ Three crises: 1. inefficacy in rats in preclinical stage, 2. suspected hepatotoxicity in preclinical stage, 3. suspected tumors in Phase 2 clinical stage (long-term toxicity studies in dogs)	1. Pathological studies in chickens and dogs (discovery of species differences) with cooperation of Mr. Kitano of the Pathology Department of the Central Research Institute 2. Determination of the route of formation of micro crystals and doctor-led clinical study on critically ill patients by Prof. Akira Yamamoto of Osaka University 3. Resumption of clinical trials of lovastatin in USA	• Discovery team undertook animal experiments by itself to address the problem of no response in rats
	Pravastatin (Mevalotin, Pravachol)			
2	Rosuvastatin (Crestor)	○ Discontinued in the early stage of the Phase II clinical study in Japan in November 1994	• License to AstraZeneca (May 1998)	
3	Leuprorelin (Leupron USA only)			
	Leuprorelin (Leuplin, Viadur)			
4	Ofloxacin (Tarivid, Floxin)/Levofloxacin (Cravit, Levaquin)	Internal opposition to development of levofloxacin (difference from ofloxacin perceived to be small relative to high production cost)		○ Ofloxacin was discovered by pursuing a discovery project along a direction different from that directed by upper management

(continued)

Table 15.4 (continued)

Group	Generic and product names (Japan, USA and Europe)	Unexpected difficulties (○: crisis that led to or could have led to discontinuation)	Overcoming difficulties	Contribution of voluntary independent research, including yami research (○: Yes)
5	Tamsulosin (Harnal, Flomax)			○ Discovery of new use of a drug candidate by the research beyond the purview of the firm's priority areas
6	Pranlukast (Onon)	Difficulty in formulating a drug for clinical trials	• Preparation difficulties resolved with cooperation of domestic startups with fine powder production technology	
7	Tacrolimus (Prograf)			
8	Pioglitazone (Actos, Glustin)	Discontinuation of ciglitazone after its early Phase II study, because of insufficient efficacy		
9	Donepezil (Aricept)	○ Low bioavailability led to decision to discontinue at preclinical stage	• Project resumed based on results of 1 year of yami research	○ Yami research for improvement of bioavailability led to resumption of the project
10	Candesartan (Blopress, Atacand, Amitas)	○ Project based on first candidate compound, CV-2973, was discontinued as a result of clinical trial but was resumed 7 years later	• Losartan developed by DuPont using the compound disclosed by Takeda was in turn used as lead compound	

(continued)

Table 15.4 (continued)

Group	Generic and product names (Japan, USA and Europe)	Unexpected difficulties (O: crisis that led to or could have led to discontinuation)	Overcoming difficulties	Contribution of voluntary independent research, including *yami* research (O: Yes)
11	Tocilizumab (Actemra)	O Three difficulties: (1) No target molecule could be identified in search for B-cell inhibitors, (2) No candidate compound could be found in drug-discovery research with IL6 as target molecule, (3) The choice of antibody as a drug candidate made production cost very high	1. Academic finding suggesting that IL6 causes autoimmune disease 2. New availability of humanization techniques for antibodies 3. Discovery of a new use of candidate drug for cancer (multiple myeloma) that could bear high production costs	
12	Nivolumab (Opdivo)	O High barrier for initiating a discovery project	• Participation of a US bio-startup	
Summary		Nine cases (O: six cases faced discontinuation crisis)		Four cases

– Donepezil (Aricept): The (formal) discovery research projects were discontinued because preclinical studies found that the candidate compound had low bioavailability.

The following three cases also encountered unexpected difficulties in the course of research and development, although they did not result in discontinuations.

– Levofloxacin (Cravit, Levaquin): There was opposition within the company to its clinical development because the difference from ofloxacin was perceived to be too small to justify the high production cost of levofloxacin.
– Pranlukast (Onon) : It was difficult to formulate the drug for clinical trials.
– Pioglitazone (Actos, Glustin): The development of discovered siglitazone was stopped in early Phase II clinical trials because of its inadequate efficacy.

These high levels of uncertainty (three-quarters of the projects faced unexpected difficulties and half of the projects faced discontinuation risk) are consistent with the results of the survey of the Japanese drug-discovery projects referred to earlier. According to the survey, about 67% of discovery projects, which yielded a registered or marketed drug (52 projects), faced unexpected difficulties, and about 57% of the same sample of discovery projects faced serious difficulties, which could have led to the discontinuation of the project over the course of their implementation (see Fig. 15.3). Among them, the projects with unknown drug targets faced significantly higher probability of unexpected difficulty (75%), although the risk of discontinuation faced by such projects was actually lower than the projects with known targets. This was presumably because the projects with known targets could face larger commercialization risk because of stronger market competition, even if their technological risks were smaller. Figure 15.3 also shows the incidence of unexpected difficulties faced by discontinued projects, which is actually lower than that of the projects with launched drugs, presumably because their durations were shorter.

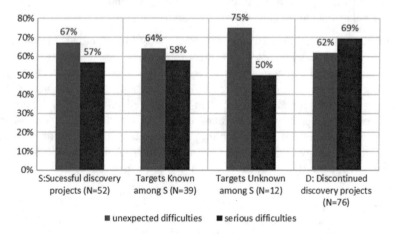

Fig. 15.3 Incidence of unexpected difficulty and discontinuation risk by type of discovery project. *Source* Nagaoka et al. (2015)

15.3.2 Overcoming Unexpected Difficulties

Many of the projects in our study could bring the innovative drugs to the market only by overcoming the unexpected difficulties. We can summarize such solutions by the following three major processes:

(1) Exploiting and responding to scientific progress.

The development of compactin by Sankyo was discontinued, and, recognizing that event, Merck also suspended the development of lovastatin. However, Merck resumed clinical development again, for which the contributions of science from the following two sources were critical. First, physicians in Japan and the United States initiated trials that strongly suggested the efficacy and safety of statins. Second, Brown and Goldstein's research clarified the mechanism of action of statins, through which it became clear that LDL receptors function to selectively lower the LDL cholesterol levels in the blood. Resumption of clinical trials of lovastatin in the United States would have significantly promoted the clinical development of Pravastatin by Sankyo.

In the case of tocilizumab (Actemra), the problem that the target molecule for B cell inhibition could not be identified was solved by the discovery of IL-6, as well as the discovery of its applicability to autoimmune disease, by the researchers at Osaka University. In drug-discovery research using IL-6 as a target molecule, the second problem was that a candidate compound for a drug could not be discovered from small molecule compounds. This was solved by using the antibody humanization technology that had been invented and was being applied by the MRC. The third problem was that price regulation in Japan might not have allowed the price to cover the high cost of the antibody drug in conventional disease. This situation was overcome by the discovery by Dr. Kawano at Hiroshima University that tocilizumab could be applied to multiple myeloma (a cancer).[6]

(2) The pursuit of possibilities through voluntary independent research by the lead researcher (including unauthorized research).

In four cases [compactin, ofloxacin (Tarivid), tamsulosin (Harnal, Flomax), and donepezil (Aricept)], voluntary independent research by the lead researcher played an important role in overcoming the crises by thoroughly pursuing technological opportunities, and exploring new uses of the drug candidate. Here, voluntary independent research is defined as research that is conducted voluntarily by the research team beyond the scope of its standard task expected from the firm. It includes *yami* research, which is defined as research that is conducted at the discretion of a laboratory unit without approval from the top management. Among the following four cases, such voluntary independent research played an essential role in overcoming the crisis.

---

[6]The prices of conventional drugs for rheumatoid arthritis were low, so that it was not clear whether the price could be set to cover the high production cost of antibody, even if it had high efficacy, under price regulation in Japan. In the case of a drug for a cancer for which there was no effective treatment, it was clear that the drug price could be set to cover the high production cost.

- Compactin: The discovery team undertook animal experiments by itself to determine the causes of inefficacy in rats. The research eventually led to the very important discovery of species differences in the efficacy of compactin.
- Tamsulosin (Harnal, Flomax) : Exploring areas that were not within the purview of the firm's priority areas led to the discovery of a new use for the drug candidate.
- Ofloxacin (Tarivid, Floxin): Pursuing a discovery project along a different direction than as directed by upper management led to a new drug discovery.
- Donepezil (Aricept): The discovery project was formally discontinued, but new chemical synthesis to improve bioavailability by structure–activity relationships was further carried out as *yami* research, which led to resumption of the project.[7]

The importance of exploiting the subsequent scientific advance as well as that of voluntary independent research in resolving these difficulties is confirmed by our large-scale survey on discovery research projects in Japan as mentioned above. Voluntary independent research, including *yami* research, enabled continuation of the projects or resumption of the discontinued projects in around 15% of the ultimately successful projects (drugs finally registered and launched in the market) that encountered unexpected difficulties. Kirsch and Ogas (2017) discuss how aspirin was also developed by *yami* research.

(3) Combining resources and knowledge across firms

Rosuvastatin (Crestor) and candesartan (Blopress, Atacand, Amitas) were cases in which discontinued projects were restarted thanks to the combination of resources or knowledges across firms. After a 4-year discontinuation period of rosuvastatin, AstraZeneca, by receiving a patent license from Shionogi, resumed clinical trials and developed a highly successful drug. AstraZeneca, unlike Shionogi, engaged in clinical developments with a view to the global launch of Crestor from the outset, and also engaged in its comparative investigations of Crestor with the most potent preceding statins (pravastatin, simvastatin, and atorvastatin) to clearly characterize rosuvastatin.

In the case of candesartan, losartan had successfully been developed by DuPont, exploiting the knowledge disclosed in the published patent of Takeda. When DuPont's hypertension research was at a deadlock, the inventions of Takeda were disclosed, and DuPont succeeded in solving the weaknesses (weak activity) of Takeda's inventions through computer simulations. In response, 7 years after the discontinuation, Takeda resumed its discovery research by exploiting the knowledge disclosed in the patent of losartan. Candesartan and losartan are examples of a success of combinatorial inventions enabled through patent disclosure.

In the case of nivolumab (Opdivo), the participation of Medarex, a US bio-startup, significantly contributed to the initiation of the discovery research to develop a humanized antibody drug. The method-of-use patent jointly filed by Ono Pharmaceutical and Prof. Honjo played a key role in attracting the participation of Medarex both as a source of information and as incentive. Medarex was developing another

---

[7]Another important case of *yami* research at Eisai is described in Itoh (2006).

antibody-based drug, ipilimumab, based on immune checkpoint inhibition. In addition, it had proprietary humanized antibody generation technology (UltiMAb) that could be applied to the generation of humanized antibody against PD-1.

15.3.3   Capturing Serendipity and Unexpected Good Luck

While research and development projects under high uncertainty often encountered unexpected difficulties, they also often encountered serendipity and unexpected luck. In 12 projects under our study, at least half of the cases had an instance of serendipity, and one-quarter of the cases captured good luck other than serendipity, as summarized in Table 15.5. Thus, our case studies show that these events were also not infrequent and capturing them played significant roles in the creation of innovative medicines.

According to Stephan (2012), serendipity is an unexpected discovery that does not meet the original research objective but is still beneficial. The following six cases are considered to have experienced important serendipities:

- Pravastatin (Mevalotin, Pravachol): A more active hydrophilic statin than compactin was discovered as a by-product of the analysis of the metabolites from compactin-treated dogs in the process of developing compactin. New drug candidates are rarely found in metabolites.
- Tamsulosin (Harnal, Flomax) : Although it was originally discovered as an antihypertensive drug, it was later found that it could be used as a drug for the treatment of urinary disorders, based on the suggestions from clinical papers and clinicians.
- Pioglitazone (Actos, Glustin) and donepezil (Aricept): Discovery research conducted for other purposes yielded lead compounds for these projects.
- Candesartan (Blopress, Atacand, Amitas): In-house basic research on chemical synthesis yielded the lead compound CV-2198. This in-house research did not target specific disease drugs. Furthermore, the discovery of the antihypertensive effect of CV-2198 occurred serendipitously in the project for a diuretic drug.
- Nivolumab (Opdivo): PD-1, the target molecule of nivolumab was discovered serendipitously in the basic research for other purposes at the Honjo laboratory, Kyoto University.

There were also four important discoveries that met the original research objectives, but were unexpected, which are defined as instances of good luck.

- Pravastatin (Mevalotin, Pravachol): Microorganisms suitable for synthesizing pravastatin was discovered at an early stage from an in-house library.
- Leuprorelin (Leuplin, Leupron, Viadur): Mistakes and repeated attempts in experiments by young researchers in synthesis research for LH-RH derivatives led to the discovery.
- Levofloxacin (Cravit, Levaquin): Optical resolution of Tarivid, which is a racemic mixture turned out to produce a compound with strong antibacterial activity and low toxicity.

Table 15.5 Unexpected good luck and serendipity

Group	Generic and product names (Japan, USA and Europe)	Capturing serendipity	Capturing good luck
1	Compactin (discontinued in Phase II)		
	Pravastatin (Mevalotin, Pravachol)	○ Discovery of pravastatin from metabolites of compactin-treated dogs	○ Early discovery of microorganisms suitable for microbial conversion from in-house libraries
2	Rosuvastatin (Crestor)		
3	Leuprorelin (Leupron USA only)		○ Mistakes and multiple attempts by young researchers in the synthesis of LH-RH derivatives led to synthesis of leuprorelin
	Leuprorelin (Leuplin, Viadur)		
4	Ofloxacin (Tarivid, Floxin)/Levofloxacin (Cravit, Levaquin)		○ Optical resolution yielded the S-form with strong activity and low side effects
5	Tamsulosin (Harnal, Flomax)	○ Compound originally discovered as an antihypertensive drug later found to be useful for treatment of urinary disorders	
6	Pranlukast (Onon)		
7	Tacrolimus (Prograf)		
8	Pioglitazone (Actos, Glustin)	○ Discovery of hypoglycemic effects from substances obtained through synthetic research for lipid-lowering drugs	
9	Donepezil (Aricept)	○ In-house hyperlipidemia research project provided a lead compound	

(continued)

Table 15.5 (continued)

Group	Generic and product names (Japan, USA and Europe)	Capturing serendipity	Capturing good luck
10	Candesartan (Blopress, Atacand, Amitas)	○ In-house basic research on reaction systems provided CV-2198 as lead compound ○ Discovery of blood pressure-lowering effect of compounds from diuretic research	○ DuPont's losartan development revealed a method for enhancing the activity and became a new lead compound
11	Tocilizumab (Actemra)		
12	Nivolumab (Opdivo)	○ Discovery of PD-1 at Honjo laboratory, Kyoto University	
Summary		Six cases	Four cases

– Candesartan (Blopress, Atacand, Amitas): Losartan developed by DuPont, which solved the weak activity of the compound synthesized by Takeda, became a new lead compound for Takeda.

15.4 Uniqueness and Competition in Discovery Research

Competition in drug discovery is often understood to occur in two phases: competition for the discovery of a new drug based on a new mechanism of action, herein referred to as "inter-mechanism competition," and competition between drugs with the same active ingredient after the expiration of the patent. However, as case studies show, it is more appropriate to understand that competition occurs in three stages. To begin with, the discovery research that sought new mechanisms of action is often unique and faces high uncertainty. Thus, competition is often rare in the initial stage in such cases. However, after the resolution of uncertainty because of the successful discovery, competition seeking early product launch and the best-in-class drug becomes very intense and can be characterized as a "patent race."[8] Such competition among drugs with the same mechanism of action but with different ingredients is referred to as "within-mechanism competition."[9] Finally, there is a competition within the drugs with the identical active ingredient through generic entries.

15.4.1 Uniqueness of Discovery Research

As shown in Table 15.6, 9 of the 12 projects included in this book (column 2 in Table 15.6) involved the discovery of a drug based on a new mechanism of action, including donepezil (Aricept), virtually the first successful drug in the world for controlling the progression of Alzheimer's disease. The other 3 projects had known mechanisms of action. It is presumed that there were only three projects out of these nine projects (column 3 in Table 15.6) in which discovery competition occurred. Thus, the discovery projects were often unique. As the following discussions suggest, the mechanisms of action were often unknown (they became clarified only after the discovery of the drugs) in these discovery projects so that high uncertainty made the research project unique and competition relatively rare, and even if the mechanism had been known, negative views against such a mechanism had prevailed.

– Compactin: The mechanism of action was established only after its discovery. While there were companies that searched for inhibitors targeting HMG-CoA

---

[8]Firms in a patent race perceive the same technological opportunity. The first firm to make the discovery acquires the patent and the race ends.

[9]If the target is different, the mechanism of action is defined as different in this paper, even if the same disease is treated, and the mechanism of action is narrowly defined.

Table 15.6 Competitive mechanisms before and after discovery

Group	Generic and product names (Japan, USA, and Europe)	Ex-ante competition for discovery of a drug with new mechanisms of action		Ex-post competition within mechanism of action	
		Discovery of a drug with new mechanism of action (○: first in the world)	Uniqueness of discovery research and existence of competition for new mechanism of action (○: likely to be unique)	Rank in launching (○: first in the world)	Competition within mechanism of action (◎: product market competition; ○: R&D competition only)
1	Compactin (discontinued in Phase II)	○ First in the world	○ (1) Mechanism of action was established only after its discovery (2) Unique discovery program of screening microorganisms	Third in the world after Merck's lovastatin and simvastatin	◎ Merck started a similar line of research, stimulated by patent publication of compactin. Subsequent entries by other firms, using the compactin as a base
	Pravastatin (Mevalotin, Pravachol)				
2	Rosuvastatin (Crestor)			Seventh in the world, second as a super statin	◎
3	Leuprorelin (Leupron USA only)	○ First in the world	• Competition is likely to have existed.	○ Both are the first in the world	
	Leuprorelin (Leuplin, Viadur)				◎ Leuplin (DDS preparation) competed with ICI (now AstraZeneca)
4	Ofloxacin (Tarivid, Floxin)/Levofloxacin (Cravit, Levaquin)			World's highest sales of antibiotics	○ Competition for optical resolution of Ofloxacin (nearly coincided with Bayer)

(continued)

Table 15.6 (continued)

Group	Generic and product names (Japan, USA, and Europe)	Ex-ante competition for discovery of a drug with new mechanisms of action		Ex-post competition within mechanism of action	
		Discovery of a drug with new mechanism of action (○: first in the world)	Uniqueness of discovery research and existence of competition for new mechanism of action (○: likely to be unique)	Rank in launching (○: first in the world)	Competition within mechanism of action (◎: product market competition; ○: R&D competition only)
5	Tamsulosin (Harnal, Flomax)	○ First in the world as an α 1-blocker	○ (1) Mechanism of action was established only after its discovery (2) Based on a serendipitous discovery from a project using in-house compound as lead compound	○ First in the world	◎ After creation of Harnal, there is competition among α1-blockers
6	Pranlukast (Onon)	○ First in the world	• Competition existed	○ First in the world (but not launched in Europe or North America)	◎ Competition with zafirlukast and montelukast

(continued)

Table 15.6 (continued)

Group	Generic and product names (Japan, USA, and Europe)	Ex-ante competition for discovery of a drug with new mechanisms of action		Ex-post competition within mechanism of action	
		Discovery of a drug with new mechanism of action (○: first in the world)	Uniqueness of discovery research and existence of competition for new mechanism of action (○: likely to be unique)	Rank in launching (○: first in the world)	Competition within mechanism of action (◎: product market competition; ○: R&D competition only)
7	Tacrolimus (Prograf)			Second immunosuppressive drug inhibiting IL2 production in the world	○ Competition for developing derivatives of tacrolimus
8	Pioglitazone (Actos, Glustin)	○ Ciglitazone is the first compound in the world to have a thiazolidinedione structure	○ (1) Mechanism of action was discovered only after drug discovery (2) Research based on a serendipitous discovery and based on development of animal models	Third in the world	◎ Following disclosure of ciglitazone, other companies also entered the R&D competition successfully

(continued)

Table 15.6 (continued)

Group	Generic and product names (Japan, USA, and Europe)	Ex-ante competition for discovery of a drug with new mechanisms of action		Ex-post competition within mechanism of action	
		Discovery of a drug with new mechanism of action (○: first in the world)	Uniqueness of discovery research and existence of competition for new mechanism of action (○: likely to be unique)	Rank in launching (○: first in the world)	Competition within mechanism of action (◎: product market competition; ○: R&D competition only)
9	Donepezil (Aricept)	○ Virtually the first in the world for controlling the progression of Alzheimer's disease	○ Project challenged the mainstream view on acetylcholine hypothesis	○ First in the world	○ When the invention of Aricept was disclosed, many companies prepared its derivatives and entered the market without success
10	Candesartan (Blopress, Atacand, Amitas)	○ First in the world discovery of nonpeptide ARB, the lead compound of losartan	• Competition is likely to have existed	Third in the world	◎ Global R&D competition after the disclosure of losartan
11	Tocilizumab (Actemra)	○ First in the world as an IL6 inhibitor	○ Mechanism of action was discovered only after drug discovery	○ First in the world as an IL6 inhibitor	• There has been no within-mechanism competition
12	Nivolumab (Opdivo)	○ First in the world as a PD-1 inhibitor	○ Negative views against cancer immunotherapy were mainstream	○ First in the world as a PD-1 inhibitor	◎ Entries of many big companies after clinical successes of Opdivo and Yervoy
Summary		New mechanism of action: nine cases	Unique discovery research: six cases	First market launch: six cases	Within-mechanism competition: 11 cases (including 3 cases of only ex-post R&D competition)

from chemical libraries, only Dr. Endo's team seems to have screened natural products at the time. The structure of the statin was complex, so that "No random chemical library would ever yield an HMG-CoA reductase inhibitor as potent as the natural statins," according to Brown and Goldstein (2004). Even though the compactin patents were disclosed and Dr. Endo reported the discoveries at international conferences, major pharmaceutical companies were not interested in them, except for Merck.

– Tamsulosin (Harnal, Flomax): The mechanism of action was established only after its discovery. It was based on a serendipitous discovery from a project for an antihypertensive drug, using an in-house compound (amosulalol) as the lead compound.

– Pioglitazone (Actos, Glustin): The novel mechanism of action (improved insulin resistance with thiazolidinediones) was discovered only after the drug discovery. The target (PPAR-gamma) was discovered only in 1995, long after the discovery by the Takeda research team of the basic molecular entity (siglitazone) for a drug in 1978. The research was based on a serendipitous discovery from substances obtained internally through the synthetic research for lipid-lowering drugs. The research was also unique and was based on the development of animal models.

– Donepezil (Aricept): When Dr. Sugimoto's team at Eisai started on the project, there had been a significant loss of support for the acetylcholine hypothesis, and the mainstream view was that drug discovery targeting it would be difficult, mainly because of toxicity issues associated with the inhibition of acetylcholine esterase.

– Tocilizumab (Actemra): The mechanism of action was discovered only after the drug discovery. After the discovery of the target IL-6 and its potential role in autoimmune diseases was presented at an academic conference, Chugai Pharmaceutical was the only firm to make an offer to Osaka University for collaborative drug development research.

– Nivolumab (Opdivo): Negative views against cancer immunotherapy had prevailed. While long-term basic research at Kyoto University clarified the target, the mechanism of action of the antibody, and its therapeutic effects in mice, the development of humanized antibody drugs was initiated only after the US bio-startup Medarex participated in the project, after refusals by many major pharmaceutical firms.

Because of high uncertainty caused by limited understanding of the mechanism of action, or because of negative views against the mechanisms, the discovery projects for drugs with new mechanisms of action were often conducted without strong support even within the firm. For this reason, pioneering researchers often needed to do research facing high uncertainty as entrepreneurial scientists. This also implies that the projects often did not attract the attention or interest of other firms. Because of this, the discovery projects for drugs with new mechanisms of action often did not face competitors until they succeeded.

These results are consistent with the finding of Cockburn and Henderson (1994), who, based on statistical analyses of discovery projects, concluded that investment into discovery projects by a pharmaceutical company is driven not by direct competitive relationships between pharmaceutical firms but by heterogeneity in company capabilities, coordination costs, and the emergence of technical opportunities. However, it is also true that once the effectiveness of a new mechanism of action is confirmed by the research of a pioneering firm, it leads to intense within-mechanism competition, as shown in Table 15.6 (columns 4 and 5).

15.4.2   Emergence of Within-Mechanism Competition and Knowledge Spillover

A pharmaceutical company that has discovered a new drug with a new mechanism of action and has acquired its patent rights has not only the first-mover advantage in clinical development, but can exclusively engage in clinical development, to the extent that protection of the patent rights is effective. At the same time, however, the new mechanism of action invented is disclosed by the patent disclosure, and as the prospect of commercial success of the new drug is confirmed by the clinical trials, drug discovery and development competition based on the same mechanism of action arises through such knowledge spillovers. This is because there often exists significant room for chemical synthesis and antibody preparation with the same mechanism of action without infringing on the preceding patents. As Table 15.6 shows, almost all of the projects under our study faced such competition within the mechanism of action, including unsuccessful R&D attempts for entries.

Within-mechanism competition was very close to a patent race in the following two cases. The discovery of levofloxacin (Cravit, Levaquin) by Daiichi Pharmaceutical was almost simultaneous with that by Bayer Germany, competing for the optical resolution of levofloxacin. The patent for candesartan (Blopress, Atacand, Amitas) issued to Takeda had been applied for in April 1990. One year later, another company applied for a patent for the same invention. In the case of donepezil (Aricept) and tacrolimus (Prograf), there were no other competitors that subsequently launched the product based on the same mechanism of action (although Prograf followed cyclosporine). However, after the disclosure of each patent, a number of companies embarked in R&D for developing a drug based on the same mechanism of action without success.

The pioneer firm in the discovery stage was not necessarily the first in launching the drug in the product market because of the rapid emergence of within-mechanism competition. Of the nine drugs with a new mechanism of action, including donepezil (Aricept), six were launched first in the world by the discoverer of the same mechanism of action, as shown in column 4 in Table 15.6. However, the following three drugs were launched later than the launch by a competitor:

- Pravastatin (Mevalotin, Pravachol): Sankyo, the first company to discover statins, lagged behind Merck's lovastatin and simvastatin in the United States and simvastatin in Europe. In this case, the fundamental reason was that Sankyo abandoned the clinical development of compactin and switched to pravastatin.
- Pioglitazone (Actos, Glustin): Takeda was the first in the world to synthesize a thiazolidinedione compound (siglitazone) as a candidate for the treatment of type 2 diabetes.[10] It was the third in the world to launch such a drug, because it took time in its discovery based on the lead compound and in starting and conducting clinical trials. However, because the discovery was carried out with an emphasis on the side effect problem, it did not come to the situation in which it was withdrawn from the market.
- Candesartan (Blopress, Atacand, Amitas): Takeda successfully discovered a class of lead compounds (e.g., CV-2198) that can lower blood pressure by a novel mechanism of action (angiotensin II receptor antagonism) but could not achieve sufficient efficacy in clinical trials and stopped the development. On the other hand, DuPont used this lead compound to create losartan, which became the first drug to be launched by this mechanism, while candesartan was the third drug in the world.

A follow-on drug with the same mechanism of action as the pioneer drug has an incentive to improve the performance of the early drug to avoid patent infringement and to differentiate them in the market. Therefore, it may make a significant contribution in terms of efficacy and side effects. For an example, Crestor's case study showed that at the time it entered the market, a number of potent statins with the same mechanism of action had been introduced into the market and their generics also entered the market. However, Crestor sales did not significantly decline after the launch of the generic drugs following the expiration of the patents for the prior statins (pravastatin, simvastatin, and atorvastatin). This indicates that the contribution of Crestor to product innovation was significant.

15.5 Implications on Management and Policy

15.5.1 Absorptive Capability and Incomplete Science

The cases described in this book strongly indicate the critical importance of exploiting the progress of science in drug discovery. Scientific advances create drug-discovery opportunities through the discovery of new targets, development of new research tools, and suggestion of new use for existing inventions. To make use of such state-of-the-art science in drug discovery, it is essential for a firm to have the ability to absorb science, the ability of corporate researchers to grasp and evaluate cutting-edge scientific advances, and to exploit them for their own drug discoveries.

---

[10]Siglitazone was not launched because of its inadequate efficacy in humans.

Coping with uncertainty is an important pillar of absorptive capability, because drug discovery often starts even if the relevant science is incomplete (Table 15.3). Universities and national research institutes rarely provide seeds in a form that can easily be transformed into drugs. In fact, the case studies suggest that the function of the target and the mechanism of action of a molecule affecting the target are often clarified only after the drug discovery is made and such drug is used as a research tool for such investigation. Incomplete science also implies high uncertainty of a drug-discovery project and the frequent emergence of unexpected difficulties. Coping with such uncertainties is necessary for successful drug discovery. Utilization and promotion of subsequent progress of science, combining resources and knowledge across firms, and thorough pursuit of possibilities through voluntary independent research (including *yami* research) played major roles.

Serendipity and good luck also often occurred in highly uncertain projects, as illustrated by our case studies. Serendipity and good luck constitute an important return for the discovery project of exploring unchartered areas and it is very important to capture them. Uncertainty is not simply a cost but also a source of gain. The case studies suggest that in the context of a firm's ability to capture serendipity and good luck, it is important to make careful observations and examinations of the data from experiments, to share information across divisions within a firm, and not to simply think of unexpected results as anomalies or failures, but to explore the mechanism for generating such data.

The key reasons why the Japanese companies under this study could make use of the scientific advancements are that they actively invested in the capability building of researchers from a long-term perspective, such as enabling them to study in universities and research institutes in Japan and abroad, and encouraging them to acquire PhDs after joining the firm. As seen in the case of Dr. Endo (statin) and Dr. Ohsugi (tocilizumab), a typical case is to give high-performing researchers a chance to undertake postdoctoral study abroad for 2 years. Working as a postdoctoral fellow in an American university requires a Ph.D. and these researchers often acquired PhDs on the basis of their corporate research completed prior to their trip to the USA.

Research fellowships at universities and research institutes in Japan and abroad were very important for transferring the advanced knowledge embodied in the researchers of universities to firms. In the case of Dr. Endo, his period of research at a U.S. university provided the opportunity for him to recognize the nature of research competition in the USA and to recognize the necessity of pursuing a research opportunity unique to Japan.

15.5.2   Encouraging Individual Initiative and Long-Term Perspective

The case studies suggest that the discovery projects leading to breakthrough drugs are often individual choices rather than organizational choices. Our case studies suggest that Japanese firms provided a significant degree of freedom to their researchers in designing research projects. The importance of individual initiatives is also revealed by the fact that voluntary independent research (including *yami* research) played an important role in overcoming the crises of initiated projects in one-third of the cases, including compactin, ofloxacin (Tarivid), tamsulosin (Harnal, Flomax), and donepezil (Aricept).

Individual initiatives are important because research projects with high uncertainty are less likely to be selected if an organizational agreement on the choice of a discovery project is the prerequisite for a launch of the project from the beginning. Indeed, the results of the survey of discovery projects in Japan as referred to earlier show that the discovery projects initiated by individuals (approximately one-quarter of the total projects) tend to be more uncertain and to have higher uniqueness than those managed by the organization from the beginning. That is, there seems to be a problem of bias of consensus-based decision-making against a discovery project with high uncertainty (e.g., the disease mechanisms and/or the target are unknown).

There is also a risk that *yami* research, which is independent and not authorized by the organization, prolongs the period of investment in unpromising projects. However, in the case of highly novel drug-discovery research, large uncertainties exist so that there are many challenges to be solved, and the return for an eventual success of such drug discovery would be large. Therefore, *yami* research may be a powerful option when a firm can ensure the consistency of long-term interests between the firm and the researchers; for example, by allowing researchers to conduct a *yami* study only after regular business hours, or by placing time limits in advance on the duration of such projects. Choosing an R&D manager in favor of transformations may also be important.[11]

For individual initiatives to work, individuals need to be able to choose projects from a long-term perspective. For example, in the case of compactin, Dr. Endo, the lead researcher, was aware that his discovery project had a high level of uncertainty and a high likelihood of failure, like gambling. He therefore set the time framework in advance that he would terminate the project if he could not obtain promising results in two years. He judged that if it were a loss of 2 years if the project failed, it would be possible to cover the loss in the framework of long-term employment. Therefore, the discovery project was designed only after the efficient screening method was constructed and the team members were gathered so that the success or failure could be decided in 2 years. This is a good illustration of the point that long-term evaluation of researchers can encourage them to choose research topics that are both highly uncertain and innovative (Azoulay et al. 2011).

The long-term employment system that is often used in Japanese firms (commitment to not dismiss until retirement age) enables a worker to take the risk of a short-term failure. For this reason, the system can be used to promote the challenge to pursue projects with high uncertainty. As pointed out by Koike (1994), the long-term employment system in Japan has the advantage of promoting innovation by

---

[11]Itoh (2015) pointed out that in organizations where innovation is important, it is important to make superiors transformative.

encouraging decision-making from a long-term perspective, and it is important to exploit such potential advantage for innovations.

15.5.3 Global Clinical Development Capabilities

In the 12 projects considered in this study, only 3 Japanese discovery firms undertook the clinical trial in the United States and Europe by itself: Yamanouchi [tamsulosin (Harnal, Flomax)], Takeda [pioglitazone (Actos, Glustin)], and Fujisawa [tacrolimus (Prograf)]. In 7 projects, major American and European pharmaceutical companies obtained the exclusive license for these markets and undertook the developments, including clinical trials.[12] This practice of international licensing is one major reason why the brands of major drug-discovery companies in Japan have not been well recognized internationally.

Licensing out the clinical development and sales of a new drug in overseas markets to major pharmaceutical firms abroad enables the exploitation of the complementary assets of such companies, such as development experience and sales networks. Such efficiency was clearly demonstrated in the case of rosuvastatin (Crestor), because AstraZeneca made a successful launch of the drug that had been given up by Shionogi. AstraZeneca undertook global development from the beginning, exploiting the economy of scale and scope in clinical trials, and also completed comprehensive comparative studies of rosuvastatin with the already approved potent statins (pravastatin, simvastatin, and atorvastatin) to demonstrate the advantage of rosuvastatin.

However, on the other hand, the transaction cost and time required to search for and negotiate with a clinical development partner and the split of property rights between drug discovery and market development can slow the clinical development process and reduce the ex-ante incentive for drug discovery. Thus, increasing the clinical development capabilities of Japanese companies in overseas markets will be very important as they expand discovery projects with high uncertainty and attempt to deal with rare diseases.

15.5.4 Patent System that Encourages Challenges to Uncertainty

The patent system protects the exclusive use of an invention that is novel, has an inventive step and utility, for a limited period of time, subject to disclosure. Case studies suggest that patent systems had a significant impact on the appropriation of the R&D investment as well as on the combination of knowledge and resources through disclosure and licensing. Both of these effects are especially important for discovery

---

[12]The drug was jointly developed in one case, and there were no sales in Europe and in the United States in one case.

projects with high uncertainty. Patent protection enhances the ex-ante incentive for R&D investment with high risks, and disclosure and licensing promote knowledge combinations for the new discovery as well as the successful commercialization of the drug candidate.

The major impact of patent protection on the appropriation[13] by a drug-discovery firm is clearly indicated by the fact that the firm's revenue from such a drug decreases dramatically after generic entries, following the expiration of the patent. For example, according to Fig. 2.2 in Chap. 2 on compactin, after the expiration of the respective patent, the sales of pravastatin, simvastatin, and atorvastatin plummeted. Furthermore, the length of the protection is limited, considering the long time necessary from the discovery to the market launch. On average for our case studies, about 11 years passed between filing of the basic patent for the discovery to the first drug approval either in Japan, the USA, or Europe. Therefore, the remaining patent protection period was less than 10 years or a half of the standard patent protection period (20 years, excluding the patent term extension) when the drug was launched. Thus, the system of patent term extension played an important role for appropriation. With the exception of pravastatin (Mevalotin), the substance patent was subject to 5 years of extension (the longest period allowed by the current patent system) for all drugs subject to the case studies in Japan. In the case of tocilizumab (Actemra), for example, it took 16 years from the basic patent to the launch, and the drug would have been protected only for 4 years if there had been no patent term extension.

The ability to secure a significant return through the creation of an innovative drug is crucial in creating incentives for a firm to invest in the environment of high uncertainty. This can be illustrated by a simple example. Assume that a new drug delivers a social benefit of S (consumer surplus and producer surplus) when the discovery project is successful with a probability $\theta (0 < \theta < 1)$ The firm appropriate $\rho (0 < \rho < 1)$ of S when the discovery is successful while it incurs the R&D cost c irrespective of the success. The necessary condition for a firm to invest in the discovery is that the ex-ante expected profit is positive:

$$\pi = \theta \rho S - c = \theta(pS - c) - (1 - \theta)c \geq 0 \qquad (15.1)$$

On the other hand, the society always gains from such investment by

$$\theta S - c = \pi + (1 - \rho)\theta S \qquad (15.2)$$

Here, $(1 - \rho)\theta S$ gives the externality to the society for undertaking such R&D. It is clear from Eq. (15.1) that when the risk of failure is high (θ is low), higher appropriability (higher ρ) is necessary to induce investment by a firm, because the firm can obtain profit with smaller possibility and the probability of incurring loss rises. If the firm can obtain only the revenue equivalent to the cost when the project is

---

[13]Patent right for the exclusive use of an invention allows the discovery firm to obtain a portion of the invention's surplus to society (e.g., benefits of patients above the cost of manufacturing and delivering the drugs) as its profit. The extent of this is called appropriation.

successful ($\rho S = c$), the ex-ante profit of the firm is negative ($-(1 - \theta)c$), so that the firm cannot invest. When appropriability is significantly limited, it will undermine drug discovery in areas of high uncertainty.[14]

Case studies also clearly show that patent systems often play an important role in facilitating combinations of knowledge and resources across organizations in the discovery and development of new drugs. First, the disclosure of inventions by the patent system enabled new inventions by combining the disclosed knowledge with new knowledge. As shown in Table 15.1, 3 of 12 drugs under investigation, rosuvastatin (Crestor), ofloxacin (Tarivid, Floxin), and candesartan (Blopress, Atacand, Amitas), are examples of the combination of the knowledge disclosed in the prior drug invention and the new research and development by the firm. It is important to note that invention disclosure through patent system takes place much earlier than that through product launch in pharmaceutical industry. In addition, as discussed earlier in Sect. 15.4.2 on emergence of within-mechanism competition and knowledge spillover, the patent disclosure promoted the follow-on inventions through knowledge spillovers. Such follow-on drugs add not only competition but also further enhance innovations since they are often differentiated. Therefore, patent disclosure facilitated the search for better and more diverse drugs with the same mechanisms of action.

Emergence of within-mechanism competition though knowledge spillover also implies that in many cases of breakthrough drugs, the patent system does not reward such drugs in a way that reflects their positive knowledge spillover effects on followers. This suggests that a system that complements the current patent system can be important, especially considering that R&D tasks involve more uncertainties. For example, the current patent term extension system focuses only on the length of the clinical trial period, but the consideration of knowledge spillover effects or the consideration of the length of the preclinical trials may also be important.

Case studies also suggest that patent protection for university inventions could be effective in promoting the transfer of expertise from university to industry and in promoting corporate investment toward the commercialization of such inventions, given high uncertainty. In the case of tocilizumab (Actemra), the basic patents for IL-6 receptor and its antibodies were acquired by Prof. Tadamitsu Kishimoto of Osaka University in the early stage of collaborative research between Chugai Pharmaceutical and Osaka University, which was carried out with the aim of drug discovery based on those patents. It is likely that the protection of these basic patents would have promoted Chugai Pharmaceutical to make sustained investments in an uncertain area, considering the fact that the project faced the risk of discontinuation even under patent protection. In the case of nivolumab (Opdivo), major pharmaceutical firms declined to participate despite the existence of patent protection. At the same

---

[14]In the case of pharmaceuticals, consumers pay only a part of their price for health insurance, but for the following reasons, the possibility of overconsumption is likely to be small for new drugs. First, the price of a new drug, even if regulated, often significantly exceeds its manufacturing cost. Second, the prescription by a doctor is necessary for a purchase.

time, the involvement of Medarex was significantly promoted by the method-of-use patent of anti-PD-1 antibodies through its disclosure and as an incentive. Thus, patenting seems to have promoted innovation significantly in these cases.

15.5.5 Science and R&D Infrastructure in Japan

Japanese science and infrastructure for industrial R&D also play a very important role in drug discovery. Although foreign science, especially that of the United States, was more often the most important knowledge source for initiating the drug-discovery projects of the cases under this study, the result of basic research at Japanese universities played key drug-discovery roles in two cases, tocilizumab (Actemra) and nivolumab (Opdivo), both of which were biological drug discoveries. The status of Japanese fundamental science in the field of immunology as a world leader seems to have contributed to two highly innovative drug discoveries. These examples demonstrate that pure basic research that aims to increase the understanding of fundamental questions, rather than to seek specific applications, can sometimes provide directly a source of highly significant innovations. Competitive grants for basic research played critical roles for these two scientific discoveries.

Many Japanese universities contributed to the development of the drugs by enhancing the drug-discovery capabilities of Japanese pharmaceutical companies, including the preparations of animal models, the introduction of new screening technologies, and cutting-edge chemical synthesis technologies. As Table 15.2 shows, new tools developed by or transferred from Japanese universities contributed to the drug discoveries and development in at least 7 cases out of 12 drugs. Japanese universities also contributed to the elucidation of the mechanism of action of the innovative medicine and to the clinical research based on physician initiative in the clinical development stage. The development and diffusion of cutting-edge research tools functions as provision of public goods by industrial research and development infrastructure, and it seems to be an important field as a science and technology innovation policy.

15.5.6 Regulations in Japan

Case studies suggest that about half of the drugs under study were launched first in the United States or Europe, rather than in Japan, despite their Japanese discovery origins. In the following three cases, the clinical study preceded in the United States, because of constraints on clinical trials in Japan.

– Donepezil (Aricept): Side effects were observed in the Phase I study in Japan, which significantly limited the dose and made it difficult to detect the efficacy of Aricept in the Japanese Phase II study. On the other hand, higher doses could be tested in the United States, allowing evaluations of drug efficacy at varying dose levels. The US Food and Drug Administration (FDA) issued an unusually quick approval of the application in 8 months.

- Leuprorelin (Leuplin, Leupron, Viadur): A clinical trial of a drug with long-term sustained-release injection was without precedent, so that it was anticipated that the trials in Japan would receive strong resistance. Takeda, therefore, decided to go ahead with the US clinical trials by TAP, a joint venture between Takeda and Abbott. The study in the USA was conducted in cooperation with the FDA, and Takeda was able to obtain useful advice from the FDA.
- Tacrolimus (Prograf): Although the preclinical tests had finished in Japan by the end of the 1980s, a large-scale clinical trial involving transplantation operations was impossible. Since 1990, all large clinical trials with more than 500 patients had been conducted both in the United States and Europe.

One of the cases considered in this study showed how the price regulation of drugs in Japan can present a hindrance to the development of innovative drugs. In the case of tocilizumab (Actemra), because the price of existing drugs for rheumatoid arthritis was low, there was a concern that even a drug with a high efficacy could not be expected to have a correspondingly high price to cover the cost of antibody production. This was apparently one major reason for why this drug was originally developed as a cancer drug (multiple myeloma).

Regulatory reforms have taken place in Japan to encourage pharmaceutical innovation in Japan, and case studies strongly suggested the importance of such innovation-oriented reforms.

15.5.7 Conclusion

This book analyzed the sources of drug innovations in Japan based on detailed case studies of 12 groups of 15 innovative drugs. It covers the first statin in the world up to the recent major breakthrough in cancer therapy, an immune checkpoint inhibitor that is based on the scientific discovery for which the 2018 Nobel Prize in Physiology or Medicine was awarded. The study showed pervasive high uncertainty in drug discovery: frequent occurrences of unexpected difficulties, discontinuations, and serendipitous discoveries and instances of good luck. Much of this uncertainty is present because drug discovery starts when the underlying science is incomplete. Thus, there exist dynamic interactions between scientific progress and drug discovery. High uncertainty also makes the value of an entrepreneurial scientist high. Such scientists fill the knowledge gaps by absorbing and combining external scientific progress and by relentless pursuit of possibilities through their own research, often including unauthorized research, to overcome the crises. Furthermore, high uncertainty and its resolution also significantly characterizes the evolution of competition in the drug industry. The patent system promotes innovation under high uncertainty not only by enhancing appropriability of such R&D investment but also by facilitating the combination of knowledge and capabilities between different organizations through mandating disclosures. We hope that this book will significantly contribute to wise innovation management and policy practices through deeper understanding of an innovation process where uncertainty, dynamic interactions between scientific progress and drug discovery, and entrepreneurial scientists play key roles.

References

Azoulay, P., Zivin, J. S. G., & Manso, G. (2011). Incentives and creativity: Evidence from the academic life sciences. *RAND Journal of Economics, 42*(3), 527–554.

Brown, M. S., & Goldstein, J. L. (2004). A tribute to Akira Endo, discoverer of a "Penicillin" for cholesterol. *Atherosclerosis Supplements, 5,* 13–16.

Cockburn, I. M., & Henderson, R. (1994). Racing to invest? The dynamics of competition in ethical drug discovery. *Journal of Economics and Management Strategy, 3,* 4819.

Itoh, M. (2006). *Drug discovery story, Shinkoh_igaku* (in Japanese).

Itoh, H. (2015). *Organizing for change: Preference diversity, effort incentives, and separation of decision and execution.* RIETI Discussion Paper Series 15-E-082.

Kirsch, D. B., & Ogas, O. (2017). *The drug hunters.* New York: Arcade Publishing.

Koike, K. (1994). *Employment system of Japan—Its universality and strengths.* Toyo Keizai Shimbun, Inc. (In Japanese).

Nagaoka, S., Nishimura, H., & Genda, K. (2015). *Drug discovery and science-survey on the science origins of pharmaceutical innovation and their economic effects* (1). IIR Working Paper 15-16, Institute of Innovation Research, Hitotsubashi University. (in Japanese).

Stephan, P. (2012). *How economics shapes science.* Cambridge: Harvard University Press.

Zucker, G. Z., Darby, M. R., & Brewer, M. B. (1998). Intellectual human capital and the birth of US biotechnology enterprises. *American Economic Review, 88*(1), 290–306.

Index

© Springer Nature Singapore Pte Ltd. 2019
S. Nagaoka (ed.), *Drug Discovery in Japan*,
https://doi.org/10.1007/978-981-13-8906-1

Printed in the United States
By Bookmasters